Andre Dallmann

Perturbed or not perturbed?

Andre Dallmann

Perturbed or not perturbed?

Structure and dynamics of fluorophore-labelled dsDNA helices probed by NMR-spectroscopy

Südwestdeutscher Verlag für Hochschulschriften

Impressum/Imprint (nur für Deutschland/ only for Germany)
Bibliografische Information der Deutschen Nationalbibliothek: Die Deutsche Nationalbibliothek verzeichnet diese Publikation in der Deutschen Nationalbibliografie; detaillierte bibliografische Daten sind im Internet über http://dnb.d-nb.de abrufbar.

Alle in diesem Buch genannten Marken und Produktnamen unterliegen warenzeichen-, marken- oder patentrechtlichem Schutz bzw. sind Warenzeichen oder eingetragene Warenzeichen der jeweiligen Inhaber. Die Wiedergabe von Marken, Produktnamen, Gebrauchsnamen, Handelsnamen, Warenbezeichnungen u.s.w. in diesem Werk berechtigt auch ohne besondere Kennzeichnung nicht zu der Annahme, dass solche Namen im Sinne der Warenzeichen- und Markenschutzgesetzgebung als frei zu betrachten wären und daher von jedermann benutzt werden dürften.

Verlag: Südwestdeutscher Verlag für Hochschulschriften Aktiengesellschaft & Co. KG
Dudweiler Landstr. 99, 66123 Saarbrücken, Deutschland
Telefon +49 681 37 20 271-1, Telefax +49 681 37 20 271-0
Email: info@svh-verlag.de
Zugl.: Berlin, Humboldt Universität zu Berlin, Dissertation, 2009

Herstellung in Deutschland:
Schaltungsdienst Lange o.H.G., Berlin
Books on Demand GmbH, Norderstedt
Reha GmbH, Saarbrücken
Amazon Distribution GmbH, Leipzig
ISBN: 978-3-8381-1767-6

Imprint (only for USA, GB)
Bibliographic information published by the Deutsche Nationalbibliothek: The Deutsche Nationalbibliothek lists this publication in the Deutsche Nationalbibliografie; detailed bibliographic data are available in the Internet at http://dnb.d-nb.de.

Any brand names and product names mentioned in this book are subject to trademark, brand or patent protection and are trademarks or registered trademarks of their respective holders. The use of brand names, product names, common names, trade names, product descriptions etc. even without a particular marking in this works is in no way to be construed to mean that such names may be regarded as unrestricted in respect of trademark and brand protection legislation and could thus be used by anyone.

Publisher: Südwestdeutscher Verlag für Hochschulschriften Aktiengesellschaft & Co. KG
Dudweiler Landstr. 99, 66123 Saarbrücken, Germany
Phone +49 681 37 20 271-1, Fax +49 681 37 20 271-0
Email: info@svh-verlag.de

Printed in the U.S.A.
Printed in the U.K. by (see last page)
ISBN: 978-3-8381-1767-6

Copyright © 2010 by the author and Südwestdeutscher Verlag für Hochschulschriften Aktiengesellschaft & Co. KG and licensors
All rights reserved. Saarbrücken 2010

To my wife

Your love and support wrote this thesis

Abstract

Structural and dynamic perturbations in DNA upon incorporation of either fluorophore, 2-Aminopurine (2AP) or 2-Hydroxy-7-nitrofluorene (HNF), are characterized by NMR spectroscopy. For this purpose the NMR solution structures of the modified DNA duplexes with the sequence 5'-GCTGCAXACGTCG-3' are solved. For X=2AP (13mer2AP) the partner base in the complementary strand is T, while for X=HNF (13merHNF) an abasic site is introduced to avoid steric strain.

By comparing results on 13mer2AP with the corresponding unmodified DNA duplex (13merRef, X=A), any perturbation can be unambiguously assigned to 2AP incorporation. For the NMR solution structure of 13merRef and 13mer2AP small but significant changes in helical parameters are found throughout the helix. Imino proton exchange measurements reveal an extended, distributed effect of 2AP incorporation on the lifetimes of the central seven base pair. However, the reduced base pair lifetime of 2AP:T cannot fully account for the rapid water exchange observed with saturation transfer experiments in the absence of base catalyst. This indicates enhanced intrinsic catalysis. As a possible catalytic site the T O4 atom opposite 2AP is discussed, which is easily accessible through the major groove and lacks a hydrogen bonding partner within the base pair.

The overall NMR solution structure is found to be B-DNA. However the NOE cross-peaks involving the HNF residue can only be accounted for by two different orientations of the HNF inside the DNA helical stack. Their population ratio is estimated to be 1:1. Dynamical perturbation is indicated by the increased linewidth and strong upfield shift of the T residue to the 5'-side of the abasic site.

Zusammenfassung

Mittels NMR-Spektroskopie werden Störungen in Struktur und Dynamik von DNA untersucht, die durch den Einbau jeweils eines der beiden Fluorophore 2-Aminopurin (2AP) und 2-Hydroxy-7-nitrofluoren (HNF) hervorgerufen werden. Zu diesem Zweck werden die NMR-Strukturen der modifizierten Duplexe mit der Sequenz 5'-GCTGCAXACGTCG-3' berechnet. Im Fall X=2AP (13mer2AP) ist die Partnerbase im Komplementärstrang ein T, während gegenüber X=HNF (13mer-HNF) eine abasische Stelle eingeführt wird.

Durch den Vergleich der Ergebnisse zum 13mer2AP mit denjenigen des entsprechenden unmodifizierten DNA Doppelstranges (13merRef, X=A) konnte jegliche Änderung eindeutig dem Einbau von 2AP zugordnet werden. Für die NMR-Strukturen von 13merRef und 13mer2AP können kleine aber signifikante, über die gesamte Helix verteilte Strukturstörungen nachgewiesen werden. Experimente zum Iminoprotonenaustausch mit Wasser ergeben, daß der Einbau von 2AP die Basenpaarlebensdauern der 7 zentralen Basenpaare erniedrigt. Die kürzere Lebensdauer des 2AP:T Basenpaares kann jedoch nicht den schnellen Wasseraustausch im Sättigungstransfer-Experiment ohne Zugabe von Basenkatalysator erklären. Als Erklärung für diese Diskrepanz wird eine effizientere intrinsische Katalyse vermutet. Als mögliche, katalytisch aktive Stelle wird das T O4 Atom diskutiert, welches über die große Furche leicht zugänglich ist und das keine Wasserstoffbrückenbindung innerhalb des Basenpaares ausbilden kann.

Die übergeordnete Struktur des 13merHNF ist eine B-Form DNA Helix. Die NOE Kreuzpeaks zu den Protonen im HNF können jedoch nur durch zwei verschiedene Orientierungen des HNFs in der helikalen Anordnung beschrieben werden. Das Verhältnis der beiden Orientierungen untereinander wird als 1:1 abgeschätzt. Störungen in der Basenpaardynamik werden durch die höhere Linienbreite und die starke Hochfeldverschiebung des T auf der 5'-Seite ausgehend von der abasischen Stelle angedeutet.

Inhaltsverzeichnis

1	**Introduction**	**1**
2	**Theoretical background**	**7**
2.1	Structural aspects of DNA	7
2.2	Base pair dynamics in DNA	11
	2.2.1 Imino proton exchange theory	11
	2.2.2 Inversion recovery experiments	16
2.3	Solution structure determination of DNA	21
	2.3.1 Nuclear Overhauser Effect spectroscopy	21
	2.3.2 Structural information from Nuclear Overhauser Enhancement effects	25
	2.3.3 Residual Dipolar Couplings	29
	2.3.4 Simulated Annealing calculations	35
3	**Experimental details**	**39**
3.1	2-Aminopurine	39
	3.1.1 Sample preparation	39
	3.1.2 Measurements	41
	3.1.3 Restraint generation	44
	3.1.4 Structure calculation	50
3.2	2-Hydroxy-7-nitrofluorene	53
	3.2.1 Sample preparation	53
	3.2.2 Measurements	54
	3.2.3 Restraint generation	55
	3.2.4 Structure calculation	56

Inhaltsverzeichnis

4 Results and discussion — 57
4.1 2-Aminopurine — 57
4.1.1 Introduction — 57
4.1.2 Results — 62
4.1.2.1 Chemical shift analysis — 62
4.1.2.2 Structural comparison — 65
4.1.2.3 Comparison of base pair dynamics — 71
4.1.2.4 Comparison of duplex melting — 75
4.1.3 Discussion — 76
4.2 2-Hydroxy-7nitrofluorene — 86
4.2.1 Introduction — 86
4.2.2 Results — 88
4.2.2.1 Chemical shift analysis — 88
4.2.2.2 NMR solution structure — 91
4.2.2.3 Duplex melting — 94
4.2.3 Discussion — 95

Summary — 99

Zusammenfassung — 101

Script code — 147
1. Force field parameter and topology files — 147
2. Xplor-NIH calculation input files — 196
3. Xplor-NIH calculation restraints files and structures — 212
4. Lua scripts written for data export from CARA — 212
5. Utility scripts — 253

Chemical shifts — 261
6. Chemical shifts of 13merHNF — 261
7. Chemical shifts of 13merRefGC — 265
8. Chemical shifts of 13merRef — 267
9. Chemical shifts of 13mer2AP — 270
10. Chemical shift differences for 13merRef and 13mer2AP — 273

Helical parameter   **275**

    11   Helical parameters for 13merRef 275

    12   Helical parameters for 13mer2AP 278

List of abbreviations   **281**

Index of figures   **283**

Index of tables   **285**

Acknowledgement

Mein besonderer Dank gilt Herrn Professor Doktor Nikolaus P. Ernsting. Über die normale Betreuung bei einer Dissertation hinaus half er mir jederzeit engagiert bei der Lösung nicht nur fachlicher Probleme. In unseren fachlichen Diskussionen hat er mich immer wieder zur Äußerung eigener Ideen ermutigt und dabei großes Vertrauen in mein Können gesetzt. Das wissenschaftliche Arbeiten war so nicht nur ein Pflichtprogramm, sondern gleichzeitig ein Hobby. Durch die Freiheit in Bezug auf Arbeits- und Zeiteinteilung, die er mir gewährte, trug er in hohem Maße dazu bei, daß ich Arbeit und Familie vereibaren konnte. Dafür bin ich Herrn Professor Ernsting außerordentlich dankbar.

Herrn Professor Doktor Christian Griesinger danke ich für seine großzügige Unterstützung und seine vielen hilfreichen Anregungen, die mir wichtige Impulse für meine Arbeit lieferten.

Herr Professor Doktor Clemens Mügge hat mir viel am und um das NMR-Spektrometer und mit vielen hilfreichen Hinweisen und Diskussionen geholfen. Ohne seinen Einsatz und seine Flexibilität bei der Meßzeiteinteilung, hätte ich meine Arbeit nicht in so kurzer Zeit durchführen können.

Mein Dank gilt ferner Frau Doktor Jennifer Tuma, die mir den Einstieg in die Thematik der NMR-Strukturanalytik sowie die Handhabung eines NMR-Gerätes sehr erleichterte.

Ich möchte mich auch ganz herzlich bei Herrn Diplom-Chemiker Lars Dehmel bedanken, der - trotz eigener Doppelbelastung durch Kind und Diplomarbeit - die Zeit fand, mit mir mehrere Tage hintereinander im Institut zu "campieren", um die NMR-Titration zu optimieren. Desweiteren machte er sich um die Auswertung dieser Messungen durch das Schreiben mehrerer Skripte sehr verdient.

Herrn Doktor Horst Hennig danke ich für die große Hilfe in vielen formalen Angelegenheiten und darüber hinaus. Frau Iris Suter bzw. Frau Sabrina Penn und Frau Heiderose Steingräber möchte ich herzlich danken, da sie stets sehr hilfsbereit waren und ein offenes Ohr für die kleinen Probleme des (wissenschaftlichen) Alltags hatten.

Inhaltsverzeichnis

Mein Dank gilt auch den restlichen Mitgliedern der Arbeitsgruppe Ernsting – Herrn Doktor Sergej Kovalenko, Herrn Doktor Alexander Dobryakov, Herrn Doktor Luis P. Lustres, Herrn Diplom-Chemiker Alexander Weigel – für die Unterstützung und die vielen Hilfestellungen beim Arbeiten.

Bei Herrn Diplom-Chemiker Patrick Wilke möchte ich mich für die Unterstützung im Rahmen seines Forschungspraktikums und darüber hinaus bedanken.

Meinen Eltern und meinen Schwiegereltern danke ich für die große Hilfe bei der Betreuung meiner Kinder, durch die ich so intensiv an dieser Dissertation arbeiten konnte. Ihre Anteilnahme und ihre (wissenschaftlichen) Ratschläge haben mich ermutigt und motiviert. Auch für die finanzielle Unterstützung bin ich ihnen sehr dankbar.

Amélie und Tristan - ihr seid meine größten Schätze. Ich liebe Euch.

1 Introduction

In 1953 Watson and Crick (WC) solved the puzzle of the deoxyribonucleic acid (DNA) structure [Watson and Crick, 1953]. Aided by the results of Chargaff [1950] and Franklin and Gosling [1953] they proposed the double helical structure. Since then the double helical arrangement of DNA has become common knowledge. But although the overall conformation had been deduced, questions remained. The idealized WC-model could not explain certain sequence-specific effects; an atomic picture was lacking.

In 1980 Dickerson and coworkers were the first to solve the crystal structure of a DNA duplex [Wing et al., 1980]. They refined the WC-model and showed that the ideal helical parameters (as theoretically predicted by Watson and Crick) are true on average but can deviate substantially dependent on sequence [Dickerson and Drew, 1981b,a, Drew et al., 1981]. Subsequently, more crystal structures of DNA sequences were solved, revealing large variations in helical parameters for B-DNA [Dickerson et al., 1982, Kopka et al., 1983, Heinemann and Alings, 1989] and giving structural insights into other conformations like A-DNA [Shakked et al., 1981, Conner et al., 1982, 1984] or Z-DNA [Drew et al., 1980, Wang et al., 1981, Drew and Dickerson, 1981, Rich et al., 1984]. A detailed atomic picture of DNA helical structure was now provided.

However it was found that crystallization can have profound effects on the conformation of DNA. In the crystalline state DNA strongly favours the A-form double helix, whereas DNA in solution occurs predominantly in the B-DNA form [Bloomfield et al., 2000]. Furthermore, the helical parameters strongly depend on the crystallization conditions [Jain and Sundaralingam, 1989, Shakked et al., 1989, Johansson et al., 2000]. Thus single crystal X-ray crystallography is of limited use when studying biological problems,

1 Introduction

particularly those involving DNA.

Parallel to the breakthrough of Dickerson and coworkers [Wing et al., 1980] the advent of 2-dimensional techniques [Jeener et al., 1979, Kumar et al., 1980a, Macura and Ernst, 1980] extended Nuclear Magnetic Resonance Spectroscopy (NMR) methods in a way that studying large biological molecules was now feasible. First efforts centered on protein structure determination [Wagner et al., 1981, Wuthrich et al., 1982, Zuiderweg et al., 1983], but eventually application to DNA structure determination in solution followed [Hare et al., 1983, Feigon et al., 1984, Clore and Gronenborn, 1984, 1985, Hosur et al., 1986]. These early studies focused on the sequential assignment of DNA or protein resonances, the structural content however was discussed only qualitatively. With the advent of powerful computers, techniques were developed that allowed for determination of 3-dimensional structures of biomolecules with NMR structural information as restraints [Williamson et al., 1985, Zuiderweg et al., 1985].

The development and subsequent refinement of the solid-phase phosphor-amidite approach for oligonucleotide synthesis [Sinha et al., 1984, Dahl et al., 1987, Schulhof et al., 1987, Caruthers et al., 1987] allowed for relatively cheap and easy access to large quantities of nucleic acids with a defined primary sequence. This marked a breakthrough for nucleic acids research since it was now possible to vary specific base positions in a predefined sequence. Thereby studying the effect of base mismatches on the helical arrangement of the duplex was facilitated [Kalnik et al., 1988, Roongta et al., 1990, Moe and Russu, 1992]. Furthermore, a means for introduction of arbitrary artificial nucleotides at any position in the duplex was provided by automated solid-phase synthesis.

Artificial DNA double strand structures have been investigated since the late 1980s [Li et al., 1987, Evans and Levine, 1988]. Several different types of modifications have to be distinguished. These involve: backbone modification [Pieters et al., 1989, Betts et al., 1995, Nielsen et al., 2009], fluorophores covalently linked to natural bases [Krugh et al., 1989, Schwartz et al., 1997, Subramaniam et al., 2001], fluorophores substituting a natural base [Nordlund et al., 1989, Guckian et al., 1998, Engman et al., 2004] or even a base pair [Matray and Kool, 1998, Guckian et al., 2000, Smirnov et al., 2002], and intercalators, which bind to DNA through stacking and/or electrostatic interactions [Fede et al., 1993, Spielmann et al., 1995, Davies et al., 1997]. All of the latter studies focus on the introduction of chromophores, since their fluorescent properties can be exploited for studying DNA [Wojczewski et al., 1999].

The spectroscopic properties of covalently attached DNA modifications are utilized in numerous ways. Over the past decades different strategies have been developed for detecting single nucleotide polymorphisms (SNPs). These strategies employ fluorophore-quencher systems (molecular beacons) [Tan et al., 2004], DNA-mediated electron transfer (DETEQ) [Wagenknecht, 2008] or forced intercalation probes (FIT) [Koehler et al., 2005]. Furthermore, fluorescent molecules are introduced at different locations into DNA in order to get long-range structural information by exploiting fluorescence resonance energy transfer (FRET) [Lilley and Wilson, 2000]. In addition it has been demostrated that transient absorption spectroscopy of fluorophore-modified DNA can be used to follow dynamics on the pico- to femtosecond timescale [Zewail, 2000].

Supramolecular vibrational modes of biological molecules are important for their function, and many have frequencies below $200\,\mathrm{cm}^{-1}$ or 6 THz. Examples are the primary event of vision ($60\,\mathrm{cm}^{-1}$) [Wang et al., 1994], oxygen acceptance of hemoglobin ($39\,\mathrm{cm}^{-1}$) [Klug et al., 2002], chemical reactions in myoglobin ($51\,\mathrm{cm}^{-1}$) [Austin et al., 1989], and conformational change of bacteriorhodopsin ($115\,\mathrm{cm}^{-1}$) [Xie et al., 2001]. For DNA transcription the double helix must be opened to expose the coding bases to chemical reactions. Thermal melting of double-stranded oligonucleotides is similar because it starts with a "denaturation bubble" [Prohofsky et al., 1979]. The latter is reached through collective modes between 60 and $140\,\mathrm{cm}^{-1}$ which compress and stretch the interbase H-bonds [Cocco and Monasson, 2000]. However, such resonances in the low-frequency region are difficult to detect due to mixing of the DNA modes with those of hydration water. Resolving such collective vibrational modes of a biological molecule by molecular THz spectroscopy is a new but potent application for chromophores in DNA.

Here the chromophore functions like a THz light source when its charge distribution is suddenly altered by femtosecond optical excitation $S_1 \leftarrow S_0$. The electric field around

1 Introduction

the probe is changed instantly and acts on nearby groups with partial charges. Most of these change their nuclear position in an overdamped fashion but some may oscillate briefly. Altogether a reaction field R(t) is created which is reported by the polar probe molecule, through an emission frequency which depends on R(t). The probe molecule is therefore not only light source but also detector. A response function can be obtained which is related to the local THz absorption spectrum. In this way the low-frequency vibrational structure of biomolecules can be accessed.

Molecular THz spectroscopy should reduce inhomogeneous broadening because the perturbing electric field and the reaction field are local. Only those modes will interact which have oscillator strength in the region, at the right direction. The obvious disadvantage is the need to embed a probe molecule inside double-stranded DNA as an artificial nucleobase. The probe has to be free of internal modes which are active below $\approx 300\,\mathrm{cm}^{-1}$, since they would mix with the macromolecular dynamics to be reported. For this reason the best-studied polarity probes, coumarins [Horng et al., 1995, Zhao et al., 2005], are not eligible. Instead one must use chromophors which have been shown to report the far infrared spectrum of pure liquids such as acetonitrile [Ruthmann et al., 1998, Karunakaran et al., 2008] or water [Lustres et al., 2005]. Required is bio-organic development of suited chromophores guided by optical femtosecond spectroscopy in the condensed phase. However, excellent suitability from an optical point of view is to no avail when the helical structure is severely disrupted, since duplex features are to be probed.

Thus, for all aforementioned spectroscopic techniques which investigate DNA features, it is advantageous or even imperative to know the exact orientation of the fluorophore inside the DNA double helix. Moreover it is instructive to have information on the stacking interactions of the fluorophore with the adjacent base pairs. Finally, it can be decisive - especially for biologically motivated hybridization studies - to be able to characterize the structural and dynamic perturbation of the DNA helical structure in terms of helical parameters and base pair lifetimes.

In order to solve above question, this work utilizes NMR spectroscopy for the structure

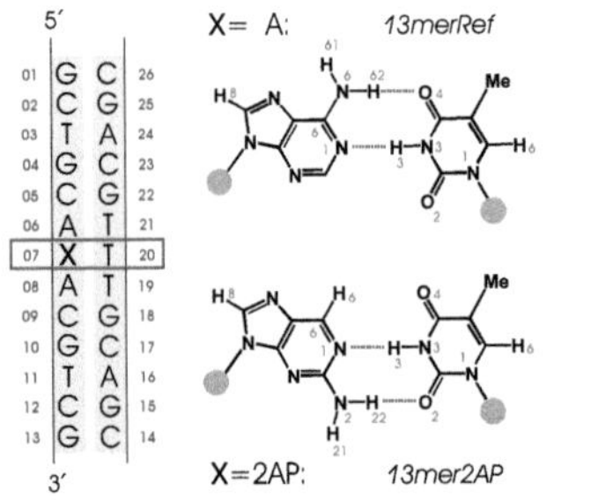

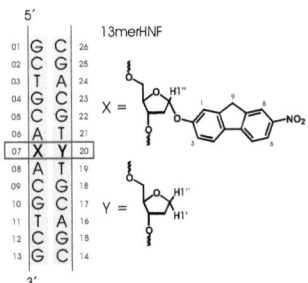

Fig. 1.1: DNA duplex sequence with chemical structure of the 2AP-T and A-T base pairs. A symmetric, nonpalindromic 13 base pair sequence was chosen to dismiss the possibility of mispairing, loop formation or fraying effects and to have a single perturbation site.

Fig. 1.2: DNA duplex sequence with chemical structure of the HNF fluorophore. Opposite to the HNF residue has been placed an abasic site (**Y**) in order to minimize steric strain on the HNF residue.

determination of the DNA sequences given in Fig. 1.1 and 1.2 in order to solve above question. In both sequences the central base or base pair is modified in order to have the single modification site and its adjacent base pairs unperturbed by fraying effects at the helix termini [Nonin et al., 1995]. Two different modifications are examined, 2-Aminopurine (2AP) and 2-Hydroxy-7-nitrofluorene (HNF).

Duplex DNA with 2-Hydroxy-7-nitrofluorene

The probe HNF is incorporated opposite to an abasic site (1'-2'-didesoxyribose) to avoid steric strain which might otherwise disrupt the overall B-DNA conformation or force the fluorophore into an extrahelical position. However, deletion of the partner base introduces increased flexibility into the DNA duplex at the modification site [Lin and de los Santos, 2001, Smirnov et al., 2002]. In conjunction with the different electronic properties of the HNF residue as compared to a natural base pair, HNF is expected to introduce large local perturbations compared to a more native modification. A first purpose of this work is to find out whether the DNA duplex adopts an overall B-DNA conformation

1 Introduction

with the HNF stacked inside the double helix. This information is essential since HNF has been designed to report on macromolecular vibrational modes in DNA via transient absorption spectroscopy, which cannot be observed in case of an extrahelical orientation of HNF.

Duplex DNA with 2-Aminopurine

The 2AP-containing duplex represents a substantially different case. 2AP causes only a slight perturbation of the directly adjacent base pairs as suggested in earlier works by Lycksell et al. [1987] and Nordlund et al. [1989]. 2AP is commonly used to monitor base stacking-unstacking events in biologically relevant sequences [Allan and Reich, 1996, Reddy and Rao, 2000, Bernards et al., 2002, Daujotyte et al., 2004, Neely et al., 2005, Lenz et al., 2007]. Therefore it is crucial to characterize the perturbation induced upon 2AP incorporation structurally (in terms of helical parameters) and dynamically (in terms of reliable base pair lifetimes). This is the second aim of this work.

Since 2AP is structurally isomeric to adenine (A), it closely resembles the latter in size and shape. It has been found that it can also form stable base pairs with thymine (T) [Ronen, 1979, Sowers et al., 1986]. The two duplex structures of 13merRef with X=A and 13mer2AP with X=2AP differ only in the location of the amino group of the central residue (see Fig. 1.1). Thus any structural or dynamic differences that are observed between the two corresponding solution structures can be directly attributed to the incorporation of 2AP. This is the major difference compared to pertinent works; these investigate a ten base pair palindromic DNA duplex with two 2AP residues incorporated [Lycksell et al., 1987, Nordlund et al., 1989].

2 Theoretical background

2.1 Structural aspects of DNA

DNA is composed of four naturally occuring nucleobases, adenine (A), guanine (G), thymine (T) and cytosine (C). While A and G are derived from purine, T and C are pyrimidine derivatives. Base pairs are formed between A:T and G:C. The structure and nomenclature of these two base pair motifs is depicted in Fig. 2.1. The higher thermodynamic stability of G:C compared to A:T base pairs [Xia et al., 1998] originates from the fact that the latter base pair forms only two instead of three hydrogen bonds (Fig. 2.1).

By attaching A, G, C, or T to the C1'-atom of a 2''-deoxy-β-D-ribose the nucleosides adenosine, guanosine, cytidine and thymidine are formed, respectively. The structure and nomenclature of 2''-deoxy-β-D-ribose, which is called sugar in the following, is shown in Fig. 2.2. All atoms of the sugar are marked with a " ' " to distinguish them from nucleobase atoms. The sugar conformation is defined by five dihedral angles $\nu_0 - \nu_4$ (Fig. 2.3). The latter are interdependent and thus can be described by only two parameters

$$\nu_j = \Psi_m * \cos(P + 144(j-2)) \quad j = 0, 1, 2, 3, 4 \qquad (2.1)$$

the pucker angle P and the pucker amplitude Ψ_m. Note that this simple relation is valid only for cyclopentane, but deviations can be accounted for by introducing correction terms [Altona and Sundaralingam, 1972].

The glycosidic torsion angle χ and the backbone dihedral angles $\alpha - \zeta$ can be used to characterize the helical structure of DNA (Fig. 2.3). While $\alpha - \zeta$ define the sugar-

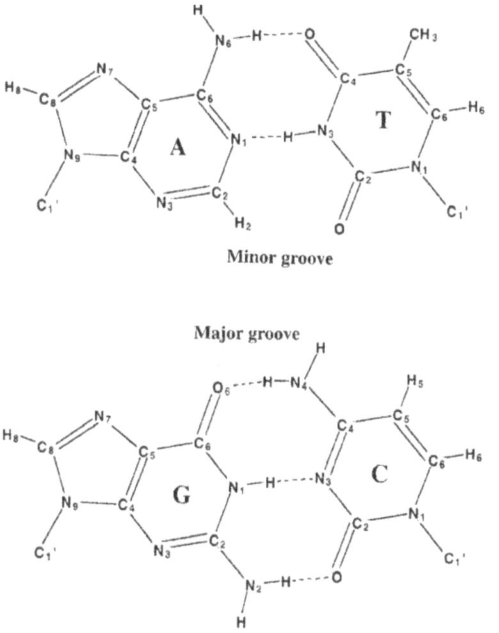

2 Theoretical background

Fig. 2.1: *Structure and nomenclature of the Watson-Crick base pairs A:T and G:C. The picture is taken from Bloomfield et al. [2000].*

phosphate backbone of DNA, χ determines the position of the nucleobase relative to the sugar. Two different orientations are sampled, the anti range which centers around $\chi = -135°$ and the syn range which samples values around $\chi = +45°$ [Bloomfield et al., 2000]. The latter is less stable as the nucleobase is located above the sugar, which leads to steric clashes. Thus the syn-orientation of nucleobases is only found for a special helical arrangement of DNA, the left-handed Z-DNA, which needs external stabilization, e.g. by high salt concentration [Rich et al., 1984].

Two main helical arrangements are found for DNA, the A-form and the B-form. While the former is more often found in crystal structures, B-DNA is the dominant conforma-

2.1 Structural aspects of DNA

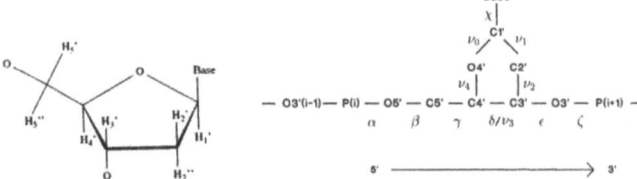

Fig. 2.2: Nomenclature and structure of 2''-deoxy-β-D-ribose [Roberts, 1993].

Fig. 2.3: Dihedral angles in the sugar-phosphate backbone of DNA. The picture is taken from Roberts [1993].

	A-DNA	B-DNA
Helix handedness	Right	Right
bp/repeating unit	1	1
bp/turn	11	10
Helix twist, (°)	32.7	36.0
Rise/bp, (Å)	2.9	3.4
Helix pitch, (Å)	32	34
Base pair inclination, (°)	12	2.4
P distance from helix axis, (Å)	9.5	9.4
X displacement from bp to helix axis, Å	−4.1	0.8
Glycosidic bond orientiation	anti	anti
Sugar conformation	C3'-endo	C2'-endo
Major groove depth	13.5	8.5
width, (Å)	2.7	11.7
Minor groove depth	2.8	7.5
width, (Å)	11.0	5.7

(a) Important parameters

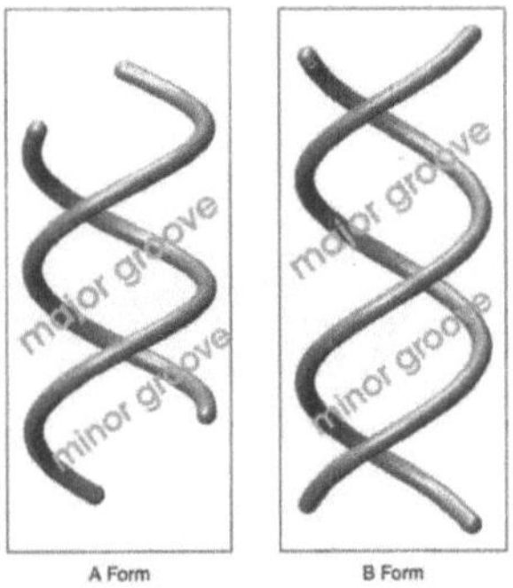

(b) Helical backbone

Fig. 2.4: Comparison of A- und B-form DNA. The pictures are taken from Bloomfield et al. [2000].

tion in solution. The most important parameters for both helical arrangements are compiled in Fig. 2.4. While the overall arrangement is similar (right-handedness, glycosidic bond orientation), differences exist. One that is commonly used to distinguish between the two helical arrangements is the sugar conformation, with C3'-endo (P around 18°) dominant for A-DNA and C2'-endo (P around 162°) for B-DNA. The widths of the minor and major grooves, which are depicted on the molecular and atomic level in Fig. 2.1 and Fig. 2.4 respectively, also differ significantly. The larger helical rise and twist values for B-DNA lead to an elongated shape of the latter, while A-DNA is much more

2 Theoretical background

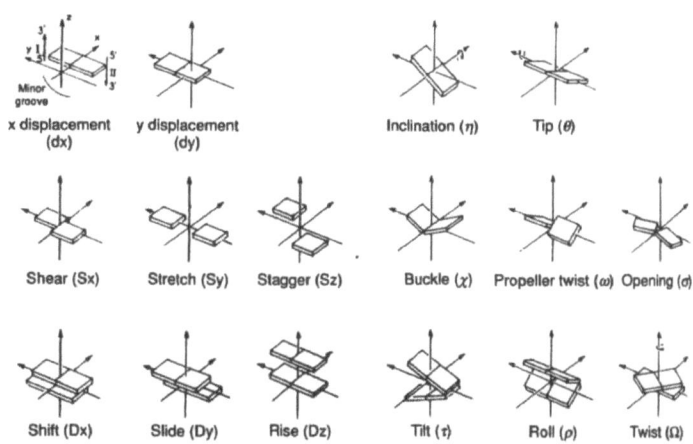

Fig. 2.5: *Helical parameters that describe the orientation of base pairs relative to the to the molecular frame (upper row), base pair partners (middle row) and conscutive base pairs (lower row) relative to each other. The picture is taken from Roberts [1993].*

compressed (Fig. 2.4).

The helical parameters, which are visualized in Fig. 2.5, describe the arrangement of oligonucleotides in a double helix. They can be divided into three subgroups. Parameters in the upper row yield information on the orientation of base pairs relative to the molecular frame. The middle row parameters report on the orientation of the two base pair partners relative to each other. The lower row is most often used to characterize the helical arrangement, since these parameters give information about the orientation of two consecutive base pairs relative to each other. Especially the rise and twist-values determine the overall shape of the double helix (Fig. 2.4).

2.2 Base pair dynamics in DNA

2.2.1 Imino proton exchange theory

The imino protons of G and T are located near the helical axis at the center of each base pair and thus are effectively shielded against solvent or catalyst attack. Consequently, the central assumption in imino proton exchange theory of DNA is that exchange with bulk water can only proceed via a transient opening of the base pair [Kochoyan et al., 1987, 1988, Leijon and Graslund, 1992]. For Watson-Crick duplexes, base pair lifetimes do not depend on the nature of the adjacent pairs. This suggests that opening involves single base pairs only; the possibility of collective opening motions is ruled out [Leroy et al., 1985, Gueron et al., 1987].

Fig. 2.6 depicts a kinetic scheme of the processes involved in imino proton exchange theory [Leijon and Graslund, 1992]. Exchange from the closed state is not possible [Nonin et al., 1995], thus the first step must be the opening of the base pair with rate constants k_{op} and k_{cl} for opening and closing, respectively. From the open state, where the imino proton is assumed to be fully accessible [Kochoyan et al., 1988], two different processes can occur: exchange via an external base catalyst (rate constant k_{cat}^{ext}) and via an intrinsic pathway (k_{cat}^{int}). As a possible intrinsic catalyst the endocyclic nitrogen of the complementary base has been proposed [Leroy et al., 1985, Kochoyan et al., 1988]. Frequently used external base catalysts are Trishydroxymethylaminomethane (TRIS) or ammonia [Kochoyan et al., 1988, Moe and Russu, 1990, Bhattacharya et al., 2002].

Under conditions of a stable structure ($k_{op} \ll k_{cl}$), which can be safely assumed for duplex lengths > 10 base pairs, the concentration of base pairs in the open state $[NH^*]$ is quasistationary, i.e. the kinetics are pseudo-first-order.

$$\frac{d[NH^*]}{dt} = 0 = k_{op}[NH^* \cdots N] - (k_{cl} + k_{cat}^{int} + k_{cat}^{ext})[NH^*] \qquad (2.2a)$$

$$(k_{cl} + k_{cat}^{int} + k_{cat}^{ext})[NH^*] = k_{op}[NH^* \cdots N] \qquad (2.2b)$$

$$[NH^*] = \frac{k_{op}}{(k_{cl} + k_{cat}^{int} + k_{cat}^{ext})}[NH^* \cdots N] \qquad (2.2c)$$

2 Theoretical background

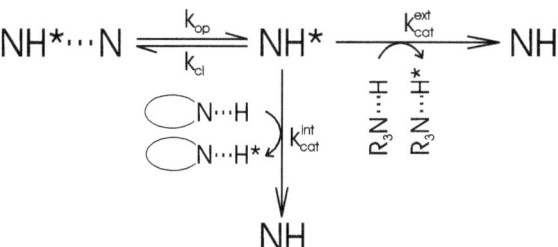

Fig. 2.6: *Kinetic scheme of imino proton exchange*

The effective imino proton exchange rate k_{ex} is defined by

$$k_{ex}[NH^* \cdots N] = (k_{cat}^{int} + k_{cat}^{ext})[NH^*] \tag{2.3}$$

where $[NH^* \cdots N]$ is the concentration of the base pair in the closed state. Substituting eq. (2.2c) into eq. (2.3) and cancelling of $[NH^* \cdots N]$ gives

$$k_{ex} = k_{op} \frac{k_{cat}^{int} + k_{cat}^{ext}}{k_{cl} + k_{cat}^{int} + k_{cat}^{ext}} \tag{2.4a}$$

$$\frac{1}{k_{ex}} = \frac{1}{k_{op}} \left(1 + \frac{k_{cl}}{k_{cat}^{int} + k_{cat}^{ext}}\right) \tag{2.4b}$$

which can be reduced to

$$\frac{1}{k_{ex}} = \frac{1}{k_{op}} + \frac{1}{K_{diss}(k_{cat}^{int} + k_{cat}^{ext})} \tag{2.5}$$

by substituting $K_{diss} = \frac{k_{op}}{k_{cl}}$. Introducing $\tau_{ex} = \frac{1}{k_{ex}}$, the imino proton exchange time, and $\tau_{op} = \frac{1}{k_{op}}$, the base pair lifetime, gives

$$\tau_{ex} = \tau_{op} + \frac{1}{k_{cat}^{int} K_{diss} + k_{cat}^{ext} K_{diss}} \tag{2.6}$$

The transfer rate of the imino proton to an external catalyst in the isolated mono-

2.2 Base pair dynamics in DNA

nucleoside (k_i) should be comparable to k_{cat}^{ext}. The former is given by

$$k_i = \frac{k_{coll}\,[B]}{1 + 10^{pK_a^n - pK_a^c}} = k_{iso}\,[B] \tag{2.7}$$

where k_{coll} is the collision rate constant and pK_a^n and pK_a^c are the pK_a-values for the nucleotide and the catalyst respectively [Eigen, 1964]. Since the latter are all constant, they can be substituted by introducing k_{iso}, the constant of proportionality between k_i and the concentration of the external base catalyst [B]. Thus k_{cat}^{ext} can be expressed as

$$k_{cat}^{ext} = \alpha\, k_{iso}\,[B] \tag{2.8}$$

where restricted accessibility of the imino proton in the open state as compared to the isolated nucleoside is taken into account by the parameter α, which ranges from 0 (not accessible) to 1 (unrestricted accessibility). For natural base pairs and all the commonly used base catalysts α was found to be approximately 1 [Kochoyan et al., 1988]. Consequently, differences in accessibility might become important for modified nucleotides, but can be safely neglected for natural ones. Substituting eq. (2.8) into eq. (2.6) gives

$$\tau_{ex} = \tau_{op} + \frac{1}{k_{cat}^{int} K_{diss} + \alpha\, k_{iso}\, K_{diss}\,[B]} \tag{2.9}$$

which is an accurate description when $k_{cat}^{int} \approx k_{cat}^{ext}$. At high external base catalyst concentrations, k_{cat}^{ext} dominates k_{cat}^{int} and eq. (2.9) simplifies to

$$\tau_{ex} = \tau_{op} + \frac{1}{(\alpha\, k_{iso}\, K_{diss})}\frac{1}{[B]} \tag{2.10}$$

and the imino proton exchange time becomes a linear function of the inverse of the external catalyst concentration. In the other extreme case, when no external catalyst is available, $k_{cat}^{int} \gg k_{cat}^{ext}$ and consequently eq. (2.9) reduces to

$$\tau_{ex} = \tau_{op} + \frac{1}{k_{cat}^{int} K_{diss}} \tag{2.11}$$

2 Theoretical background

which means that changes in the imino proton exchange can be either due to different base pair lifetimes (which can be determined via eq. (2.10)) or altered intrinsic exchange.

Early works on imino proton exchange considered the opening of the base pair to be rate-limiting, i.e. $k_{cl} \ll (k_{cat}^{int} + k_{cat}^{ext})$ [Teitelbaum and Englander, 1975b,a, Patel and Hilbers, 1975, Hilbers and Patel, 1975]. In that case, eq. (2.4a) simplifies to

$$k_{op} \frac{k_{cat}^{int} + k_{cat}^{ext}}{k_{cat}^{int} + k_{cat}^{ext}} = k_{op} = k_{ex} \tag{2.12}$$

which implies that imino proton exchange with water directly measures the lifetime of the closed base pair (τ_{op}) since the latter is no longer a function of [B]. Exchange occurs every time the base pair opens, thus τ_{ex} and consequently τ_{op} are maximal. As a result, base pair lifetimes published before 1985 were overestimated by approximately one order of magnitude. In that year Gueron and coworkers showed that τ_{ex} depends on the external catalyst concentration [Leroy et al., 1985, Gueron et al., 1987], demonstrating that imino proton exchange in polynucleotides is not opening-limited. Instead, eq. (2.10) is validated by their results. By introducing the apparent dissociation constant αK_d (eq. (2.13a)) [Kochoyan et al., 1988] one obtains the commonly used expression for τ_{ex} (eq. (2.13b)):

$$\alpha K_d = \alpha k_{iso} K_{diss} \tag{2.13a}$$

$$\tau_{ex} = \tau_{op} + \frac{1}{(\alpha K_d)} \frac{1}{[B]} \tag{2.13b}$$

A plot of $[B]^{-1}$ vs τ_{ex} allows for determination of the apparent dissociation constant from the slope of the linear fit. However, interpretation of αK_d is not straightforward and hampered by the approximations detailed above. Much more informative of base pair dynamics is τ_{op}, which can be determined from the intercept with the ordinate (in the limit of infinite [B]) of the plot $[B]^{-1}$ vs τ_{ex}.

Based on the reaction $[BH^+] + [H_2O] \rightleftharpoons [B] + [H_3O^+]$ and the definition of the acidity constant (K_s)

$$K_s = K_{eq}[H_2O] = \frac{[B][H_3O^+]}{[BH^+]} \tag{2.14}$$

2.2 Base pair dynamics in DNA

an expression is obtained, which relates the concentration of free external base catalyst $[B]$ and the total added base concentration $[B_0]$.

$$\frac{1}{[B]} = (1 + 10^{(pK_a^c - pH)}) \frac{1}{[B_0]} \qquad (2.15)$$

It follows that $[B]$ is a function of $[B_O]$ and the pH-value.

The imino proton exchange time τ_{ex} can be determined by measuring the line broadening of the imino proton resonance due to addition of base catalyst [Lycksell et al., 1987].

$$\frac{1}{\tau_{ex}} = \pi \Delta \qquad (2.16)$$

Line broadening provides an easy and time-efficient way to experimentally determine τ_{ex}. However, since line broadening can also have other sources apart from the exchange process (e.g. different shim settings for each titration point) results obtained with this method tend to be inaccurate.

An alternative way is to calculate τ_{ex} from the difference of the spin-lattice relaxation time with (T_1^{ext}) and without external base catalyst (T_1^{int}) [Bhattacharya et al., 2002].

$$\frac{1}{\tau_{ex}} = \frac{1}{T_1^{ext}} - \frac{1}{T_1^{int}} \qquad (2.17)$$

T_1^{ext} and T_1^{int} can be measured by NMR spectroscopy with high precision - errors below 2 % can be achieved [Sass and Ziessow, 1977] - and thus allow for more precise values of τ_{ex} to be determined.

2.2.2 Inversion recovery experiments

Base pair dynamics of DNA can be followed by NMR spectroscopy. As detailed above (section 2.2.1), the key observable is chemical exchange of imino protons with bulk water as a function of external catalyst concentration. According to eq. (2.17) τ_{ex} can be measured by determining T_1^{ext} and T_1^{int}. However, a difference in magnetization between the imino proton and water protons has to be created in order to make the exchange observable by NMR.

This difference can be created by employing two alternative experimental approaches, each of which targets either the water or the imino proton signals. Consequently, four different approaches can be distinguished in the literature: *selective saturation of the water signal* [Leroy et al., 1985, Nonin et al., 1995], *selective saturation of the imino proton region* [Lycksell et al., 1987, Kochoyan et al., 1987, 1988, Leroy et al., 1988, Moe and Russu, 1990, 1992, Leijon and Graslund, 1992, Leroy et al., 1993, Folta-Stogniew and Russu, 1994, Moe et al., 1995, Folta-Stogniew and Russu, 1996, Leijon, 1996], *selective inversion of the water signal* [Snoussi and Leroy, 2001, Mihailescu and Russu, 2001, Chen and Russu, 2004, Coman and Russu, 2005] and *selective inversion of the imino proton region* [Dornberger et al., 1999, Bhattacharya et al., 2002].

Although the majority of the former studies used the saturation recovery method, more recent works rely on inversion recovery experiments. Saturation recovery might be preferable for measurements of long T_1-values ($> 5\,\mathrm{s}$) [Levy and Peat, 1975] or when the choice of optimal delay times is difficult due to a large range of T_1-values of interest [Roscher et al., 1996]. But these conditions are clearly not fulfilled for T_1-values of imino protons which are on a ms-timescale [Moe and Russu, 1990]. Becker et al. [1980] compared the efficiency of inversion and saturation recovery in determining T_1-values with a certain precision. They report inversion recovery to be much more efficient, with saturation recovery requiring 8-times more scans to achieve comparable signal-to-noise ratio [Weiss et al., 1980, Becker et al., 1980]. Thus inversion recovery should generally be preferred over saturation recovery experiments.

As to the question whether to invert the water or the imino proton region, no com-

2.2 Base pair dynamics in DNA

prehensive study has yet been made. Regarding water inversion, Mihailescu and Russu [2001] point out that in order to obtain values for τ_{ex}, the magnetization of the imino protons as a function of the exchange delay τ has to be fitted to the equation

$$M(\tau) = M(0) - \frac{[M(0) - M^0]}{(T_1 + k_{ex})^{-1}} - \frac{k_{ex} M_0 T_1^{water} \tau}{(q-1)^{-1}} \quad (2.18)$$

where the values of T_1^{water} and q (efficiency of inversion for the water signal) have to be determined in a separate experiment. Furthermore, the value of $(T_1 + k_{ex})$ is determined for each imino proton resonance of interest by selective saturation [Mihailescu and Russu, 2001]. These additional experiments are time-consuming and thus inversion of the water signal is not to be preferred. In consequence, the method of choice for measuring base pair lifetimes is selective inversion of the imino proton region.

In the standard inversion recovery experiment the z-magnetization of one or multiple spins is inverted by an initial 180°-pulse ($+M_z \rightarrow -M_z$). After a variable delay time (τ) a 90°-pulse is used to detect the recovered magnetization (Fig. 2.7). The plot of signal intensity vs delay time is fitted exponentially with three instead of two parameters (A,B and T_1) to account for imperfect inversion of the signal

$$I(\tau) = A + B\, exp(-\tau/T_1) \quad (2.19)$$

[Sass and Ziessow, 1977, Kowalewski et al., 1977]. To create differences in magnetization between water and the imino protons, the initial 180°-pulse has to be applied selectively to the imino proton region.

Selective inversion can be achieved in numerous ways. In principle, the commonly used constant-amplitude (rectangular), unselective (hard) pulse can be applied selectively by varying the pulse length and adjusting the pulse power. Rectangular pulses have the advantage of relatively straightforward implementation and optimization due to their analytical solution to the Bloch equations [Hajduk et al., 1993]. However, the excitation profile, which can be calculated by the latter equations, is imperfect. While there is a central lobe of strong excitation, the sharp leading and trailing edges of the pulse give rise

2 Theoretical background

to a set of sidelobes (cf. Fig. 2.8), whose amplitude decreases with offset from the centre frequency [Freeman, 1992]. The sidelobes lead to considerable off-resonance excitation. Similar reasoning can be applied to inversion. In order to reduce off-resonance excitation, the transition region at the edges of the pulse must be smoothed.

A number of shaped pulses for selective bandwidth inversion were created when - with the advent of the computer - numerical calculation of the Bloch equations became possible. Among the first shapes to be proposed was the Gaussian pulse envelope, which is still popular [Bhattacharya et al., 2002]. The advantage of the latter is that in the linear response region (pulse length $\tau_p \ll T_1, T_2$) the excitation function is another Gaussian (see Fig. 2.8). Thus sidelobes are effectively avoided [Bauer et al., 1984]. However, a Gaussian excitation function is far from the ideal of a rectangle, consequently the completely inverted region is quite small [McDonald and Warren, 1991]. Other pulse shapes like Hermitian [Warren, 1984], Gaussian cascades [Emsley and Bodenhausen, 1990] and quaternion pulses [Emsley and Bodenhausen, 1992] try to extend the region of complete inversion with increasing success.

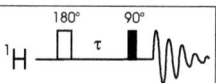

Fig. 2.7: *Scheme of a general inversion recovery pulse sequence*

However, the advantages, which were outlined above are achieved at the cost of longer pulse durations. But when the the relaxation times T_1 and T_2 become comparable to (or even shorter than) the pulse duration, relaxation during the 180°-pulse is no longer negligible. This can have a profound effect on the excitation profile and hamper applicability of shaped pulses for selective excitation [Hajduk et al., 1993]. Furthermore, most shaped pulses were derived on the basis of the Bloch equations neglecting such effects as radiation damping, relaxation and coupling [McDonald and Warren, 1991]. While each of these effects can be accounted for separately [Warren et al., 1989, Hajduk et al., 1993, Ewing et al., 1990], to account for all of them at once is difficult.

The adiabatic sweep is a completely different approach to selective bandwidth inversion. The usually employed pulsed NMR experiment is operated at a static magnetic field B_o and uses a pulsed magnetic field B_1 to simultaneously invert all frequencies of

2.2 Base pair dynamics in DNA

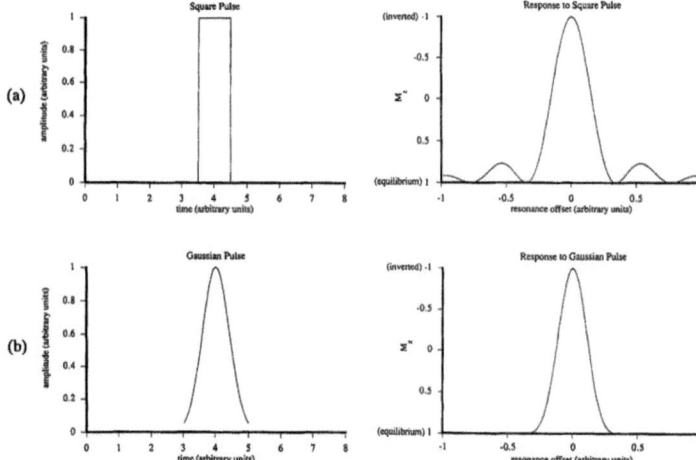

Fig. 2.8: Pulse shapes and excitation forms of a) rectangular and b) Gaussian pulse. This picture is taken from McDonald and Warren [1991].

interest with a constant, bandwidth-centered carrier frequency. In the adiabatic passage experiment, the carrier frequency is modulated over the whole bandwidth and thus the frequencies of interest are inverted successively [Tannús and Garwood, 1997]. An alternative, equivalent approach is to modulate the phase of the center frequency [Garwood and DelaBarre, 2001]. Adiabatic pulses have two main advantages over standard pulses. They are rather insensitive to B_1 field inhomogeneity [Bohlen and Bodenhausen, 1993] and require much less radiofrequency power for inverting nuclear spins over a wide range of chemical shifts [Kupce and Freeman, 1995]. Many different pulse shapes, which enhance one or the other advantage, were created: the hyperbolic secant pulse [Silver et al., 1984, 1985], the chirped pulse [Bohlen and Bodenhausen, 1993, Fu and Bodenhausen, 1995] or the WURST pulse [Kupce and Freeman, 1995, 1997]. Some are now utilized in broadband heteronuclear spin decoupling where radiofrequency power is limited [Kupce and Freeman, 2007] or in *in vivo* NMR where surface coils with a spatially inhomoge-

2 Theoretical background

neous B_1 field are used [Tannús and Garwood, 1997]. Yet there are also disadvantages. One of them is a distinct phase distortion which is introduced for large bandwidths due to the different times at which the frequencies are inverted. This disadvantage can be circumvented, however, by introducing a second adiabatic pulse which cancels the phase distortion introduced by the first [Kupce and Freeman, 1997]. Further disadvantages are the requirement that adiabatic rotations must be accomplished rapidly relative to T_1 and T_2 of the nuclear spins of interest [Garwood and DelaBarre, 2001]. This limits application of adiabatic pulses to spins with quite long relaxation times and those which span a large number of chemical shifts, e.g. heteronuclei like ^{13}C.

2.3 Solution structure determination of DNA

2.3.1 Nuclear Overhauser Effect spectroscopy

NMR spectroscopy has become the method of choice for structure determination of biomolecules in solution. This success is primarily based on the possibility to directly extract distance information utilizing the Nuclear Overhauser Enhancement effect (NOE). Overhauser [1953] was the first to discover that the intensity of the signal of one resonance is changed upon perturbation of another due to cross-relaxation. Initially the NOE was observed by selective saturation of one line and subsequent recording of the 1D spectrum over the whole spectral region of interest. This approach proved to be very useful for the structure elucidation of small molecules [Colson et al., 1967, Woods et al., 1968, Schirmer et al., 1970, Schirmer and Noggle, 1972]. Its applicability to biomolecules however, was limited by the long accumulation time and severe spectral overlap which leads to poor selectivity of saturation [Kumar et al., 1981]. The development of the 2D-NMR spectroscopy finally allowed to aquire the complete set of NOE effects for a macromolecule with a single experiment [Macura and Ernst, 1980]. Furthermore, the 2D Nuclear Overhauser Enhancement Spectroscopy (NOESY)-experiment could be readily repeated in H_2O [Kumar et al., 1980b].

The pulse sequence of a general 2D NOESY experiment is shown in Fig. 2.9. With an initial 90°-pulse transverse magnetization is created. The latter is allowed to precess freely during the evolution time t_1, thereby frequency-labelling the magnetization components. A second 90°-pulse rotates the magnetization onto the negative z-axis.

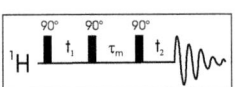

Fig. 2.9: *Scheme of a general NOESY pulse sequence*

During the subsequent mixing period with variable length τ_m, z-magnetization components exchange through dipole-dipole cross-relaxation. A third 90°-pulse again creates transverse magnetization which is finally detected. All three pulses are applied non-selectively [Macura and Ernst, 1980].

In the absence of scalar spin-spin-interactions, cross-relaxation of the longitudinal

2 Theoretical background

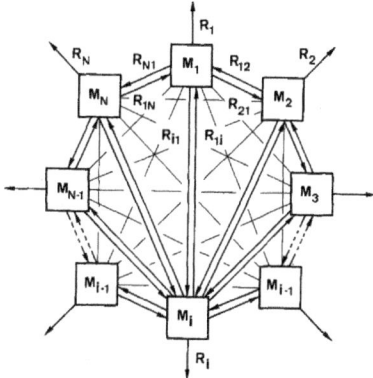

Fig. 2.10: *The cross-relaxation network of N groups of nuclei. The cross relaxation rates R_{ij} lead to a distribution of magnetization within the system, while the leakage relaxation rates R_i lead to a loss of magnetization towards the environment. This picture is taken from Macura and Ernst [1980].*

magnetization components M_{ij} can be described with the equation:

$$\dot{\mathbf{m}} = \mathbf{R} \times \mathbf{m} \tag{2.20}$$

$\mathbf{R}$ is the relaxation matrix comprising the cross relaxation rates R_{ij} and the external relaxation (leakage) rates R_i (Fig. 2.10). The vector $\mathbf{m}$ comprises the deviation of M_{zi} from thermal equilibrium for i spins

$$m_i = M_{zi} - \frac{n_i}{N} M_o \tag{2.21}$$

where M_o is the total equilibrium magnetization of the N nuclei. After the evolution period the initial z-magnetization components are encoded by the precession frequencies

2.3 Solution structure determination of DNA

of the components (ω_i)

$$m_i(0) = M_o \frac{n_i}{N} \left[cos(\omega_i t_1) \, exp\left(-\frac{t_1}{T_{2i}}\right) - 1 \right] \tag{2.22}$$

such that the cross-relaxation pathway can be traced back to its origin. The recovery of the magnetization back to equilibrium during the mixing period can be described by the following solution to eq. (2.20) [Macura and Ernst, 1980]

$$\mathbf{m}(\tau_m) = exp[-\mathbf{R}\,\tau_m]\,\mathbf{m}(0) \tag{2.23}$$

where $\mathbf{m}(\tau_m)$ is the matrix comprising the magnetization components after the mixing period, $\mathbf{m}(0)$ represents the intensities of the diagonal peaks at $\tau_m = 0$. The diagonal (R_{ii}) and off-diagonal (R_{ij}) relaxation matrix elements are given as [Roberts, 1993]:

$$R_{ii} = q_{ij} \sum_{i,j} \left\{ J_{0,ij}(\omega_i - \omega_j) + 3\left[J_{1,ij}(\omega_i) + J_{1,ij}(\omega_j)\right] + \right.$$

$$\left. 6\,J_{2,ij}(\omega_i + \omega_j) + R_{1i} \right\} \tag{2.24a}$$

$$R_{ij} = q_{ij} \left[6\,J_{2,ij}(\omega) - J_{0,ij}(\omega) \right] \tag{2.24b}$$

where R_{1i} is the leakage rate, which can usually be neglected in the absence of paramagnetic nuclei. The factor q_{ij} comprises all constant values

$$q_{ij} = \frac{\gamma_i^2\,\gamma_j^2\,\hbar^2\,\mu_0^2}{160\,\pi^2} \tag{2.25}$$

where γ_i and γ_j are the gyromagnetic ratios for spins i and j respectively, $\hbar$ is the Planck constant (divided by 2π) and μ_0 is the magnetic constant or vacuum permeability. $J_{n,ij}(\omega)$ represents the spectral densities for the zero, single and double quantum transitions (n=0,1,2 resp.)

$$J_{n,ij}(\omega) = \frac{\tau_c^{ij}}{1 + \left(n\,\omega_0\,\tau_c^{ij}\right)^2} \frac{1}{r_{ij}^6} \tag{2.26}$$

2 Theoretical background

where τ_c^{ij} is the rotational correlation time of the vector between spins i and j and r_{ij} is the distance between the latter. Since the differences in resonance frequency for various spins are negligibly small compared to the value of the resonance frequency itself, ω_i and ω_j are approximated by the center frequency ω_0. The factor n in the denominator is given by the number of quanta involved in the transition. With the help of some assumptions the intensities of NOE cross-peaks can be directly related to distances in the molecule under investigation, as is seen in the next section.

2.3.2 Structural information from Nuclear Overhauser Enhancement effects

The NOE effect is an invaluable tool for structure elucidation, since it is correlated with the distance between the interacting nuclei. There are different methods to extract structural information out of NOE cross-peak intensities. A very popular method is to classify peaks according to their peak intensities into strong, medium and weak and setting up loose distance restraints of 1.8-2.8 Å, 1.8-3.3 Å and 1.8-5.0 Å, respectively. This approach has the clear advantage of simplicity and abolishes the need to integrate NOE cross-peaks in the 2D-spectrum. However, intensities are only representative of peak volumes when the lineshape for all peaks is the same. This is true only when the digital resolution is larger than the peakwidth [Roberts, 1993], which is clearly not fulfilled for modern high-resolution spectra of macromolecules. A second disadvantage is that such loose distance restraints do not have much restraining power and consequently the structure is ill-defined. In particular for nucleic acids, the classification on the basis of peak intensities is unsuitable, since the proton density is much less than in proteins (0.35 protons/atom versus 0.52 protons/atom in proteins) and thus yields fewer distance restraints. Additionally, long-range peaks as detected in proteins cannot be observed in DNA due to its rod-like shape. The need for more accurate distance restraints in DNA structure determination by NMR is thus intensified [MacDonald and Lu, 2002].

Very accurate distance restraints can be extracted using the Full Matrix Relaxation Approach. Here distances are calculated directly from the off-diagonal elements of the relaxation matrix, which in turn is obtained via

$$\mathbf{R} = \frac{ln[\mathbf{m}(0)] - ln[\mathbf{m}(\tau_m)]}{\tau_m} \qquad (2.27)$$

[Roberts, 1993]. To obtain the complete intensity matrix **m** is not possible in practice. Thus two different algorithms have been proposed to fill the "gaps" in **m**. Schematic representations for both algorithms are given in Fig. 2.11. Both algorithms rely on a reasonable starting structure from which an intensity matrix is calculated. The latter is then combined with the intensity matrix derived from the experimental NOEs and

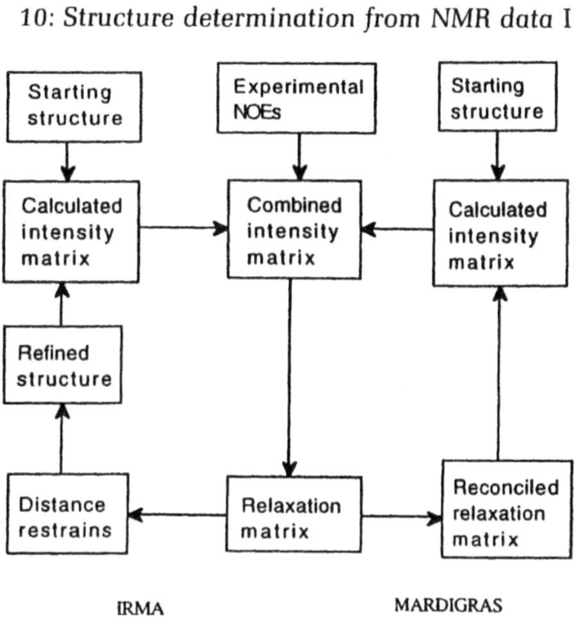

10: Structure determination from NMR data I

Fig. 2.11: *Two different approaches to obtain distances from NOE experiments. While one approach is based on the self-consistency of the relaxation matrix (right part), the other relies on external structure calculation (left part). Popular implementations of these approaches are the programs MARDI-GRAS [Borgias and James, 1990] and IRMA [Boelens et al., 1988]. This picture is taken from Roberts [1993].*

the relaxation matrix is calculated using eq. (2.27). At this step the two algorithms differ. One approach checks the relaxation matrix for self-consistency and produces a reconciled relaxation matrix (MARDIGRAS [Borgias and James, 1990]) from which the intensity matrix is in turn calculated. The other approach (IRMA [Boelens et al., 1988]) produces a set of distance restraints from the relaxation matrix. The restraints are used to refine the starting structure. The refined structure is then used to calculate an intensity matrix and the process starts all over again. The main advantage of either approach

2.3 Solution structure determination of DNA

is that by measuring NOESY spectra at different mixing times, spin diffusion can be accounted for and very accurate distance restraints are produced. However, as pointed out by Lane [1996] and Tonelli and James [1998] error limits are often too small since conformational averaging leads to considerable errors in NOE intensiy. Furthermore, aquiring and especially accurately integrating NOE data for several mixing times is exceedingly time-consuming.

A third possibility to derive distance restraints from NOE cross-peak intensities is the Isolated Spin Pair Approximation (ISPA). Several assumptions are made. First, a single correlation time (τ_c) for the whole molecule is introduced, with which τ_c^{ij} in eq. (2.26) is replaced.

$$J_{n,ij}(\omega) = \frac{\tau_c}{1 + n^2\omega_0^2\tau_c^2} \frac{1}{r_{ij}^6} \qquad (2.28)$$

In some cases local mobility of residues must be taken into account and a modified spectral density function has to be used [Lipari and Szabo, 1982a,b]

$$J_{n,ij}(\omega) = \left(\frac{S^2\tau_c}{1 + n\omega_0^2\tau_c^2} + \frac{S^2\tau_e}{1 + n\omega_0^2\tau_e^2}\right) \frac{1}{2\,r_{ij}^6} \qquad (2.29)$$

where τ_e is the effective correlation time of the local mobility site and S^2 is the generalized order parameter, which is a measure for the flexibility of the site with values ranging from 0 (unrestricted motion) to 1 (fully restricted motion). In proteins, order parameters range from 1 to as low as 0.6 for flexible side chains [Flynn et al., 2001], while S^2 in DNA is on the order of 0.8 for all proton pairs [Lane, 1993, 1996]. In the ISPA approach all distances are referenced to a fixed distance (both of which have $S^2 \approx 0.8$). Thus the contribution of local mobility is cancelled. The assumption of a single correlation time for the whole molecule is valid since correlation times for base and sugar protons are comparable [Reid et al., 1989] and oligonucleotides shorter 15 base pairs in length can be assumed isotropic rotors [Birchall and Lane, 1990].

In the ISPA approach the matrix exponential of eq. (2.23) is approximated with a Taylor expansion

$$exp[-\mathbf{R}\,\tau_m] = 1 - \mathbf{R}\,\tau_m + \frac{1}{2}\mathbf{R}^2\,\tau_m^2 - \ldots \qquad (2.30)$$

2 Theoretical background

whereby the peak intensities in the NOESY spectrum are given as

$$A_{ij} = \delta_{ij} - R_{ij}\,\tau_m + \frac{1}{2}\sum_k R_{ik}\,R_{jk}\,\tau_m^2 - \ldots \qquad (2.31)$$

The central assumption in ISPA is that for short mixing times, the Taylor expansion can be truncated after the linear term. Thus any effects of spin diffusion, which is magnetization transfer over third atoms (represented by the quadratic term), are neglected and the cross-peak intensity ($i \neq j$) becomes a linear function of r_{ij}^{-6}.

$$A_{ij} = R_{ij}\,\tau_m = q_{ij}\,\tau_c\,\tau_m \left(\frac{5 - 4\,\omega_0^2\,\tau_c^2}{1 + 4\,\omega_0^2\,\tau_c^2}\right) \frac{1}{r_{ij}^6} \qquad (2.32)$$

Consequently, distances can be derived by referencing to a known, fixed distance whereby all constant terms are cancelled.

$$\frac{A_{ref}}{A_{ij}} = \left(\frac{r_{ij}}{r_{ref}}\right)^6 \qquad or \qquad r_{ij} = r_{ref}\,\sqrt[6]{\frac{A_{ref}}{A_{ij}}} \qquad (2.33)$$

A commonly used reference distance is the C H5-H6 distance which is fixed at appr. 2.5 Å [Reid et al., 1989]. While many cross-peaks can be referenced with this distance, it is necessary to introduce several others to account for fast rotation in methyl groups (e.g. C7-H7) or solvent exchange with amino and imino protons (e.g. H42-H5).

2.3 Solution structure determination of DNA

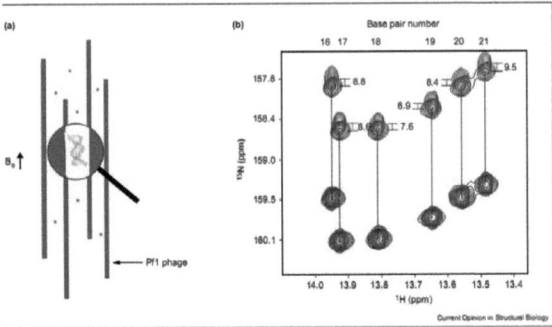

Fig. 2.12: Panel (a) shows the steric interaction of DNA with Pf1 which prevents the DNA from tumbling isotropically and thus induces residual order. The latter gives rise to Residual Dipolar Couplings (RDC) which can be determined by measuring the difference in dipolar coupling of i.e. an N-H bond vector with and without alignment (panel (b)). This picture is taken from MacDonald and Lu [2002].

2.3.3 Residual Dipolar Couplings

In combination with NOE data, Residual Dipolar Couplings (RDC) are now routinely employed for the NMR structure determination of macromolecules as they compensate for the drawbacks of NOE data. Due to the r_{ij}^{-6}-dependence of the NOE (cf. section 2.3.1) only information on closely spaced spins ($r_{ij} \approx 5\,\text{Å}$) is provided. In proteins, long-range peaks between residues far apart in the primary sequence can be observed owing to the tertiary fold. Due to the rod-like shape of short and medium-size DNA, only information on directly adjacent base pairs is available. Long-range effects like kinking in A-tract DNA could thus not be described prior to the development of RDC measurement [MacDonald et al., 2001, Stefl et al., 2004].

RDCs yield information on the orientation of bond vectors relative to the molecular frame. Thus also distant parts in a macromolecule can be characterized relative to each other, which significantly improves results from structure calculations, especially with regard to the global fold [Mauffret et al., 2002].

The first account of measuring anisotropic interactions was published more than 45

29

2 Theoretical background

years ago by Saupe and Englert [1963]. Only after the advent of high-resolution NMR spectrometers and methods, this technique could be applied to biological macromolecules with success. Bax and Tjandra [1997] were able to demonstrate that dissolving the protein ubiquitin in a very dilute solution of bicelles induces residual order while retaining the high resolution of NMR spectra. This marked a breakthrough in the field of biomolecular NMR since measurement of residual order effects such as RDCs, were not limited to molecules with natively high magnetic suceptibility anisotropy anymore [Tolman et al., 1995, Tjandra et al., 1996]. Initially application of RDCs to NMR structure determination centered on proteins [Bax and Tjandra, 1997, Tjandra and Bax, 1997, Tjandra et al., 1997], but subsequently the importance of RDCs for the structure determination of DNA was demonstrated extensively in theory [Vermeulen et al., 2000, Mauffret et al., 2002] and in practice [Tjandra et al., 2000, Zidek et al., 2001, MacDonald and Lu, 2002, Wu et al., 2004, Stefl et al., 2004].

Inducing just enough alignment for reliable measurement of RDCs, while retaining the high resolution and unambiguousness of the spectra is central to the success of RDC measurements. Tjandra and Bax [1997] showed that the degree of alignment can easily be adjusted by varying the concentration of the bicelles. In the following, diverse alignment media were developed in order to extend the applicability of the approach: bicelle-based alignment [Tjandra and Bax, 1997, Tjandra et al., 1997, Ottiger and Bax, 1999a, Barrientos et al., 2000, Al-Hashimi et al., 2000, Ruckert and Otting, 2000], filamentous phage [Hansen et al., 1998], stretched gels [Chou et al., 2001, Ma et al., 2008] or paramagnetic tagging [Wohnert et al., 2003]. The bacteriophage Pf1 proved to be particularly suited for aligning oligonucleotides since it is stable over a wide range of temperatures and Pf1-DNA interaction is minimized due to electrostatic repulsion between the two negatively charged macromolecules [Hansen et al., 1998, Prestegard et al., 2000]. The steric interaction of Pf1 with DNA is illustrated in Fig. 2.12 a.

RDCs (D_{ij}) are determined by measuring the difference of the dipolar coupling in the

2.3 Solution structure determination of DNA

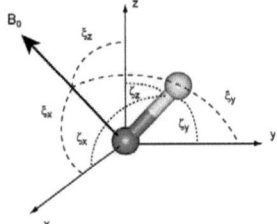

Fig. 2.13: Orientation of the magnetic field, defined by angles ξ_x, ξ_y, ξ_z and the internuclear vector, described by angles $\zeta_x, \zeta_y, \zeta_z$, in the macromolecular frame. This picture is taken from Blackledge [2005].

presence ($^1J_{ij}^{ani}$) and absence of molecular alignment ($^1J_{ij}^{iso}$), as is shown in Fig. 2.12 b.

$$^1J_{ij}^{ani} = {}^1J_{ij}^{iso} + D_{ij} \tag{2.34}$$

The RDC term can be described by the following equation

$$D_{ij} = -\frac{\gamma_i \gamma_j \mu_0 \hbar}{4\pi^2} \left\langle \frac{3\cos^2\alpha_{ij}(t) - 1}{2 r_{ij}^3(t)} \right\rangle \tag{2.35}$$

where α denotes the angle of the internuclear vector between atoms i and j with the applied magnetic field and $\langle \rangle$ designates the time and ensemble average. When no preferred alignment is induced, all values of α are sampled with equal probability over time and thus D_{ij} is averaged to zero. If residual order is induced by aligning the molecule, information on the orientation of the internuclear vector can be extracted. However, in order to have a fixed reference system, it is desirable to relate the orientation of the internuclear vector to the macromolecular frame rather than the magnetic field. To achieve this, two conditions must be met: 1) the distance between the atoms which give rise to D_{ij} does not change significantly over time (or else its distribution is known), such that r_{ij} can be substituted in eq. (2.35) with an effective distance r_{ij}^{eff} which is known and where averaging is already taken into account. 2) It is assumed that for macromolecules the time average of α can be expressed by two convoluted motions, the macromolecule tumbling with respect to the magnetic field vector (ξ_x, ξ_y, ξ_z) and the internuclear vector

2 Theoretical background

moving inside the macromolecular frame ($\zeta_x, \zeta_y, \zeta_z$) as illustrated in Fig. 2.13.

$$cos\,\alpha_{ij} = \begin{pmatrix} cos\,\xi_x \\ cos\,\xi_y \\ cos\,\xi_y \end{pmatrix} \begin{pmatrix} cos\,\zeta_x \\ cos\,\zeta_y \\ cos\,\zeta_y \end{pmatrix} = \sum_{k=}^{x,y,z} cos\,\xi_k\,cos\,\zeta_k \qquad (2.36)$$

To a first approximation, the internuclear vector is considered to be fixed within the macromolecular frame and thus time-averaging has no effect on ζ. Introducing the alignment tensor **A** with dimensionless units

$$A_{kl} = \langle cos\,\xi_k\,cos\,\xi_l \rangle \qquad (2.37)$$

eq. (2.35) can then be rewritten as

$$D_{ij} = -\frac{3\,\gamma_i\,\gamma_j\,\mu_0\,\hbar\,S_{flex}}{8\,\pi^2\,(r_{ij}^{eff})^3} \sum_{k,l=}^{x,y,z} \left(A_{kl}\,cos\,\zeta_k\,cos\,\zeta_l - \frac{1}{9}\delta_{kl} \right) \qquad (2.38)$$

where S_{flex} represents an order parameter to account for local flexibility of the internuclear vector. Here the assumption is that the local motion does not influences the overall alignment of the macromolecule, which is reasonable in the absence of large amplitude fluctuations. In most cases, the local motion can be approximated with the diffusion-in-a-cone model, where the order parameter S_{flex} is related to the generalized order parameter S, which scales down the measured RDCs linearly [de Alba and Tjandra, 2002].

In the present framework the alignment tensor **A** has all elements non-zero. It is however desirable to find a specific molecular frame in which all off-diagonal elemtents of **A** are zero. Such a frame, called the principal axis system (PAS) can be found by a 3D Euler rotation of the current molecular frame with parameters α, β and γ [Blackledge, 2005]. Eq. (2.38) can then be rewritten in terms of the polar angles θ, ϕ (Fig. 2.14, left panel), which describe the orientation of the internuclear vector in the eigenframe of the

2.3 Solution structure determination of DNA

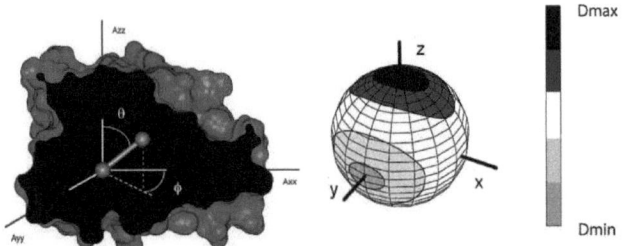

Fig. 2.14: *Orientation of the internuclear vector with polar coordinates θ and φ in the eigenframe of the alignment tensor with eigenvalues A_{xx}, A_{yy} and A_{zz} (left panel). The right panel illustrates the orientational degeneracy of RDCs. As can be clearly seen a large number of different orientations can be sampled when the RDC takes intermediate values. Only extreme RDC value define unique orientations of the internuclear vector. This picture is taken from Blackledge [2005].*

alignment tensor with eigenvalues $|A_{xx}| \leq |A_{yy}| \leq |A_{zz}|$ [Lipsitz and Tjandra, 2004]

$$D_{ij} = -\frac{\gamma_i \, \gamma_j \, \mu_0 \, \hbar \, S_{flex}}{4 \, \pi^2 \, (r_{ij}^{eff})^3} \left[A_a \left(3 \cos^2\theta - 1\right) + A_r \, sin^2\theta \, cos \, 2\phi \right] \quad (2.39)$$

where A_a and A_r are the axial and rhombic components of the alignment tensor **A** [Prestegard et al., 2000], which are given by

$$A_a = \frac{1}{2} A_{zz} \quad and \quad A_r = \frac{1}{2}(A_{xx} - A_{yy}) \ . \quad (2.40)$$

In consequence, the alignment tensor **A** is determined by 5 independent parameters: the three Euler angles needed for rotation of the reference frame (α, β, γ) and the two components of the alignment tensor **A** (A_a, A_r). The alignment tensor can then be unambiguously determined from a minimum set of 5 experimental RDCs by singular value decomposition [Losonczi et al., 1999].

With **A** at hand, the orientation of any internuclear vector with respect to the macromolecular frame can be calculated. Unfortunately, there exists a multitude of orientations of an internuclear vector that is consistent with a given intermediate RDCs value, as is illustrated in Fig. 2.14, right panel. Only extreme values correspond to unambiguous

2 Theoretical background

orientations. The orientational degeneracy for intermediate RDC values clearly limits their value for structure determination. Thus it is desirable to remove this degeneracy by either measuring more RDCs per residue, or RDCs for the same internuclear vectors but with a different alignment tensor [de Alba and Tjandra, 2002, Blackledge, 2005].

2.3.4 Simulated Annealing calculations

The main idea behind Simulated Annealing (SA) calculations is coupling simple energy minimization to Molecular Dynamics. Thus the problem that molecules converge to a local minimum instead of the global can be circumvented. The Newtonian equation of motion

$$F_i(t) = -\frac{\partial V}{\partial r_i} = m_i \, a_i(t) \qquad (2.41)$$

relates the acceleration $a_i(t)$ of each atom i at time t to the derivative of the potential energy (V) with respect to the atom position r_i. $F_i(t)$ represents the force which acts upon the atom i with mass m_i [Bloomfield et al., 2000]. The atoms are moved according to the force that is exerted upon them for a given time-step (typically between 1-5 fs). From the knowledge of the last and current atom positions and velocities, a new force is calculated which in turn acts on the atoms. This cycle is repeated until a convergence criterion (e.g. a minimum change in the gradient of the potential energy) is met. Initially, atom velocities are computed using a Gaussian or Maxwell distribution. The atom coordinates are derived from a starting structure. Since the initial coordinates and velocities determine all subsequent ones, it is important to start from a reasonable structure. This is usually achieved by starting from already known crystal or NMR structures.

Temperature-coupling of Molecular Dynamics is achieved by introducing an average kinetic energy which is computed via Boltzmann statistics

$$<E_{kin}> = <\frac{1}{2} m \, \nu^2> = \frac{3}{2} k_b T \qquad (2.42)$$

where k_b is the Boltzmann factor, m the atom mass and ν the atom velocity. The higher the temperature (T) is chosen, the higher the kinetic energy of the system. Thus at high temperatures kinetic barriers can be overcome, and the global minimum should be accessible.

The number of atoms in macromolecules such as DNA or proteins is on the order of thousands with three times as much cartesian coordinates to be calculated at each step. In order to make computation of macromolecules feasible, Molecular Dynamics calculati-

2 Theoretical background

ons rely on the predefinition of atom types. For these atom types many parameters such as bond lengths, bond angles, dihedral angles, partial charges etc. are assumed to be fixed and are comprised in the force field. In the present work the program XPLOR-NIH [Schwieters et al., 2003] was used, which employs the CHARMM force field [Weiner et al., 1984, MacKerell et al., 2000]. The total potential energy V_{tot} consists of two components [Brünger, 1996]

$$V_{tot} = E_{emp} + E_{eff} \tag{2.43}$$

where the empirical (E_{emp}) and the effective energy term (E_{eff}) are given as [Brünger, 1996].

$$E_{emp} = E_{bond} + E_{angle} + E_{dihe} + E_{vdW} + E_{Coulomb} \tag{2.44a}$$

$$E_{eff} = E_{noe} + E_{rdc} + E_{plan} + E_{cdih} \tag{2.44b}$$

E_{bond}, E_{angle}, E_{dihe} and all energy terms of E_{eff} are calculated as the product of a scaling factor (the force constant) and the deviation of the observed value from the equilibrium one, e.g.

$$E_{bond} = k_{bond} \left(r_{ij}^{obs} - r_{ij}^{equ} \right) \tag{2.45}$$

The equilibrium values for E_{noe} and E_{rdc} are taken from experiment while these of E_{cdih} and E_{plan} are averages from the literature [Brünger, 1996]. The corresponding scaling factors are defined in the calculation input and thus can be used to increase the restraining power of selected energy terms. Equilibrium values and force constants of E_{bond}, E_{angle}, E_{dihe} constitute one part of the force field. E_{vdW} is given as

$$E_{vdW} = \sum_{i,j} \left(\frac{A_{ij}}{r_{ij}^{12}} - \frac{B_{ij}}{r_{ij}^{6}} \right) \tag{2.46}$$

with $A_{ij} = 2\sqrt{\varepsilon_{ii}\varepsilon_{jj}}(\sigma_{ii} - \sigma_{jj})$ and $A_{ij} = 2\sqrt{\varepsilon_{ii}\varepsilon_{jj}}(\sigma_{ii} - \sigma_{jj})$. The atomic permitivities ($\varepsilon_{ii/jj}$) and van-der-Waals radii ($r_{ii/jj}$) set up another part of the force field. The partial

2.3 Solution structure determination of DNA

atomic charges (q_i, q_j) in the $E_{Coulomb}$-term constitute the last part of the force field.

$$E_{Coulomb} = \sum_{ij} \frac{q_i \, q_j}{\varepsilon_0 \, r_{ij}} \qquad (2.47)$$

When explicit treatment of water is not feasible due to restrictions on the calculation time, the solvent screening effect can be approximated by introducing a distance dependent permittivity of free space $\varepsilon_0(r_{ij})$.

Ultimately, force field parameters are based on crystal structures and ab-initio calculations of small model molecules (bond lengths, bond angles and dihedral angles), infrared spectroscopy data (force constants) and empirical testing (where no experimental source is available) [Weiner et al., 1984]. When chemically modified nucleotides are to be incorporated into calculations, parameters for this nucleotide have to be added to the existing force field. In the absence of experimental data, these parameters are calculated using ab-initio methods.

3 Experimental details

3.1 2-Aminopurine

3.1.1 Sample preparation

NMR sample preparation DNA strands were obtained (from BIOTEZ, Berlin, and BIO-SPRING, Frankfurt/M, Germany) already purified by reverse-phase high-pressure liquid chromatograpy (HPLC). After hybridization they were subjected to size exclusion chromatography (SEPHADEX G25) and ammonia treatment (lyophylization with $\approx 3\,\%$ NH_3-solution) to remove residual, low molecular weight impurities (mainly Et_3N-buffer from HPLC). Equivalent amounts of complementary single strands were hybridized by rapid heating to $90,°C$ and subsequent gradual cooling to room temperature at a rate of $1\,°C$ per minute. NMR samples were prepared at 3 mM duplex concentration in D_2O (D_2O 99.98 %) and H_2O ($H_2O:D_2O/90{:}10$) at pH 6.5 in 10 mM Na_2HPO_4/NaH_2PO_4 and 150 mM NaCl solution.

RDC sample preparation Samples for measuring residual dipolar couplings (RDC) were prepared in D_2O as described above with the addition of 20 mg/ml Pf1 (obtained from ASLA BIOTECH, Riga) suspended in the same buffer. This necessitates rebuffering of Pf1 since it is obtained in a different and non-deuterated buffer. Rebuffering is achieved by ultracentrifuging 100 μl of Pf1 two times with 600 μl phosphate buffer (as described above) and subsequently two times with 600 μl deuterated phosphate buffer at 60000 rpm. Each ultracentrifuging step is performed at 4 °C for 2 h. Next the sediment is suspended in the DNA sample. The high viscosity of Pf1 complicates sample handling and thus the suspension has to be stirred until a viscose, clear, gel-like sample is obtained. After

3 Experimental details

transferring it into the NMR tube, bubbles have to be removed by slow centrifugation of the NMR tube (max. 500 rpm). The degree of orientation can be checked by measuring the quadrupolar deuterium splitting of the HOD signal [Lipsitz and Tjandra, 2004]. In case of degradation or non-complete suspension of Pf1 in the sample, the expected symmetric doublett with splitting of 10±5 Hz is asymmetric, extremely broadened or even non-observable.

3.1 2-Aminopurine

3.1.2 Measurements

Duplex melting Melting of the 13mer2AP duplex was monitored by optical absorption at 260 nm and by the fluorescence quantum yield (due to the 2AP nucleobase) from 310 nm excitation [Evans et al., 1992]. For comparison, melting of the 13merRef duplex was measured by absorption only. The solutions had a total single-strand concentration of 23.7 mM and the optical path length was 1 cm. Temperature was varied between 25 and 85 °C; the standard error of transfer between absorption and emission temperatures was ±0.011 °C. Following changes of typically 1 °C, equilibration was allowed for at least 15 minutes. The relative fluorescence yield of pure 2AP in buffered water was measured in the same manner for reference. Measurements were corrected for density changes.

Imino proton exchange Exchange rates of the imino protons were obtained from inversion recovery experiments at 298 K on a BRUKER AVANCE 600 NMR spectrometer. For this purpose the standard 1D 1H-WATERGATE (water suppression by gradient tailored excitation) pulse program was modified to start with a selective inversion pulse. As discussed in section 2.2.2 the pulse shapes least sensitive to relaxation effects are rectangular and gaussian pulses. Dornberger et al. [1999] and Bhattacharya et al. [2002] use millisecond gaussian-shape pulses for inversion, but such long pulse durations again facilitate relaxation effects. Additionally, the imino protons are shifted upfield ($\approx$ 13 ppm) of the bulk of the DNA (1-9 ppm) and especially the HOD ($\approx$ 4.80 ppm) NMR signals and thus off-resonance excitation effects can be ruled out. In consequence, a selective rectangular inversion pulse of 327 μs duration centered on the imino proton region was employed in this work. After a variable delay with 21 settings ranging from τ=20 μs up to 2 ms, a 1D spectrum was recorded. The settings for the variable delay have been optimized to give accurate results for slowly as well as fast exchanging imino protons.

Samples were prepared in H$_2$O with the buffer described earlier. In order to cancel the effect of intrinsic catalysis (cf. section 2.2.1), the spin-lattice relaxation time T_1^{int} of all imino protons was measured prior to titration with base catalyst. Subsequently, each duplex was titrated with 1 M TRIS buffer as base catalyst for proton exchange. Test runs

3 Experimental details

with ammonia as base catalyst gave inferior results, since the latter is highly volatile. This makes high-concentration salt solutions of ammonia, which are necessary in order to keep the pH stable upon addition of the catalyst, inaccessible. Another set of test runs, where pure ammonia and HCl (for pH compensation) were added separately to the NMR sample, resulted in temporary but extreme pH changes (pH values ranging from 1 to 11). These changes led to gradual degradation of the DNA sample. Furthermore, the error in concentration introduced by evaporation of ammonia is intolerable when high precision results are to be obtained. In order to keep the pH as stable as possible, TRIS-hydrochloride at pH 7.4 was used. At every titration point the pH was measured in order to determine the amount of free base catalyst precisely (cf. eq. (2.15) in section 2.2.1). All pH values centered around 7.1. To estimate the error in T_1-determination, three sets of measurements were carried out at each titration point, resulting in 693 1D spectra (21 delay settings, 11 titration points).

Each spectrum was fitted (with the MATHEMATICA program environment) to give the integrals of the imino proton peaks. Complete spectral fitting became necessary since standard region integration as implemented in TOPSPIN is sensitive to peak overlap. Due to continued broadening of the imino proton signals upon base catalyst addition, integrals obtained by the latter method were inconsistent and thus gave rise to large errors (up to 100 %) in T_1-values. In contrast, complete spectral fitting in MATHEMATICA yielded T_1-values with errors of typically 0.5-5 %.

Structural NMR NMR measurements for structure calculation were carried out on a BRUKER AVANCE 600 NMR spectrometer. The optimum temperature of 298 K was determined by monitoring the imino proton signal intensity (cf. Fig. 3.1). For each duplex, NOESY-, DQF-COSY- (double quantum filtered correlated spectroscopy), TOCSY- (total correlation spectroscopy) and HMQC-spectra (heteronuclear multiple quantum coherence spectroscopy) in D_2O, a WATERGATE-NOESY-spectrum in H_2O and an HMQC-spectrum in D_2O/Pf1 were recorded. The quadrupolar splitting of the ^2H NMR signal after addition of Pf1-phage was 15.72 Hz and 13.65 Hz for 13merRef and 13mer2AP, re-

3.1 2-Aminopurine

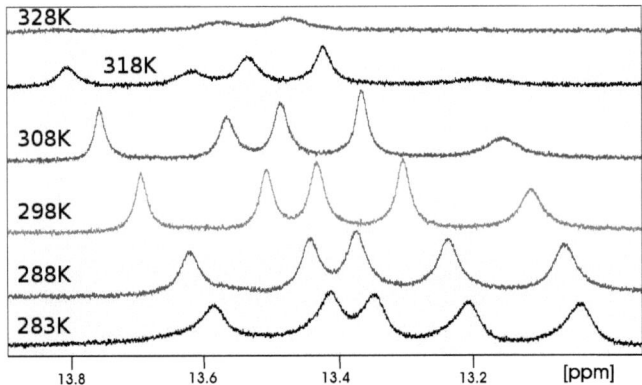

Fig. 3.1: *Imino proton signal intensities for the temperature range 283-328 K. in 13mer2AP.*

spectively. Upon addition of Pf1, the sample became very viscose and standard shimming procedures failed to produce reasonable linewidths. Two workarounds were investigated. One was to optimize the shim for a BRUKER standard sample and use these shim settings on the DNA/Pf1 sample. The second workaround involved opimization of the lineshape of the HOD signal. This procedure demands that after every change of the shim settings, a 1D spectrum of the sample had to be aquired. Although the latter procedure is more time-consuming than the other one, it yielded supreme results and thus was employed.

For DQF-COSY- (TOCSY-) spectra, 16 (32) transients were acquired, with 2048×256 points in F2 and F1 dimensions. For NOESY-spectra in both solvents, 16 transients were acquired with 4096×2048 points at a mixing time of 150 ms. For the HMQC-spectra with and without Pf1, 192 transients were acquired with 8192×512 points. The high number of data points was necessary to obtain sufficient resolution in the proton dimension in order to determine the RDC-values with a precision below 1 Hz. The optimal d2-delay, where both the non-aromatic and aromatic region are equally enhanced, was determined to be 2.5 ms. All spectra were processed with the BRUKER TOPSPIN-software, and signals were assigned with the help of CARA [Keller, 2004].

3 Experimental details

3.1.3 Restraint generation

Force field parametrization Density functional theory (DFT) calculations of the 2AP-nucleotide were performed with GAUSSIAN03 employing a triple zeta valence plus polarisation (TZVP) basis set. Due to the structural proximity of 2AP and A, force field parameter and topology entries of the latter were adopted and changed where necessary. Special emphasis was placed on the charge distribution since this was shown to be the main difference between A and 2AP [Broo, 1998, Mishra et al., 2000, Jean and Krueger, 2006]. For this purpose several methods for obtaining point charges were compared. The fastest method is the Mulliken population analysis, which, however, gives unrealistic results [Breneman and Wiberg, 1990]. This is due to the equal distribution of overlap populations between the two involved atoms. Hence another approach has been developed, which tries to derive atomic charges via fitting of point charges to the Molecular Electrostatic Potential (MEP) [Connolly, 1983, Singh and Kollman, 1984]. Crucial to the success of this technique is the algorithm which is used for fitting. Three different algorithms were tested; the CHELP- [Chirlian and Francl, 1987], the CHELPG- [Breneman and Wiberg, 1990] and the MK-algorithm [Besler et al., 1990]. Out of these three the MK- and CHELPG-algorithms gave consistent and similar results. Partial charges derived with the CHELPG-algorithm were employed in the force field. Although the sugar moiety was simulated as well, only the partial charges for the 2AP residue were integrated into the force field. Due to the similarity of A and 2AP the influence of the latter on the partial charges of the sugar can be considered negligible (as the same values for partial charges of sugar atoms are also used for G,C and T residues). Thus the partial charge values for the sugar were taken as already defined for the native residues in the force field. To retain the neutrality of the 2AP residue the partial charge at the N7 atom of 2AP was increased from -0.61 to -0.45 (cf. section 3.1.3).

Distance restraints The assigned NOE cross-peaks were converted to distance restraints by referencing their integrals to the integrals of known distances employing the Isolated Spin Pair Approximation. The NOE cross-peaks were integrated with the program CARA [Keller, 2004] using the sum-over-rectangle method. As reference distances Me-H6 T (3.1 Å) for all NOE cross-peaks involving methyl protons, H42-H5 C (2.4 Å) for all NOE cross-peaks involving exchangeable protons and H5-H6 C (2.5 Å) for the remaining NOE cross-peaks were used (bond lengths adapted from the force field parameters). For the purpose of exporting the integral values obtained by CARA [Keller, 2004] to an XPLOR-NIH [Schwieters et al., 2003] restraints file, a LUA script was written (see section 4). This script classifies the intregrated peaks according to the overlap with other peaks and scales their volume integrals accordingly. Additionally, uncertainties for the NOE restraints are automatically calculated from the standard deviation of the reference peaks' volume integrals. The estimated uncertainty is then increased according to the classification of each peak. Furthermore, this classification is printed into a separate file, which can be used to assess whether or not peak overlap might prevent a reliable estimation of the peak volume.

Residual Dipolar Coupling restraints The experimentally determined RDCs (see tables 3.1 and 3.2) of C-H bond vectors can be input into the structure calculations with the measurement precision as error bounds. From the experimentally determined RDC-values the orientation of the corresponding internuclear vector is determined via eq. (2.39) in section 2.3.3. All constants in the latter equation are comprised within the factor D_a

$$D_a = -\frac{\gamma_i \gamma_j \mu_0 \hbar S_{flex}}{4 \pi^2 (r_{ij}^{eff})^3} \quad (3.1)$$

for which only one value is used throughout the calculation. This necessitates scaling of different sets of RDCs, for example when C-H as well as C-C RDCs have been measured. Scaling is achieved by introducing prefactors to the D_a-factor, which are defined as the ratio of the two D_a-factors involved. For example, when C-C RDCs are to be scaled to a set of C-H RDCs, the corresponding prefactor would be calculated as:

$$D_a^{pre}(CC) = \frac{D_a(CH)}{D_a(CC)} = \frac{\gamma_H}{\gamma_C}\left(\frac{r_{CC}}{r_{CH}}\right)^3 = \frac{42.576}{10.705}\left(\frac{1.496}{1.090}\right)^3 \approx 10.28 \quad . \quad (3.2)$$

3 Experimental details

where $\frac{\gamma_H}{\gamma_C}$ is the change in gyromagnetic ratio when working with C-H or C-C RDCs and r_{CC} and r_{CH} are the C-C and C-H bond lengths between the atoms defining the internuclear vector. Although in this work only C-H RDCs were measured, scaling issues become important for the implementation of the experimentally determined C-H RDCs of the T methyl groups.

Due to the fast rotation of the latter, only an averaged value for all three C-H bond vectors can be measured. This fast rotation scales the corresponding RDC values by

$$P_2(cos\beta) = \frac{3}{2} cos^2\beta - \frac{1}{2} \qquad (3.3)$$

where β is the C5-C7-H7* angle. The latter is usually assumed to be the ideal tetrahedral angle of 109.5°, but was experimentally determined for methyl groups to be 110.9° by Ottiger and Bax [1999b]. Although small, the deviation from the ideal angle has a strong impact on methyl RDC scaling since $P_2(cos\beta)$ is a steep function β. Thus with $P_2(\cos 109.5)$=-0.3329 and $P_2(\cos 110.9)$=-0.3089, conversion factors between C5-C7 and C7-H7* methyl RDCs can be calculated as -3.42 and -3.17, respectively, according to

$$\frac{D_{CH\_Me}}{D_{CC\_Me}} = P_2(cos\beta)\, D_a^{pre}(CC) \quad . \qquad (3.4)$$

The value of -3.17 was determined by measuring the correlation of experimentally determined C-H and C-C methyl RDCs [Ottiger and Bax, 1999b].

In order to implement the methyl C-H RDCs into the structure calculations, they have to be associated with a unique internuclear vector. Thus it is necessary to convert the C7-H7* RDCs into the corresponding C5-C7 RDCs. Since the latter must now be input as C-C RDCs, a seperate restraints file has to be used. There are in principle two ways how this implementation can be achieved in XPLOR-NIH v2.20 [Schwieters et al., 2003]. One way is to convert all experimentally determined methyl C-H RDC values by hand to the corresponding C-C values with the factor 1 / -3.17 = -0.3155. This is done prior to implementing them into the calculation. In that case, the prefactor $D_a^{pre}(CC)$ can be used to scale the C-C RDC input file to the C-H RDC input. According to the manual, this scaling is handled automatically by the *scale_toCH* routine of the program XPLOR-NIH v2.20 [Schwieters et al., 2003]. Due to a bug however, the automatic scaling is only effective when the methyl C-C RDCs are input as type "CACO". The other way to input C-H methyl RDCs into structure calculations will be implemented in the new version of XPLOR-NIH (v2.24) but was made available to the author in advance (personal communication of Dr. Charles Schwieters). In that new version, the methyl RDCs are input as a separate restraints file too, but their values do not have to be scaled by hand. They are automatically recognized by the program as methyl RDCs and scaled to the C-H RDC input by

$$D_{Me}^{pre} = -3.17 \, D_a^{pre}(CC) = 32.59 \quad . \tag{3.5}$$

However, scaling of the D_a-factor and scaling the RDC values is not equivalent, since the energy for the RDC potential term is given by [Brünger, 1996]

$$E_{RDC} = k_{RDC} \, (D_{calc} - D_{obs})^2 \tag{3.6}$$

where k_{RDC} is the scale factor for the RDC energy term. The latter must be modified for different sets of RDCs according to their respective errors with a weighting factor ω_{ij}. When using the prefactor D_{Me}^{pre} for implementation of methyl RDCs, the energy of the C-C methyl RDCs has to be scaled by

$$\omega_{CC} = \frac{1}{(-3.17)^2} \, \omega_{CH} = 0.0995 \, \omega_{CH} \tag{3.7}$$

due to dependence of the energy on the square of the difference between calculated (D_{calc}) and observed (D_{obs}) RDC.

3 Experimental details

Tabelle 3.1: *Experimentally determined RDCs for 13merRef. The RDCs were measured with a precision of ±0.6 Hz.*

Res	Vector	$J_{(CH)}$ (Hz)	$J_{(CH)}$(aligned) (Hz)	RDC (Hz)
A6	C2-H2	201.6	225.6	24.0
A7	C2-H2	203.4	224.4	21.0
A8	C2-H2	202.2	223.8	21.6
A16	C2-H2	202.8	214.8	12.0
A24	C2-H2	201.6	224.4	22.8
C9	C5-H5	171.6	190.2	18.6
C17	C5-H5	170.4	183.6	13.2
T3	C7-H7	126.6	122.4	-4.2
T11	C7-H7	126.0	118.8	-7.2
T19	C7-H7	123.0	112.8	-10.2
T20	C7-H7	127.2	118.2	-9.0
T21	C7-H7	126.6	118.2	-8.4
T3	C1'-H1'	155.4	167.4	12.0
C9	C1'-H1'	162.0	169.2	7.2
C17	C1'-H1'	162.6	166.8	4.2
T20	C1'-H1'	163.8	178.2	14.4
T21	C1'-H1'	153.6	174.0	20.4
T3	C6-H6	177.0	195.0	18.0
C5	C6-H6	180.0	201.0	21.0
T20	C6-H6	177.0	192.6	15.6
T21	C6-H6	175.8	190.8	15.0
C23	C6-H6	175.8	201.0	25.2
A7	C8-H8	213.0	235.8	22.8
A16	C8-H8	213.0	230.4	17.4

3.1 2-Aminopurine

Tabelle 3.2: *Experimentally determined RDCs for 13mer2AP. The RDCs were measured with a precision of ±0.6 Hz.*

Res	Vector	$J_{(CH)}$ (Hz)	$J_{(CH)}$(aligned) (Hz)	RDC (Hz)
A6	C2-H2	202.2	222.0	19.8
2AP7	C6-H6	179.4	200.4	21.0
A8	C2-H2	202.8	218.4	15.6
A16	C2-H2	201.6	213.6	12.0
A24	C2-H2	201.6	219.0	17.4
C5	C5-H5	170.4	184.2	13.8
C9	C5-H5	171.0	189.6	18.6
C17	C5-H5	170.4	183.0	12.6
T3	C7-H7	127.2	121.2	-6.0
T11	C7-H7	127.2	121.2	-6.0
T19	C7-H7	126.6	118.2	-8.4
T20	C7-H7	127.2	120.6	-6.6
T21	C7-H7	126.6	119.4	-7.2
T3	C1'-H1'	160.8	172.2	11.4
2AP7	C1'-H1'	156.6	170.4	13.8
C9	C1'-H1'	160.8	168.6	7.8
C17	C1'-H1'	162.6	166.2	3.6
T21	C1'-H1'	161.4	182.4	21.0
T3	C6-H6	175.2	193.8	18.6
C5	C6-H6	175.2	195.6	20.4
T20	C6-H6	176.4	190.2	13.8
T21	C6-H6	176.4	192.0	15.6
C23	C6-H6	174.0	199.2	25.2
2AP7	C8-H8	214.2	234.6	20.4
A16	C8-H8	213.6	229.8	16.2

3 Experimental details

3.1.4 Structure calculation

Calculation input Structure calculations were performed with XPLOR-NIH V2.20 [Schwieters et al., 2003]. A total of 340 (333) NOE distance restraints and 24 (25) Residual Dipolar Couplings were used for 13merRef (13mer2AP). The experimental data were supplemented with 144 backbone dihedral restraints, 78 hydrogen bond distance restraints and 28 planarity restraints (see Table 3.3).

Initial MD-calculations were performed with dihedral restraints allowing both A-form and B-form conformations (with error bars of $\pm 50°$). B-form conformation was experimentally confirmed by ^{3}J coupling constants for H1'-H2' derived by P.E.COSY (primitive exclusive correlated spectroscopy) and NOESY-cross-peak intensities characteristic of B-DNA. Consequently, regular dihedral values from the literature [Roberts, 1993] were included in the calculations.

Tabelle 3.3: *Overview of structural statistics for 13merRef and 13mer2AP calculations*

	13merRef	13mer2AP
NOE restraints		
- total	340	333
- interresidue	196	188
- intraresidue	144	145
- 2AP to DNA	-	9
RDC restraints	24	25
Dihedral angle restraints	144	144
H-bond restraints	78	78
Base pair planarity restr.	28	28
NOE viol. (> 0.5 Å)	0	0
RDC viol. (> 0.4 Hz)	0	0
Dihe viol. ($> 5°$)	0	0
RMSD to ave. struct. in Å	0.30	0.33

The structures were calculated in two steps. First, a reasonable starting structure with well defined local conformation was computed. To ensure that no bias is introduced towards local energy minima, the calculation started from an elongated and equilibrated structure. The resulting structure, which is mainly defined by NOE restraint data, was used as input for Simulated Annealing calculations including RDC data. The need for locally well defined starting structures in order to calculate reasonable structures which

3.1 2-Aminopurine

satisfy NOE as well as RDC data is documented in the literature [Vermeulen et al., 2000, Mauffret et al., 2002].

Simulated Annealing protocol The two complete MD protocols are given in the Appendix, section 2. Only the protocols for the 13mer2AP calculation is given, since the other calculations were carried out with the same protocol (with the exception of the input file names). The input scripts are based on the example files of the XPLOR-NIH package (refine_full.py and sa.inp) but were substantially modified in the course of this work.

The MD protocol with only NOE restraints as experimental input consisted of an initial minimization (50 steps) followed by 48 ps of high-temperature cartesian coordinate dynamics at 3000 K, subsequent gradual cooling to 25 K in 120 steps of 0.05 ps length and a final minimization (3000 steps).

The MD protocol including the experimental RDC restraints consisted of an initial cartesian coordinate minimization (1000 steps) followed by 50 ps of high-temperature cartesian coordinate torsion angle dynamics at 20000 K, subsequent gradual cooling to 25 K in 154 steps of 0.5 ps length (34 steps to cool down to 3000 K, followed by 120 steps to reach the end temperature) and a final minimization (3000 steps). The alignment tensor values were allowed to float during the calculations, as implemented in XPLOR-NIH (v2.20) [Schwieters et al., 2003].

For each run an ensemble of 100 structures was computed. The 10 minimum energy structures without violation of restraints were chosen to compute an averaged structure which was energy-minimized to yield the final structure. The root-mean-squares-deviation (RMSD) of the 10 minimum energy structures to the average, minimized structure is a measure for the precision of the calculation.

Structure validation To check the accuracy of the structures, NOESY-spectra were back-calculated from the average structures with the Full Matrix Relaxation Approach implemented in XPLOR-NIH (v2.20) [Schwieters et al., 2003]. The back-calculated spectra were visualized with the program GIFA [Pons et al., 1996] and overlayed with the

3 Experimental details

experimental ones. Furthermore, RDCs were predicted from the average structure using the program PALES [Zweckstetter and Bax, 2000] and compared to the experimental ones. Finally, helical parameters were calculated with the help of the program 3DNA [Lu and Olson, 2003]. These were checked for consistency with values for regular B-DNA helices.

A number of convenience scripts were written to automate data conversion between PALES [Zweckstetter and Bax, 2000] and XPLOR-NIH (V2.20) [Schwieters et al., 2003], the latter program and GIFA [Pons et al., 1996] and for quick access to energies or number of restraints in structure calculations output files. These utility scripts are comprised in the Appendix in section 5.

3.2 2-Hydroxy-7-nitrofluorene

Since many experimental details for HNF are equivalent to the ones for 2AP, the description for the latter is referenced where appropiate in order to avoid redundance.

3.2.1 Sample preparation

2-Hydroxy-7-nitro-fluorene was synthesized (Matthias Pfaffe) in four steps. The 2'-deoxyriboside dRi-HNF was prepared by reaction with 1'α-Chloro-3',5'-di-O-toluoyl-2'-deoxy-D-ribose in the presence of activated molecular sieve. Quantification of the H1'-H2'/H2" and H3'-H2'/H2" NOE cross-peaks in the NOESY-spectra of the main product revealed that the α-glycoside was formed predominantly. After purification of the latter by column chromatography, the corresponding phosphor-amidite was reached by standard methods. Fixed-phase synthesis of the labelled DNA strand (BIOTEZ) required a fourfold increase over the normal reaction time for coupling dRi-HNF. Further sample preparation is equivalent to the one described in section 3.1.1.

3 Experimental details

3.2.2 Measurements

Duplex melting Melting of the 13merHNF duplex was monitored by optical absorption at 260 nm and by the red-shift of the weak absorption band of HNF at 380 nm. The solution had a total single-strand concentration of 23.5 mM and the optical path length was 1 cm. The following procedures are equivalent to the ones described in section 3.1.2.

Structural NMR NMR measurements for structure calculation were carried out on a BRUKER AVANCE 600 NMR spectrometer at 298 K. For each duplex, NOESY-, DQF-COSY-, TOCSY- and HMQC-spectra in D_2O, a WATERGATE-NOESY-spectrum in H_2O and an HMQC-spectrum in D_2O/Pf1 were recorded. The quadrupolar splitting of the ^2H NMR signal after addition of Pf1-phage was 10.55 Hz. The following procedures and settings are equivalent to the ones described in section 3.1.2.

3.2.3 Restraint generation

Distance restraints, RDC restraints and force field parameters were generated as described in section 3.1.3. The experimental RDC values that were used for restraint generation are given in Table 3.4. In the case of HNF however, no topology information for a structurally similar compound was available. Thus the corresponding parameter and topology input files had to be generated from the structure calculated by DFT methods as described in section 3.1.3. The latter can be achieved utilizing the program XPLO2D [Kleywegt and Jones, 1998].

The influence of the HNF on the partial charges of the sugar residue was considered to be non-negligible for C1' and the directly bonded atoms due to the strong structural and electronic differences between native nucleobases and HNF. Thus, in contrast to 2AP, for the atoms C1', C4', H4', H1", and O4' of HNF the partial charges derived from the DFT-calculations were used. The charge at O4' was increased by +0.13 in order to retain neutrality of the residue (cf. section 3.1.3).

Tabelle 3.4: *Experimentally determined RDCs for 13merHNF. The RDCs were measured with a precision of $\pm 0.6\,Hz$.*

Res	Vector	$J_{(CH)}$ (Hz)	$J_{(CH)}$(aligned) (Hz)	RDC (Hz)
A6	C2-H2	202,2	228,0	25,8
A8	C2-H2	201,0	223,8	22,8
A16	C2-H2	201,6	220,8	19,2
A24	C2-H2	201,0	229,8	28,8
T3	C7-H7	124,8	120,0	-1,6
T11	C7-H7	125,4	117,0	-2,8
HNF7	C1'-H1"	169,2	182,4	13,2
C9	C1'-H1'	163,2	172,2	9,0
C17	C1'-H1'	161,4	166,2	4,8
T3	C6-H6	176,4	198,0	21,6
A6	C8-H8	213,6	244,2	30,6
A8	C8-H8	215,4	238,8	23,4
A16	C8-H8	213,6	234,0	20,4
A24	C8-H8	214,8	240,6	25,8
HNF7	C6-H6	157,8	189,0	31,2
HNF7	C8-H8	162,0	191,4	29,4
ABA20	C1'-H1'	147,0	147,0	0,0
ABA20	C1'-H1"	145,8	148,2	2,4

3 Experimental details

3.2.4 Structure calculation

The Simulated Annealing protocol and Structure validation were carried out as described in section 3.1.4.

Calculation input Structure calculations were performed with XPLOR-NIH v2.20 [Schwieters et al., 2003]. As the HNF was found to exist in two different orientation with in the duplex, two calculation runs were performed with the HNF methylene group pointing towards the major or minor groove for face-up and face-down orientations, respectively. A total of 401 (403) NOE distance restraints and 19 Residual Dipolar Couplings were used for the calculation of the face-up (face-down) orientation. The experimental data were supplemented with 124 backbone dihedral restraints, 72 hydrogen bond distance restraints and 24 planarity restraints (see Table 3.5).

Initial MD-calculations were performed with dihedral restraints, allowing both A-form and B-form conformations (with error bars of $\pm 50°$). B-form conformation was experimentally confirmed by NOESY-cross-peaks intensities characteristic of B-DNA. Consequently, regular dihedral values from the literature [Roberts, 1993] were included in the calculations for all but the HNF residue.

Tabelle 3.5: *Overview of structural statistics for 13merHNF calculations*

	face-up	face-down
NOE restraints		
- total	401	403
- interresidue	225	226
- intraresidue	176	177
- 2AP to DNA	33	31
RDC restraints	19	19
Dihedral angle restraints	124	124
H-bond restraints	72	72
Base pair planarity restr.	24	24
NOE viol. (> 0.5 Å)	0	0
RDC viol. (> 0.4 Hz)	0	0
Dihe viol. ($> 5°$)	0	0
RMSD to ave. struct. in Å	0.63	0.46

4 Results and discussion

4.1 2-Aminopurine

4.1.1 Introduction

2AP is known since the 1950s for its role in the mutagenic transition of A:T to G:C base pairs [Freese, 1959, Rogan and Bessman, 1970, Ronen, 1979]. As a structural isomer of A it can form stable WC-type base pairs with T (see Fig. 4.1) [Ronen, 1979, Sowers et al., 1986]. Base pairs with C, A or G are much weaker [Sowers et al., 1986, Fazakerley et al., 1987, Hochstrasser et al., 1994, Law et al., 1996]. Thus 2AP is incorporated instead of A by DNA polymerase opposite to T [Freese, 1959], though at a

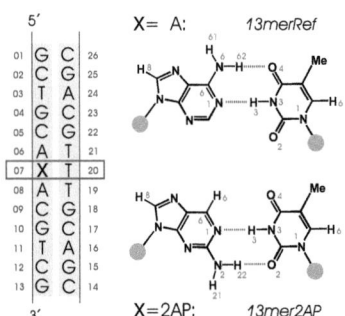

Fig. 4.1: DNA duplex sequence with chemical structure of the 2AP-T and A-T base pairs.

lower frequency than A [Rogan and Bessman, 1970, Watanabe and Goodman, 1982]. The mutagenic transition of A:T to G:C occurs during replication of 2AP:T base pairs [Bernstein et al., 1976, Goodman et al., 1977]. By adjusting the type of polymerase [Grossberger and Clough, 1981] or the concentration of nuclease [Clayton et al., 1979] the mutagenic rate can be tuned. The influence of nucleotide sequence on the mutagenic rate has no discernable logical pattern [Pless and Bessman, 1983]. Another way to stimulate the mutagenic transition is to increase the dCTP- or decrease the TTP-level

4 Results and discussion

in the cell [Caras et al., 1982]. In 1981 Watanabe et al. [Watanabe and Goodman, 1981] demonstrated that 2AP:C mispairs form at 320-fold higher frequency than A-C mispairs. They concluded that 2AP:C mispairing is central to the mutagenicity of 2AP.

The stabilization of the 2AP:C base pair in solution therefore had to be investigated. Four models have been proposed for the geometry of this mispair. Ronen [1979] suggested a WC geometry with one hydrogen bond, while Goodman and Ratliff [1983] presented evidence for two hydrogen bonds involving one of the bases in the rare tautomeric imino form. Sowers et al. put forth the other two pairing schemes. Based on NMR imino proton data and pH-titration experiments, they proposed WC geometry through protonation of the mispair at neutral pH [Sowers et al., 1986]. A ^{15}N-enriched NMR-study later suggested a wobble geometry at high pH [Sowers et al., 1989], corroborated by fluorescence anisotropy measurements which showed pH-dependent conformational changes in the geometry of the 2AP:C mispair [Guest et al., 1991]. ^{15}N-enriched NMR-studies by Fagan et al. [1996] and Sowers et al. [2000] reproduced this pH-dependence and could finally demonstrate that the wobble geometry is predominant at neutral pH while protonation occurs sequence-dependently at lower pH. This conclusion is supported by a theoretical study [Sherer and Cramer, 2001].

The fluorescence properties of 2AP differ markedly from those of the natural nucleobases [Longworth et al., 1966, Callis, 1979, Serrano-Andres et al., 2006]. This was recognized originally by Ward et al. [1969] who observed a decrease in quantum yield of fluorescence by two orders of magnitude upon stacking of 2AP in a DNA helix. The effect was first exploited to determine thermodynamic parameters of stacking associations by Bierzynski et al. [Bierzynski et al., 1977b,a, Gajewska et al., 1982].

Fluorescence quenching of 2AP can also be used to study structural transitions in biological systems [Ward et al., 1969]. At the beginning such applications were thwarted by the lack of site-specific incorporation techniques. After chemical synthesis of 2AP-containing DNA-duplexes was realized [Eritja et al., 1986, McLaughlin et al., 1988], fluorescence quenching of 2AP was employed to study structural transitions in DNA [Lycksell et al., 1987, Patel et al., 1992]. In 1993 Bloom et al. [Bloom et al., 1993, 1994] were the

4.1 2-Aminopurine

first to use 2AP fluorescence quenching to follow the kinetics of polymerase-catalyzed 2AP-insertion on a millisecond timescale. But on the whole, fluorescence quenching of a 2AP nucleobase surrogate is used to detect base stacking-unstacking transitions.

The local dynamics in the vicinity of 2AP, and its variation with sequence, can be studied by monitoring the fluorescence anisotropy decay [Nordlund et al., 1989]. This technique was used by Guest et al. [1991] to characterize the dynamic behaviour of mismatched base pairs. Hochstrasser et al. [1994] combined both techniques - fluorescence quenching and anisotropy decay - to study DNA double strand melting upon binding to the Klenow fragment. Xu et al. [1994] utilized 2AP fluorescence to show that DNA duplex melting proceeds via a highly flexible, yet B-DNA type transition state.

Numerous studies - facilitated by improved synthesis [Fujimoto et al., 1996] and subsequent commercial availability of 2AP-containing oligonucleotides - exploit 2AP fluorescence to investigate various problems in structural biology and biophysics: methyltransferase-induced base flipping [Allan and Reich, 1996, Allan et al., 1998, Holz et al., 1998, Allan et al., 1999, Stivers et al., 1999, Gowher and Jeltsch, 2000, Reddy and Rao, 2000, Bernards et al., 2002, Daujotyte et al., 2004, Su et al., 2004, Neely et al., 2005, Lenz et al., 2007], conformational changes and enzymatic cleavage of the hammerhead ribozyme [Menger et al., 1996, 2000, Kirk et al., 2001], promoter binding and clearance of T7 RNA polymerase [Jia et al., 1996, Ujvari and Martin, 1996, Bandwar and Patel, 2001, Liu and Martin, 2002], binding and strand separation of primer-template DNA by T4 DNA polymerase [Marquez and Reha-Krantz, 1996, Beechem et al., 1998, Fidalgo da Silva et al., 2002, Mandal et al., 2002, Hariharan and Reha-Krantz, 2005, Hariharan et al., 2006, Tleugabulova and Reha-Krantz, 2007], and charge transfer mechanisms together with polar solvation in DNA [Kelley and Barton, 1999, Wan et al., 2000, Fiebig et al., 2002, Jean and Hall, 2002, O'Neill and Barton, 2002a,b, Jean and Hall, 2004, O'Neill and Barton, 2004, Hardman and Thompson, 2006, Jean and Krueger, 2006, Hardman and Thompson, 2007]. All of these studies utilize 2AP fluorescence.

A new possibility to study structural transitions in biomolecules has recently been devised which makes use of unique properties of 2AP. Johnson et al. [2004] demonstrated

4 Results and discussion

that a low-energy Circular Dichroism band is observed for 2AP dinucleotides incorporated into DNA or RNA double strands. They used this method to monitor structural changes in RNA hairpin loops [Johnson et al., 2005b] or breathing fluctuations at replication forks [Johnson et al., 2005a]. This new technique and the high number of 2AP-related publications show the significance of 2AP for the study of biological macromolecules.

Structure perturbation upon incorporation of 2AP into a DNA-helix was investigated in detail only once, by Nordlund et al. [1989]. A palindromic sequence containing two 2AP:T base pairs was investigated by NMR spectroscopy, fluorescence decay and Molecular Dynamics simulations. Base pairing is observed in the duplex, but the melting temperature of the duplex was considerably lower than in the reference with WC A:T base pairs. The missing of NOE-crosspeaks to the 5'-side of the 2AP-substitution was interpreted as a local disturbance of the overall B-type helix [Nordlund et al., 1989]. Studies on minor groove binding drugs support this conclusion. They indicate that 2AP incorporation into A-T tracts reduces the binding affinities of these drugs to the level of G-C tracts [Loontiens et al., 1991, Patel et al., 1992]. Minor groove distortions by 2AP have also been proposed by a theoretical study [Lankas et al., 2002].

Dynamical effects due to incorporation of 2AP were examined even more scarcely. Nordlund et al. [1989] reported efficient recognition and cleavage of 2AP-containing duplexes by EcoRI, and concluded that dynamics are not perturbed. On the contrary, Lycksell et al. [1987] observed a 1 ms lifetime for the 2AP:T base pair, lower than for the corresponding A:T base pair, but state that the accuracy is correct probably only within a factor of 2. A destabilizing effect on the neighbouring pairs was not observed.

When structural transitions in biological systems are studied with a molecular probe, the assumption is that the modified system behaves like the natural one. Consequently the introduction of the probe must leave the structure unchanged. Disturbance of the original helix when 2AP occupies an A site would limit or even prevent the use of 2AP-fluorescence in some studies of biological systems. Therefore the above-mentioned hypothesis of a local disturbance has to be tested.

For this purpose a detailed analysis of the solution structure of a DNA-duplex in which

4.1 2-Aminopurine

an A is substituted by 2AP is presented. The only change introduced into the helix is the difference in the position of the amino group of A and 2AP (cf. Fig. 4.1). In contrast to previous studies [Lycksell et al., 1987, Nordlund et al., 1989] the nonpalindromic nature of the sequence allows to have a single perturbation site. Thus any change in structure or base pair dynamics can be directly attributed to the 2AP incorporation. The central base pair was modified in order to detect possible long-range effects (up to three base pairs) of 2AP incorporation, which otherwise might be rendered ambiguous due to base pair fraying effects [Nonin et al., 1995]. A symmetric 13 base pair sequence was chosen to dismiss the possibility of mispairing, loop formation and fraying effects.

4 Results and discussion

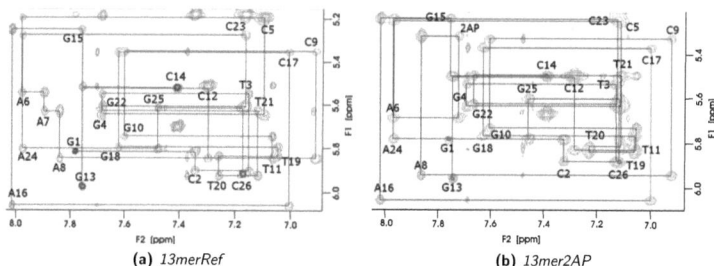

Fig. 4.2: *Sequential assignment in the sugar H1' base H6/H8 region of the NOESY-spectrum in D_2O for 13merRef (a) and 13mer2AP (b). Only the intraresidual cross-peaks H1'-H6/H8 are marked for clarity. Starting points are marked with blue, end points with red circles. All cross-peaks expected for a regular B-DNA helix are observed for both duplexes.*

4.1.2 Results

4.1.2.1 Chemical shift analysis

The NOESY-spectra were assigned with the sequential approach [Roberts, 1993, Bloomfield et al., 2000]. The sequential assignments of the sugar H1' and base H6/H8 protons for 13merRef and 13mer2AP are shown in Fig. 4.2a and 4.2b, respectively. Cross peaks expected for a regular B-DNA helix are present in the NOESY-spectra of both samples. Based on the assignment of the H1' and H6/H8 protons, the remaining base and sugar protons could be assigned by combining the information from the COSY-, TOCSY-, HMQC-, and NOESY-spectra. Assignment of the exchangeable protons was done in the WATERGATE-NOESY-spectrum. The imino proton of T20 was easily identified, because of the symmetry of the sequence, but the remaining imino, amino, and H2 protons had to be referenced to the already assigned non-exchangeable C H5 protons via the H42/H41-H5 NOE cross-peaks.

^{1}H Chemical Shift Differences (CSDs) between 13merRef and 13mer2AP are negligible (< 0.10 ppm) for all but the H1', imino, and H2 protons of the central three base pairs (see Table 4.1 and Table 13 in the Appendix, section 10). There is no significant trend that either 3'- or 5'-neighbours exhibit larger CSDs. No CSDs for the H6 and H8 protons

4.1 2-Aminopurine

Tabelle 4.1: *Selected ^1H Chemical shift differences (CSD) (reference HOD at 4.80 ppm, calculated as X(13merRef)-X(13mer2AP)) for 13merRef and 13mer2AP. Only the inner 7 base pairs are shown. In the case of X7 H2 the difference between the chemical shifts of H2 (A) and H6 (2AP) is taken.*

Residue	H1'	H1/H3/H41/H42	H2/H5/H7	H6/H8
G4	-0.01	-0.01	-	-0.03
C5	-0.08	0.03/-0.01	-0.02	-0.04
A6	-0.17	-	-0.16	-0.01
X7	0.29	-	-0.67	0.16
A8	-0.12	-	-0.13	-0.04
C9	0.00	-0.03/0.02	-0.06	-0.04
G10	-0.01	-0.05	-	-0.02
C17	-0.04	-0.01/-0.02	-0.02	-0.02
G18	0.00	0.03	-	-0.03
T19	-0.03	0.21	-0.04	-0.04
T20	0.09	0.53	-0.03	0.01
T21	0.11	0.13	0.00	-0.02
G22	-0.04	-0.05	-	0.00
C23	0.01	-0.02/-0.02	-0.02	0.01

for purine and pyrimidine bases with the exception of residues A7/2AP7 are observed.

In Fig. 4.3 the sum of the absolute values of the CSDs of all protons belonging to one residue is given for 13merRef to 13mer2AP (blue columns) and 13merRef to 13merRefGC (red columns). In 13merRefGC the central base pair of 13merRef is substituted by G:C. With the exception of the modification site, absolute per-residue CSDs from 13merRef to 13mer2AP are smaller than to 13merRefGC. Per-residue CSDs are significant two bases in each direction from the modification site in the 2AP-containing strand. In the unmodified counterstrand only the central three T bases exhibit significant CSDs. The chemical shifts of all assigned protons and carbons for 13merRef and 13mer2AP are given in the Appendix, sections 8 and 9, respectively.

4 Results and discussion

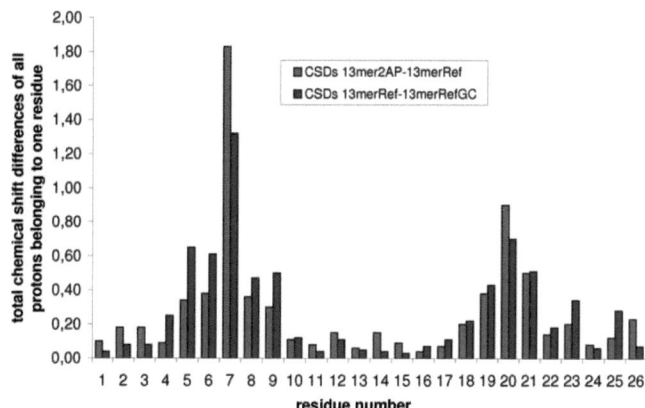

Fig. 4.3: *The sum of the absolute values of the CSDs of all protons belonging to one residue is given for: 13mer2AP to 13merRef (blue columns) and 13merRef to 13merRefGC (red columns) for comparison.*

4.1 2-Aminopurine

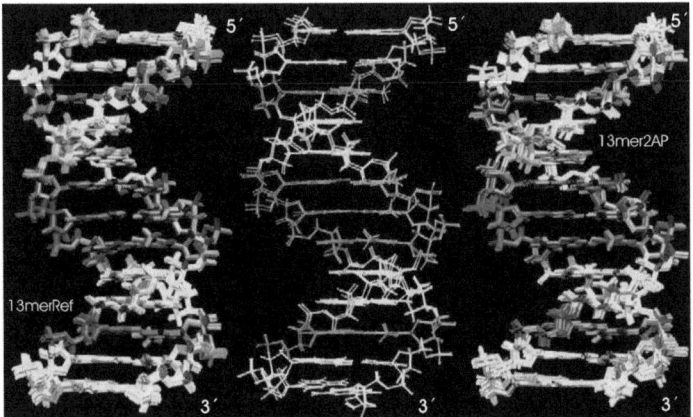

Fig. 4.4: *Overlay of the 10 best-energy, minimum-violation structures for 13merRef and 13mer2AP. The RMSD among each set of structures is 0.30 Å and 0.33 Å, respectively. The average structures for the two duplexes are compared in the middle; their RMSD is 0.46 Å.*

4.1.2.2 Structural comparison

From a family of 100 calculated structures the 10 minimum-energy, violation-free structures are shown in Fig. 4.4 for 13merRef (left) and 13mer2AP (right). The latter were used to calculate an average structure each. The accuracy of the calculations was checked by back-calculating the NOESY-spectrum from the average structure, followed by comparison with the experimental spectrum (Fig. 4.5a and 4.5b). Additionally, the RDCs were back-calculated from the average structure with the program PALES [Zweckstetter, 2008]. The correlation plots of experimentally determined vs predicted RDCs are shown in Fig. 4.6a and 4.6b and yielded correlation factors (R) of 1.000 and q-factors of 0.002 and 0.003 for *13merRef* and *13mer2AP*, respectively. The precision of the calculations can be assessed by the RMSD of the 10 best structures from their average, 0.30 Å for 13merRef and 0.33 Å for 13mer2AP. An overlay between the averaged structures is shown in Fig. 4.4 (middle part); their RMSD is 0.46 Å.

Helical parameters were calculated for the 10 minimum-energy, violation-free structu-

4 Results and discussion

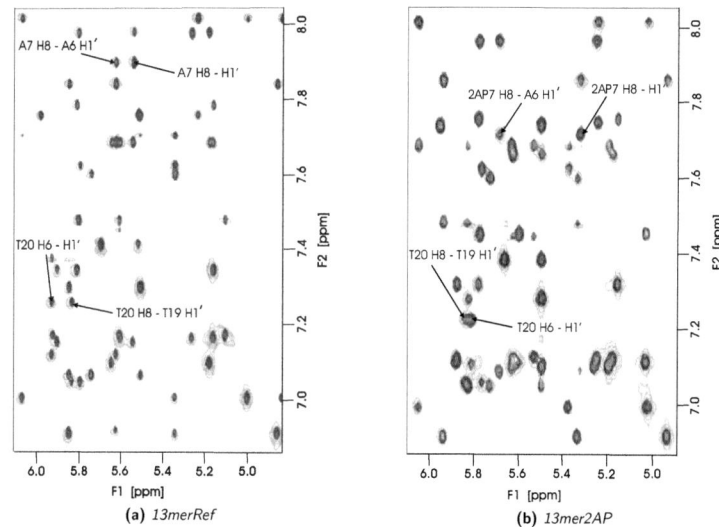

Fig. 4.5: *Overlay of the experimental NOESY-spectrum at 150 ms mixing time (green) and the NOESY-spectrum back-calculated from the average structure (red). Arrows point to NOE cross-peaks involving the modification site.*

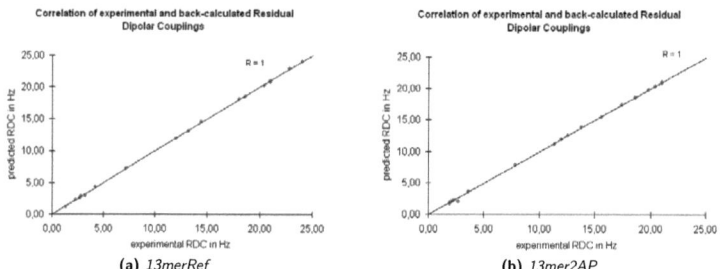

Fig. 4.6: *Plot of the experimental vs predicted RDCs for 13merRef and 13mer2AP.*

4.1 2-Aminopurine

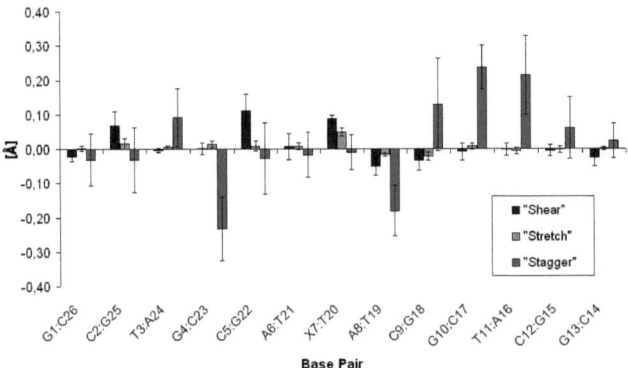

Fig. 4.7: *13merRef and 13mer2AP: Differences (X(13merRef)-X(13mer2AP)) in translational helical parameters between base pair partners [Å]. Blue columns represent "Shear"-values, green ones "Stretch" and red ones "Stagger". Error bars indicate the estimated uncertainty.*

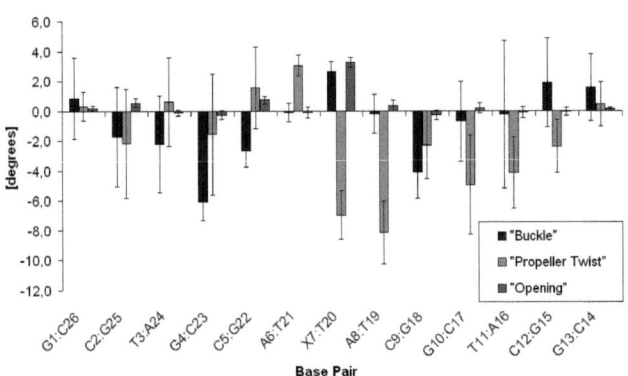

Fig. 4.8: *13merRef and 13mer2AP: Differences (X(13merRef)-X(13mer2AP)) in rotational helical parameter between base pair partners [°]. Blue columns represent "Buckle"-values, green ones "Propeller Twist" and red ones "Opening". Error bars indicate the estimated uncertainty.*

4 Results and discussion

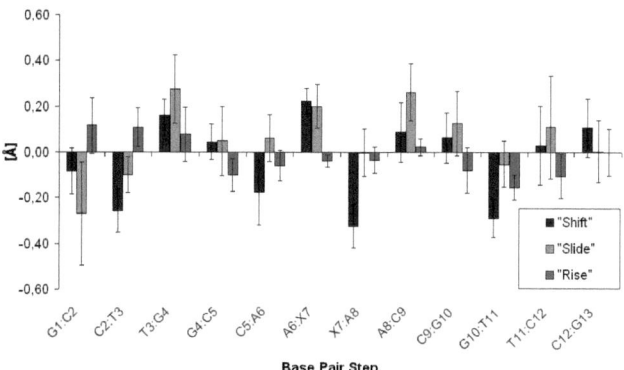

Fig. 4.9: *13merRef and 13mer2AP: Differences (X(13merRef)-X(13mer2AP)) in translational helical parameter between base pairs [Å]. Blue columns represent "Shift"-values, green ones "Slide" and red ones "Rise". Error bars indicate the estimated uncertainty.*

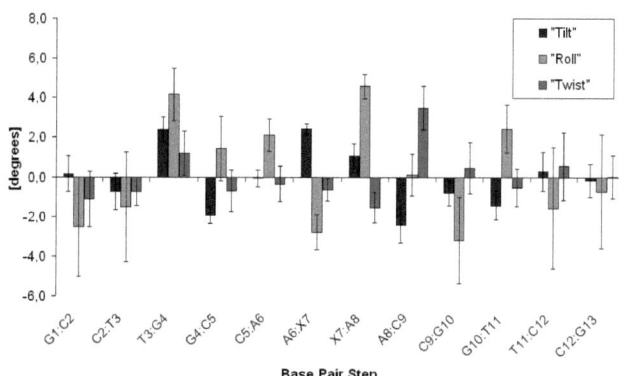

Fig. 4.10: *13merRef and 13mer2AP: Differences in rotational helical parameter between base pairs [°]. Blue columns represent "Tilt"-values, green ones "Roll" and red ones "Twist". Error bars indicate the estimated uncertainty.*

4.1 2-Aminopurine

res of 13merRef and 13mer2AP with the program 3DNA [Lu and Olson, 2003]. Fig. 4.7 and 4.8 depict differences between 13merRef and 13mer2AP in translational and rotational helical parameters between base pair partners, while Fig. 4.9 and 4.10 display differences in translational and rotational helical parameters between base pairs. The value (X) represented by a column in Fig. 4.7- 4.10 is calculated as:

$$X = \frac{1}{10} \left(\sum_{n=1}^{10} X_n(13merRef) - \sum_{n=1}^{10} X_n(13mer2AP) \right) \qquad (4.1)$$

where $X_n(13merRef)$ and $X_n(13mer2AP)$ represent the helical parameter values calculated for the 10 minimum-energy, violation-free structures of 13merRef and 13mer2AP, respectively. The RMSD of these values to their average is shown as an error bar. Thus a measure for the precision of a given helical parameter is obtained. In general, for translational parameters only values which deviate by at least 0.1 Å (with the corresponding uncertainties taken into account) and for rotational parameters $> 5°$ are interpreted. Deviations below these thresholds are too weak to be reliably described by the force field (deviations of 0.1 Å and 5° from the equilibrium value are allowed for bond lengths and angles respectively).

In the following, the observed perturbations in helical parameters are presented. For definition of the helical parameters please see section 2.1, Fig. 2.5.

- The "Stagger"-values (red columns in Fig. 4.7) differ significantly for base pairs G4:C23 (-0.23 ± 0.09 Å), A8:T19 (-0.18 ± 0.07 Å) and G10:C17 (0.24 ± 0.06 Å), while values for "Shear" and "Stretch" (blue and green columns in Fig. 4.7) do not show significant deviations.

- Of the rotational helical parameters within base pairs only the "Propeller Twist"-values (green columns in Fig. 4.8) of base pairs X7:T20 (-7.0 ± 1.6°) and A8:T19 (-8.1 ± 2.1°) exhibit significant deviations, while "Buckle"- and "Opening"-values are equal within the estimated error (blue and red columns in Fig. 4.8 respectively).

- The rotational helical parameters between base pairs "Tilt", "Roll" and "Twist" (blue, green and red columns in Fig. 4.10 respectively) do not show significant variations within

4 Results and discussion

the estimated error.

- The "Shift"-values (blue columns in Fig. 4.9) differ significantly for base pair steps C2:T3 (-0.26 ± 0.10 Å), A6:X7 (0.20 ± 0.03 Å) and A8:C9 (0.26 ± 0.03 Å). The "Slide"-values (green columns in Fig. 4.9) differ significantly for base pair steps T3:G4 (0.28 ± 0.15 Å), A6:X7 (0.22 ± 0.11 Å) and A8:C9 (-0.26 ± 0.12 Å). The "Rise"-values (red columns in Fig. 4.9) do not differ within the estimated confidence interval.

In summary, significant deviation in helical parameters are observed throughout the duplex, with the central three base pairs exhibiting the largest deviations. The average helical parameters calculated from the values for the 10 minimum-energy, violation-free structures of 13merRef and 13mer2AP are listed in the Appendix in sections 11 and 12, respectively.

4.1 2-Aminopurine

4.1.2.3 Comparison of base pair dynamics

The T20 imino proton resonance behaves substantially different depending on the counter base 2AP or A. In 13mer2AP this resonance can be observed in the 1D-spectrum, though it is broadened substantially (cf. Fig. 4.12). The corresponding diagonal signal in the NOESY-spectrum however vanishes with increasing mixing time beyond 100 ms. Fig. 4.11 illustrates this unusual decay behaviour by comparing signal intensities for the diagonal and representative cross-peak signals of T20 (•) and T3 (▲) in 13mer2AP. Diagonal signals T20 H3 and T3 H3 (solid lines) are normalized at 50 ms to show their different decay behaviour.

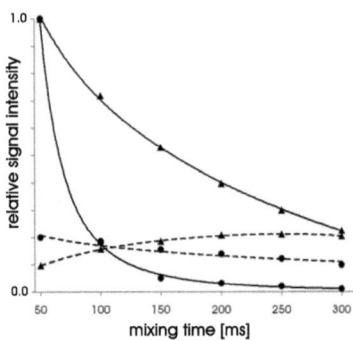

Fig. 4.11: T20 (•) and T3 (▲) imino proton intensities for 13mer2AP at 298 K. Solid lines depict diagonal signals, while dashed lines represent cross-peak intensities of T20H3–2APH6 and T3H3–A24H2.

Without this normalization, the diagonal peak of T20 H3 reaches only 20 % of the T3 H3 signal intensity at 50 ms. Cross-peaks T20H3–2APH6 and T3H3–A24H2 (dashed lines) are given relative to the corresponding diagonal signal at 50 ms. The diagonal T20 H3 peak almost vanishes when going from 50 to 100 ms mixing time. The intensity of the NOE cross-peak to 2AP H6 decays slowly, in contrast to the T3H3–A24H2 cross-peak where the intensity increases for the first 200 ms.

To test whether H_2O exchange is responsible for the unusual decay behaviour of T20 H3 we performed water saturation transfer experiments. In the latter experiments 1D-spectra in H_2O are aquired with two different methods of attenuating the water signal. With the presaturation method the water signal is saturated before the actual pulse sequence (Fig. 4.12 (B,D): irradiation time 3 s, strength 55 dB). Consequently, exchange of now saturated water protons with unsaturated imino protons leads to attenuation of the latter signal. This cosaturation can be circumvented by using the WATERGATE

4 Results and discussion

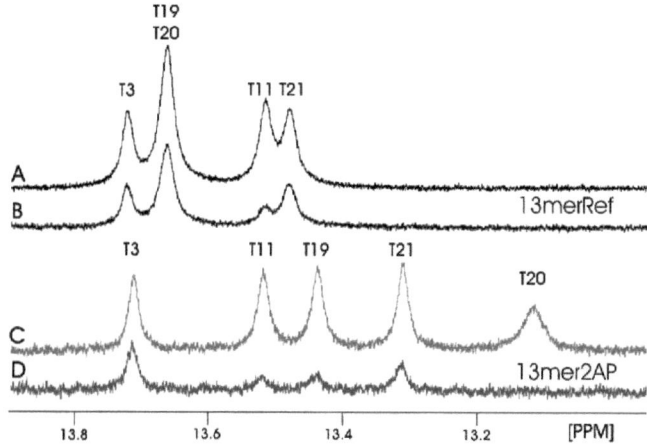

Fig. 4.12: *Saturation transfer experiments in H_2O. Resonances from exchanging protons are reduced in the 1D presaturation spectra (lower black lines, B,D), reflecting the relative rates by which imino protons exchange with the solvent. Water saturation is avoided in 1D WATERGATE spectra which are shown for comparison (upper gray lines, A,C).*

pulse sequence (Fig. 4.12 (A,C)), which does not excite the water protons and thus prevents their detection. Here water exchange with imino protons leaves their signal unperturbed. Fig. 4.12 depicts the imino proton region with both attenuation techniques, presaturation (B,D) and WATERGATE (A,C) for 13merRef, 13mer2AP respectively. In 13mer2AP the cosaturation of the T20 imino proton is so strong that the signal is not observable in presaturation experiments. The signals of the directly adjacent bases T19 and T21 are reduced in intensity when compared to the corresponding signals in 13merRef.

Base pair dynamics can be studied by measuring the base pair lifetime τ_{op}. The latter is obtained by extrapolating a plot of imino proton exchange τ_{ex} vs inverse base concentration 1/[B] to infinite catalyst concentration (for explanation see section 2.2.1). Fig. 4.13 shows the corresponding data and linear fits for the central 7 base pairs and the extrapolated lifetimes for 13merRef and 13mer2AP. Lifetimes in green refer to overlap-

4.1 2-Aminopurine

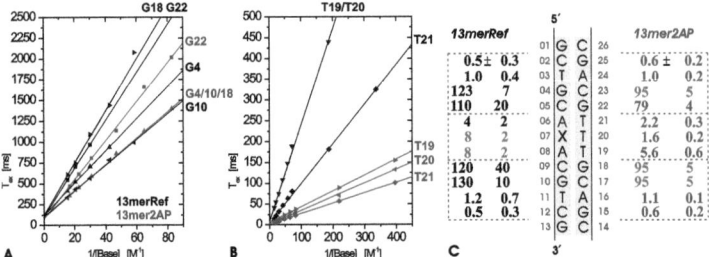

Fig. 4.13: *Basepair lifetime determination. The inversion recovery of imino proton signals is affected by TRIS base which catalyzes exchange with water. The exchange time τ_{ex} depends on the inverse base concentration $1/[B]$, with different ranges for G:C and A:T pairs. Extrapolation to infinite concentration gives the lifetimes which are collected in the right panel. Lifetimes in green could not be determined separately due to spectral overlap.*

ped resonances in the imino proton region. Their overlap is complete and signal recovery identical; therefore only averaged lifetimes can be given. The R^2-values for all linear fits are above 0.996. Confidence intervals for each lifetime are also determined from the fits. The lifetimes of the terminal base pairs could not be measured due to "base pair fraying" which is commonly observed at the helical termini. Here base pair lifetimes are considerably shortened, broadening the signal from terminal imino protons to the point of vanishing and weakening the signal from neighbouring ones [Leijon and Graslund, 1992, Nonin et al., 1995]. Lifetimes for the semiterminal G:C (0.5 ms) and A:T base pairs (1.1 ms) are found to be identical for 13merRef and 13mer2AP within the estimated error (the corresponding linear plots are shown separately in Fig. 4.15 for reasons of clarity). Contrary to the outer base pairs, lifetimes for the inner seven pairs differ between 13merRef and 13mer2AP. The substitution A→2AP reduces τ_{op} severely for the central A:T pairs (by factors 0.55, 0.20 and 0.70 for T21, T20, T19) and less so for the more remote G:C pairs (0.8-0.7).

4 Results and discussion

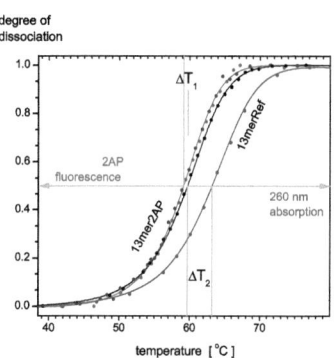

Fig. 4.14: *Melting curves via 13mer2AP absorbance (black) and 2AP fluorescence yield (blue points - after heating, red points - after cooling).*

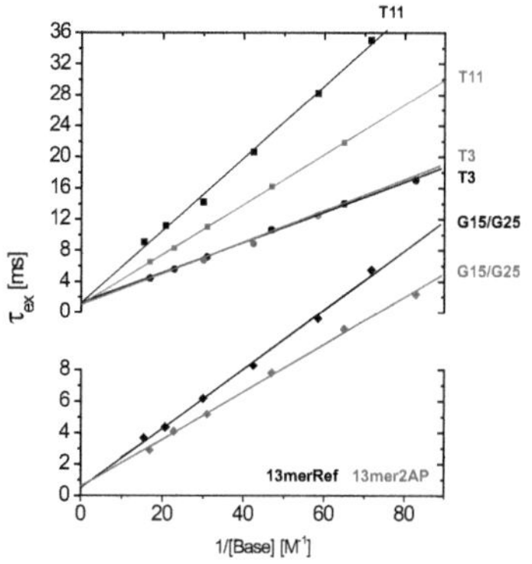

Fig. 4.15: *Linear fits for the imino proton exchange times of T3, T11, G15 and G25 vs the inverse base catalyst concentration.*

4.1.2.4 Comparison of duplex melting

Melting curves of 13mer2AP are shown in Fig. 4.14. Black points depict the melting point as determined by following the temperature-dependent absorption at 260 nm. Blue and red points represent duplex melting as determined by monitoring temperature-dependent 2AP fluorescence yield for heating and cooling, respectively. With UV absorption the melting behavior of the entire duplex is monitored whereas the 2AP fluorescence yield is sensitive to the local environment only. Melting points of 59.7 and 59.2 °C for the duplex as a whole and the centre are measured respectively. Thus $\Delta T_1 = 0.5$ K can be determined for duplex vs centre melting. The melting point of 13merRef was measured as 63.2 °C, with $\Delta T_2 = 3.5$ K for 13merRef vs 13mer2AP melting.

4 Results and discussion

4.1.3 Discussion

Structure perturbation

Nordlund et al. [1989] studied 2AP in a palindromical decamer sequence. They reported that NOE cross-peaks expected for a regular B-DNA helix, were missing to the 5'-side of the 2AP incorporation site. From that they conclude that the structure is perturbed locally with the perturbation extending only to the adjacent base pairs [Nordlund et al., 1989]. In contrast, all NOE cross-peaks expected for a regular B-DNA helix have been identified for 13mer2AP. This indicates that 2AP incorporation induces no pronounced perturbation in 13mer2AP. A possible explanation might be the higher detection limit of the 600 MHz NMR spectrometer (instead of 300 MHz) and the 5-fold higher duplex concentration employed in this work. Another reason might be the low stability of the decamer duplex, whose melting point decreases by 8.6 K (to 32.8 °C) upon introduction of the two 2AP residues [Nordlund et al., 1989].

The per-atom CSD data analysis for 13merRef and 13mer2AP (cf. section 4.1.2.1) supports the above conclusion that no pronounced perturbation is induced upon 2AP incorporation. Significant per-atom CSDs are observed exclusively between protons spatially adjacent to 2AP in 13mer2AP or A in 13merRef. This suggests that the observed CSDs are due to the different ring current effects induced by 2AP and A [Wijmenga et al., 1997]. Thus per-atom CSD data does not hint at structural differences between the two samples, but instead points to electronic differences between 2AP and A, which are evidenced by quantum chemical calculations [Broo, 1998, Mishra et al., 2000, Jean and Krueger, 2006].

On the contrary, absolute per-residue CSDs indicates that also structural differences exist between 13merRef and 13mer2AP which propagate at least two base pairs in each direction from the modification site. Comparison with the total CSDs between 13merRef and 13merRefGC suggests that the substitution of an A:T by a G:C base pair has a stronger impact on the helical structure than substitution of A by 2AP since total CSDs for the latter are equal or less for the base pairs adjacent to the modification site (cf. Fig. 4.3). This indicates that the helical structures of 13merRef and 13mer2AP both

4.1 2-Aminopurine

adopt a regular B-DNA conformation. Yet very subtle structural changes between these two structures exist, which effect the significant CSDs observed for the bases C5 and C9.

A comparison to CSD data on other single mismatch sites is instructive (Note that ^{1}H-CSD data for A→2AP substitution, other than reported here, is not available.). Bhattacharya et al. [2002] examined the effect of single mismatches A:A, G:G, C:C on the imino proton chemical shift. CSDs for the adjacent base pairs were significantly more perturbed, than in our case, reflecting the disturbed helical structure resulting from the non-WC-geometries of the mismatch sites. Klewer et al. [2000] studied incorporation of 3-nitropyrrole into DNA. They report CSDs to the unperturbed structure that are comparable to these of 13merRef/13mer2AP. This is surprising given the fact that 3-nitropyrrole is in a syn conformation. Despite this perturbation, the overall helical arrangement is found to be the B-form. This is in line with the results for 13mer2AP, where a B-type helix with small but significant deviations to the structure of 13merRef is suggested by the CSD data.

The overall B-DNA helical structure of 13merRef and 13mer2AP is supported by the calculation results. Slightly smaller RMSDs among the 10 best-energy structures (0.30 and 0.33 Å for 13merRef and 13mer2AP) as compared to the RMSD between the corresponding average structures (0.46 Å) are observed. This indicates that while the overall conformation is identical, minor but significant differences exist. These differences can be visualized by analysing and comparing the helical parameters for 13merRef and 13mer2AP. All of the helical parameters are within the range typically observed for B-DNA (cf. section 2.1). However, significant deviations between 13merRef and 13mer2AP structures exist throughout the duplex (see section 4.1.2.2) for the parameters "Stagger" and "Propeller Twist" between base pair partners, and "Slide" and "Shift" between two base pairs. The largest deviations are observed for the central three base pairs. This is shown in Fig. 4.16 which displays the latter pairs enlarged and from a tilted angle in order to improve visualization of the helical parameters of interest. In conclusion, the structural differences between 13merRef and 13mer2AP are small but detectable.

The structural perturbation induced upon 2AP incorporation is very weak compared

4 Results and discussion

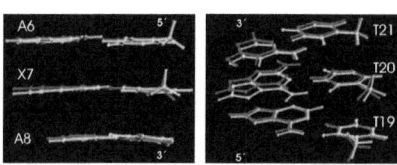

Fig. 4.16: *Comparison of the three central base pairs (average structures, center of Fig. 4.4 enlarged) for 13mer2AP (yellow) and 13merRef (red). The backbone was omitted for clarity. In the right panel the view was tilted to better visualize the differences in "Shift"- and "Slide"-values.*

4.1 2-Aminopurine

to other base analogues. Engman et al. investigated the structure of tC containing DNA. They find that the overall B-DNA conformation is preserved. However, the DNA is slightly bent and several NOE cross-peaks involving tC are not consistent with a B-DNA conformation. Furthermore, the second ring of tC, which is not involved in base pairing, extends into the major groove, thus preventing protein docking [Engman et al., 2004]. An even larger fluorophore, pyrene, was used by Smirnov et al. [2002] as a base pair analogue opposite to an abasic site. The overall B-DNA conformation is again preserved, but local mobility or even an alternative orientation of the 5'-adjacent base pair is introduced. This is indicated by the presence of two imino proton signals and very weak or lacking NOE cross-peaks for this base pair [Smirnov et al., 2002]. 2,4-Difluorotoluene - a steric mimic of T devoid of hydrogen bonding sites - was introduced by Guckian et al. [1998]. Although a strong thermodynamic destabilization is found ($\Delta T=11$ K), no significant deviation of B-DNA conformation is observed as indicated by uninterrupted sequential connectivities. However, a detailed analysis of helical parameters and base pair dynamics is lacking. A whole base pair devoid of hydrogen bonding was used in a related study [Guckian et al., 2000]. It was demonstrated that overall B-DNA conformation is retained. Local mobility however is increased as sensed by broadening of the imino proton resonance at position 5. None of these works use a joint (structure, thermodynamics and base pair kinetics) approach to study perturbation upon incorporation of fluorophores. Moreover, a detailed analysis of helical parameters and a comparison with a control with WC base pairs only is lacking. Thus the present work constitutes a novel approach to the topic of DNA perturbation studies.

Basepair lifetimes

Although the surrounding structure in 13merRef and 13mer2AP is almost identical, large dynamic differences are indicated by the unusually fast decay of the T20 imino proton diagonal signal in the NOESY-spectrum of 13mer2AP (cf. Fig. 4.11). Inspection of the diagonal signal in NOESY spectra was used by Engman et al. [2004] to qualitatively check for increased water exchange in a modified DNA duplex. The lack of any uncommon phenomena led to the conclusion that base pair dynamics are not perturbed by

4 Results and discussion

incorporation of the fluorophore tC. Following that reasoning, the fast decay observed with the T20 imino proton may indicate increased chemical exchange with water. This interpretation is corroborated by the results of the water saturation transfer experiments. Here the signal of the T20 imino proton is completely cosaturated in 13mer2AP, while the signals from T3 in the same duplex and T20 in the reference duplex are only slightly reduced (Fig. 4.12). Interestingly though, intensities of the adjacent imino protons (T19,T21) are more reduced compared to 13merRef, which indicates faster water exchange for these protons, too. This in turn suggests that 2AP incorporation also effects the dynamics of the adjacent base pairs.

The latter hypothesis has been validated by the results from base pair lifetime measurements. Upon 2AP incorporation the lifetimes of the central seven base pairs seem to be reduced. But the lifetimes of the G4, G10 and G18 imino protons for 13mer2AP and T19 and T20 for 13merRef (shown in green in Fig. 4.13) could not be determined separately due to complete spectral overlap. However their lifetimes should be similar since their recovery behaviour is identical (as assessed by inspection of the 21 1D-spectra resulting from the 21 delay settings after the inversion pulse). Thus the averaged lifetimes can be interpreted for the overlapped protons. Interpretation of differences in base pair lifetimes for the G18, T19 and T21 imino protons of 13merRef and 13mer2AP is complicated as the estimated confidence intervals overlap slightly. However, differences between these lifetimes can be assumed due to the observed reduction of imino proton signal intensity in the water saturation transfer experiment. Additionally, significantly reduced lifetimes for the G4 and G10 imino protons indicate an extended effect of 2AP incorporation on the base pair dynamics of the three adjacent base pairs in each direction. An alternative explanation for the reduced base pair lifetimes of G4 and G10 might be different sample conditions for 13merRef and 13mer2AP. This can be a significant source of errors, when comparing two different samples. However, lifetimes of the base pairs C2:G25, T3:A24,T11:A16, and C12:G15 are equal within the estimated error although these base pairs are influenced by base pair fraying effects. Since the latter effect is sensitive to solution properties (pH, buffer, temperature), different sample conditions

4.1 2-Aminopurine

can be ruled out as an explanation. Additionally, the excellent agreement between these lifetimes shows that the lifetime measurements are reliable even below 1 ms. In summary, one can conclude that 2AP incorporation has an extensive, distributed effect on the base pair dynamics of the three adjacent pairs in each direction.

Base pair lifetimes in 2AP-containing DNA have been examined by Lycksell et al. [1987]. Their result for the modification site, of 1 ± 2 ms, agrees qualitatively with the one reported here but is less precise. In contrast, these authors do not observe an extensive, distributed effect on base pair dynamics due to the incorporation of 2AP into the DNA helix. In their palindromic decamer sequence, the melting point is lowered upon incorporation of 2AP from 41.4 °C to 32.8 °C [McLaughlin et al., 1988]. This suggests that at 25 °C, the temperature at which base pair lifetimes were measured, the latter are already affected by duplex melting. This is corroborated by the fact that the opening motion of some inner base pairs is too rapid to be observed. Further support stems from the fact that the central two A:T pairs in their reference sequence constitute an A-tract motif, which would suggest a base pair lifetime for the core A:T pair of over 10 ms [Leroy et al., 1988, Moe and Russu, 1990] rather than 6 ms as observed by Lycksell et al. [1987]. Its length of 13 base pairs makes the non-palindromic duplex studied in the present work thermodynamically more stable. Thus it is less susceptible to destabilization due to base pair fraying. Influence of non-native modifications on the base pair dynamics of adjacent base pairs has been reported in the literature. Moe and Russu [1992] investigated a palindromic dodecamer sequence, with a G:T mismatch introduced at position 4. An increased base pair opening rate is demonstrated for position 5. A similar effect at position 3 could not be resolved due to the high error for the obtained base pair lifetime (60 %) in the reference and signal overlap in the mismatch sequence [Moe and Russu, 1992]. These observations support the finding that 2AP incorporation effects the base pair lifetimes of the three adjacent pairs in each direction.

The extensive, distributed effect of 2AP incorporation on base pair opening dynamics can be explained with transition state theory. For base pair opening to occur, the stabilizing enthalpy contribution from the hydrogen bonds between the partners and the

4 Results and discussion

stacking interaction with the adjacent base pairs have to be overcome when the transition state is formed. Thus a reduction in strength of either interaction would favor base pair opening and thereby lead to a reduced base pair lifetime. Similar reasoning is used to explain the severely reduced base pair lifetimes of the helical termini [Nonin et al., 1995].

Reduced stacking interactions and hydrogen bonding energies throughout the duplex are indicated by the 3.5 K reduction in overall duplex melting temperature upon 2AP incorporation. This is supported by the results of the structural analysis. The higher melting point suggests that stacking interactions and hydrogen bonding in 13merRef are stronger than for 13mer2AP. Thus any change of the position of base pairs (stacking) or base pairing partners (hydrogen bonding) relative to each other must lead to the reduction of either stabilizing energy term. Consequently, the observed deviations in "Stagger" and "Propeller Twist" (defined between base pair partners) on the one hand and "Shift" and "Slide" (defined between base pairs) on the other hand indicate reduced hydrogen bonding and stacking interactions, respectively. Locally, the strong impact of 2AP incorporation on the central three base pairs is indicated by the 0.5 K reduction in duplex melting temperature when monitored via 2AP fluorescence. Since the latter is sensitive only to the directly adjacent base pairs, the observed reduction implies considerable premelting and thus destabilization of the duplex structure around the 2AP modification site as a whole. Premelting around the incorporation site has also been observed by Law et al. [1996] in a 2AP modified undecamer, where the destabilization was found to be even stronger (1.6 K). As pointed out before, deviations in helical parameters are strongest for the central three base pairs, which corroborates the results of the melting study. The latter results are further supported by the fact, that base pair lifetimes are considerably shortened and water saturation transfer is increased for the central three pairs. Thus one can conclude that 2AP incorporation destabilizes the duplex structure, by reducing the stacking and hydrogen bond interactions. This in turn leads to a lower activation enthalpy for the transition state of the base pair opening reaction. Thereby the extensive distributed effect of 2AP incorporation on the base pair lifetimes can be

4.1 2-Aminopurine

explained.

The effect of 2AP incorporation on the base pair dynamics of remote pairs can be used to explain results from DNA-protein interaction studies. The latter indicate that although protein activity is reduced, binding affinities are enhanced upon incorporation of 2AP into DNA [Brennan et al., 1986, Petrauskene et al., 1995, Malygin et al., 1999, Reddy and Rao, 2000]. Methyltransferases bind to DNA by flipping the target base out of the helix [Allan and Reich, 1996, Allan et al., 1998, Malygin et al., 1999, Reddy and Rao, 2000]. Thus higher base pair opening rates would enhance this binding process. The latter assumption is supported by the observation of binding enhancement upon introduction of mismatches [Moe and Russu, 1992]. Reddy and Rao [2000] investigated EcoP15I DNA Methyltransferase binding to DNA. They found that binding was enhanced for 2AP-containing DNA as compared to native DNA, regardless, whether 2AP replaces A in the recognition sequence or outside. Similar results were reported by Malygin et al. [1999] for the T4 DAM Methyltransferase. Enhancement of the binding process when 2AP is substituted for an A just outside the recognition center can only be explained by the influence of 2AP substitution on the base pair opening dynamics of several pairs in each direction as was observed for 13mer2AP. Similar observations were made with 2AP-containing DNA binding to restriction endonucleases [Petrauskene et al., 1995, Brennan et al., 1986], which can be explained analogously.

However, the reduced base pair lifetime of 2AP:T cannot account for the increased water exchange in the absence of added catalyst. Despite the fact that in 13mer2AP the base pair lifetimes of A16:T11 (1.1 ms) and A24:T3 (1.0 ms) are significantly smaller than for 2AP:T (1.6 ms), the corresponding imino resonances are still detectable in the presaturation experiment, while that of 2AP:T is missing (see Fig. 4.12). This observation suggests that the short base pair lifetime of 2AP:T accounts only in part for the increased water exchange. Since τ_{op} is measured in the limit of high catalyst concentration, but the presaturation experiments were carried out in the absence of catalyst, intrinsic catalysis seems to be involved. One possible catalytic site could be the N1 atom in 2AP, which may have higher basicity compared to A. Electronic differences are indeed

4 Results and discussion

indicated by the CSD data, but it is unlikely that the pK is raised sufficiently high to account for the results. An alternative mechanism involves the T20 O4 atom, which lacks a hydrogen bond to its partner base and can be easily accessed by solvent or base catalyst via the major groove. Thus, by forming a strong hydrogen bond with a potential, effective catalyst, it brings the latter within reach of the imino proton. In this way the catalytic ensemble is preorganised, increasing the probability of the catalytic exchange. Additionally, lack of the sterically demanding amino group in the major groove suggests that for the imino proton to be accessed by the catalyst, full opening of the 2AP:T base pair is not required. Thus exchange catalysis is more effective for 2AP:T as compared to A:T base pairs. Support for this explanation comes from a comparison of T_1-values in the absence of added catalyst. The latter is shorter by roughly a factor three or more for T20 in 13mer2AP (29.5 ms) when compared to the other T imino protons in the duplex (T3: 199.7 ms, T11: 83.6 ms, T19: 98.8 ms, T21: 141.0 ms).

Another possible explanation for our observations would be exchange from the closed state. Arguments against this possibility are given by Nonin et al. [1995] who find that exchange from the closed state is negligible even for terminal base pairs, since the imino protons are not accessible sideways. The increased exchange rates of terminal imino protons are explained by reduced stacking interactions [Nonin et al., 1995]. Leroy et al. [1993] find that in $C:C^+$ mispairs the amino protons of C have different exchange times, depending on whether or not they are hydrogen bonded. This supports the conclusion that exchange from the closed state is strongly inhibited when the imino protons are not accessible sideways. The latter is true however for G:T mismatches, where possible exchange from the closed state was discussed by Moe and Russu [1992]. Since sideway accessibility is not the case for 2AP:T base pairs, exchange from the closed state can be ruled out.

The conclusion that the hydrogen bond of the T O4 atom with bulk water or base catalyst increases the efficiency of imino proton exchange is supported by results of Strekowski et al. [1987]. They studied the interaction of DNA with different intercalators and minor groove binders. Regardless of G:C or A:T binding specificity, all molecules contai-

4.1 2-Aminopurine

ning an amino side chain substituent selectively catalyzed A:T imino proton exchange with bulk water. Analogously to the 2AP:T case, this A:T specificity of can be explained by the formation of a hydrogen bond between the amino side chain of the drug and the T non-hydrogen-bonded O2 atom located in the minor groove. The possibility for this interaction is lacking in G:C base pairs since the formation of three hydrogen bonds leaves no hydrogen-bonding partner available. Thus the formation of this hydrogen bond preorganizes the catalytic ensemble, analogously to what was described for 2AP:T.

4 Results and discussion

4.2 2-Hydroxy-7nitrofluorene

4.2.1 Introduction

HNF is a fluorene derivative which has been synthesized and incorporated into a DNA double strand only recently [Dallmann et al., 2009]. Consequently, no reports on HNF or HNF-DNA constructs can be found in the literature. For 2AP subtle changes in structure and dynamics had to be analyzed in order to answer the question of applicability to biological systems. In contrast, HNF has been designed for the proof-of-principle that macromolecular vibrational modes can be measured via transient absorption spectroscopy with polarity probes [Dallmann et al., 2009]. Thus the focus of this part of the work is on the position of the fluorophore inside the DNA helix. A comparison of helical parameters with a native structure (as in case of 2AP) is not instructive since HNF is not as close to a natural base pair as 2AP:T.

Compared to 2AP, HNF is much larger in size and thus functions as a base pair surrogate. Due to synthetic reasons the HNF could only be obtained in sufficient purity in the α-glycosidic form. Preliminary molecular modelling of the HNF-containing duplex with the program HyperChem v7.5 indicated that the hydroxyl linkage introduces orientational flexibility of the HNF moiety. Thus with both, α- and β-glycosidic form of the HNF nucleotide, stacking of the HNF with its adjacent base pairs can be realized. The same test calculations suggested that the introduction of an abasic site imposes less steric strain on HNF and prevents the latter (or its partner base) from being flipped out of the helical stack. Thus the same sequence as for 13mer2AP was studied, with the central base pair substituted by **X**=HNF in the left-

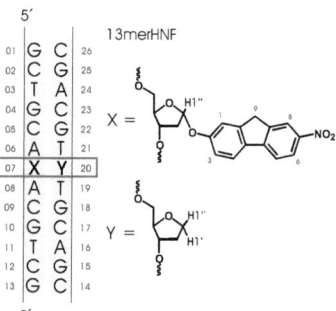

Fig. 4.17: *DNA duplex sequence with chemical structure of the HNF (**X**) and abasic site (**Y**).*

4.2 2-Hydroxy-7nitrofluorene

hand strand and **Y**=ABA (abasic site) in the other one (cf. Fig. 4.17). A symmetric, nonpalindromic 13 base pair sequence was chosen to dismiss the possibility of mispairing, loop formation and fraying effects. As in 13mer2AP, the central base pair was chosen for modification, in order to avoid fraying effects. The structures and nomenclature of HNF and the abasic site are also depicted in Fig. 4.17. For clarity only hydrogens which are important for the definition of the structure are shown. The nomenclature of the sugar protons follows the one described in section 2.1, with the hydrogen in the β-position of C1' symbolized by H1". Grey numbers indicate the numbering scheme of the carbon atoms and the corresponding hydrogens of the fluorene body (Fig. 4.17).

4 Results and discussion

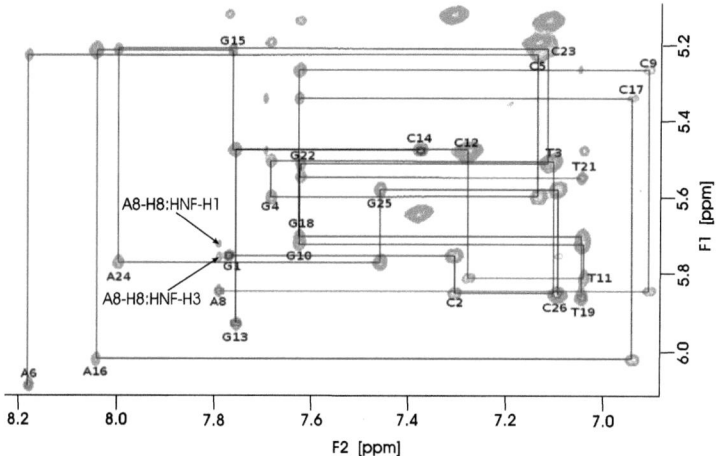

Fig. 4.18: *Sequential assignment in the sugar H1' base H6/H8 region of the NOESY-spectrum in D_2O for 13merHNF. Only the intraresidual cross-peaks H1'-H6/H8 are marked for clarity. Starting points are marked with blue, end points with red circles. Termination points of the sequential assignment within one strand due to the modification site are marked with purple circles.*

4.2.2 Results

4.2.2.1 Chemical shift analysis

The assignment of the DNA signals was achieved following standard procedures as described in Roberts [1993] and Bloomfield et al. [2000]. In contrast to 13mer2AP however, the sequential assignment was interrupted at the modification site due to the lack of base protons at the abasic site (cf. Fig. 4.18, indicated by purple circles). Assignment at the modification site was however possible through several inter- and intrastrand NOE cross-peaks from the HNF H5, H6, H8 and H1, H3, H4 protons to residues T19, ABA20, T21 and A6, A8, respectively. As an example, the NOE cross-peaks A8-H8:HNF7-H1 and A8-H8:HNF7-H3 are marked in Fig. 4.18.

Based on the assignment of the H1' and H6/H8 protons, the remaining base and sugar protons (including HNF and ABA) could be assigned by combining the information from

4.2 2-Hydroxy-7nitrofluorene

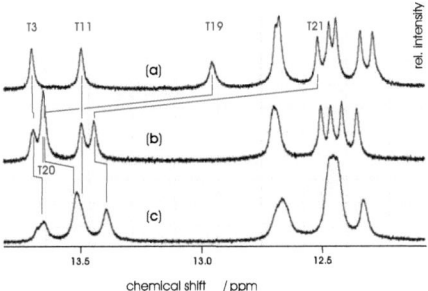

Fig. 4.19: *Comparison of the imino proton region of 13merRef, 13merRefGC and 13merHNF. Red solid lines follow the chemical shift assignments of a given imino proton resonance starting from 13merHNF (a) through 13merRef (b) to 13merRefGC (c). The solid green line indicates the chemical shift of the T20 imino proton which does not exist in the 13merHNF duplex.*

the COSY-, TOCSY-, HMQC-, and NOESY-spectra. Assignment of the exchangeable protons was done in the WATERGATE-NOESY-spectrum. Because of the symmetry of the sequence the imino, amino, and H2 protons had to be referenced to the already assigned non-exchangeable H5 protons via the H42/H41-H5 NOE cross-peaks. Assignment of the imino protons of T19 and T21 was complicated by the strong upfield shift of both protons when compared to either corresponding protons in 13merRef and 13merRefGC (see Fig. 4.19).

CSD analysis of 13merHNF and the corresponding native structures 13merRef (with a central A:T base pair) and 13merRefGC (with a central G:C base pair) is shown in Fig. 4.20, red and blue columns respectively. For comparison the CSDs between the to native structures 13merRef and 13merRefGC are shown (green columns). The sum of the absolute values of the CSDs for all protons belonging to one residue is given in order to avoid cancelling of negative and positive shifts. This results in a loss of information regarding the direction of the chemical shift differences, but allows for comparison between the different residues. With the exception of the three central base pairs, absolute per-residue CSDs upon introduction of HNF are comparable or only slightly larger than for the exchange of an A:T vs a G:C base pair (Fig. 4.20). Residues that are more than

4 Results and discussion

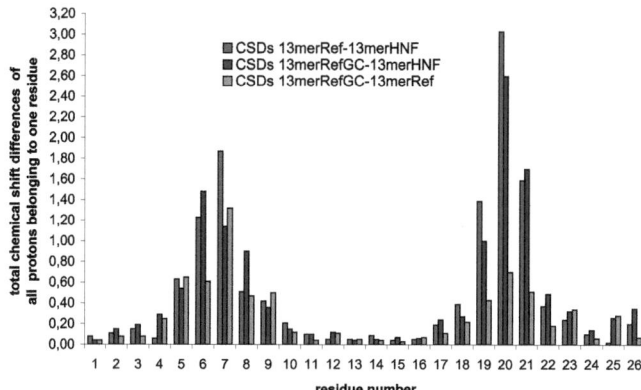

Fig. 4.20: *The sum of the absolute values of the CSDs for all protons belonging to one residue is given for: 13merRef to 13merRefGC (green columns), 13merHNF to 13merRef (red columns) and 13merHNF to 13merRefGC (blue columns). 13merRefGC has the same sequence as 13merRef but with a central G:C base pair.*

two base pairs removed from the modification site do not exhibit significant chemical shift perturbations. The chemical shifts of all assigned proton and carbon atoms for 13merHNF, 13merRefGC and 13merRef are given in the Appendix, sections 6, 7 and 8.

4.2 2-Hydroxy-7nitrofluorene

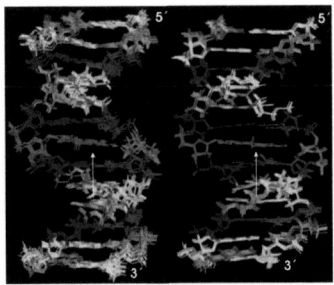 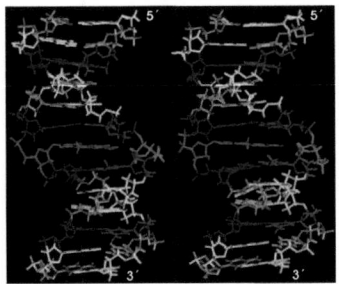

Fig. 4.21: The 10 minimum-energy, violation-free structures of 13merHNF in face-up (left panel) and face-down (right panel) orientation are depicted. White arrows indicate the view direction of Fig. 4.23

Fig. 4.22: The averaged, minimized structures of 13merHNF in face-up (left panel) and face-down (right panel) orientation are depicted.

4.2.2.2 NMR solution structure

The NMR solution structure is determined from experimental NOE and RDC data as described in section 3.1.4. While we see only one NOE data set for the duplex as a whole, the subset relating to the HNF chromophore cannot be described by a single orientation. Instead two structures of the same duplex are needed with different orientations of the chromophore: one where the fluorene methylene group points towards the major groove (face-up) and one where it points to the minor groove (face-down). Interestingly, the RDC restraints allow both orientations equally well. Simulated Annealing calculations for the two orientations produced two families of structures. The best (minimum-energy, violation-free) 10 of each, which are shown in Fig. 4.21, are used for generating the average structures. The latter are depicted in Fig. 4.22. The HNF probe fits into the helical fold (Fig. 4.21 and 4.22) and stacks with residues 6, 8, 21 and partly 19. This is illustrated in Fig. 4.23, where the central three base pairs of each orientation are zoomed and rotated to visualize stacking interactions.

The calculation results were validated by back-calculation of the NOESY-spectra (red and blue for face-up and face-down orientations respectively) and comparison with the experimental one (green). Fig. 4.24a and 4.24b differ only in the displayed region. While

4 Results and discussion

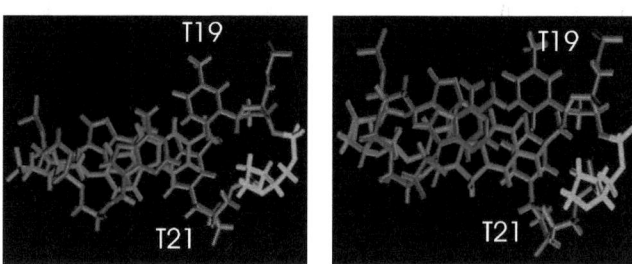

Fig. 4.23: *The central three base pairs of the average structures of 13merHNF in face-up (left panel) and face-down (right panel) orientation are depicted.*

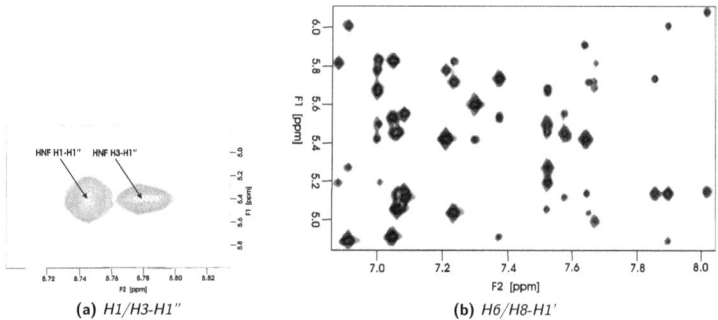

(a) H1/H3-H1'' (b) H6/H8-H1'

Fig. 4.24: *Overlay of the experimental NOESY-spectrum at 150 ms mixing time (green) and the NOESY-spectra back-calculated from the averaged, minimized structures of the HNF in the face-up (blue) and face-down (red) orientation. Arrows point to NOE cross-peaks involving the modification site. Panel (a) depicts the H1'-H1/H3 NOE cross-peaks, which can be only accounted for with the two different orientations of HNF within the double helix. Panel (b) shows the H1'-H6/H8 region, where predicted spectra for both orientations fit the experimental one equally well.*

4.2 2-Hydroxy-7nitrofluorene

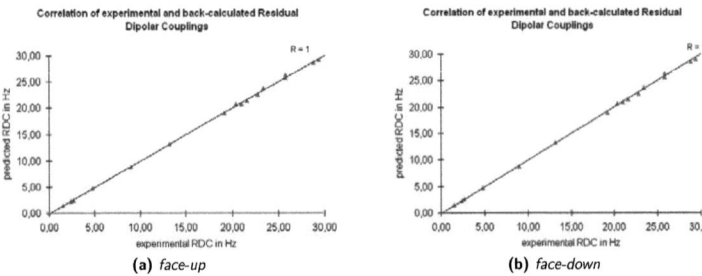

Fig. 4.25: *Plot of the experimental vs predicted RDCs for 13merHNF with HNF in the face-up (a) and face-down (b) orientations respectively.*

the latter shows the H1'-H6/H8 proton region (the same as in Fig. 4.18), Fig. 4.24a centers on two characteristic intraresidual NOE cross-peaks, namely H1/H3-H1" of the HNF residue. The latter peaks can only be accounted for by the two different orientations of the HNF inside the double helix, as is demonstrated by the back-calculated spectra. As either peak would be non-observable with only one orientation sampled, the population ratio of the face-up and face-down orientation can be estimated from the integral ratio of these two peaks to be 1:1. The accuracy of the calculations has been further validated by back-calculation of the RDCs from the average structure. The correlation plots of experimentally determined vs predicted RDCs are shown in Fig. 4.25a (4.25b) and yielded correlation factors (R) of 1.000 and q-factors of 0.004 (0.005) for the face-up (face-down) orientations. The precision of the calculations is assessed by the RMSDs of 10 minimum-energy, violation-free structures to their average structure, which are 0.66 Å and 0.47 Å for the face-up and face-down structures, respectively. An analysis of the helical parameters of the duplex structure was not possible since the modification of the central base pair made the definition of the helical axis impossible for the commonly used programs 3DNA [Lu and Olson, 2003] and CURVES+ [Lavery et al., 2009].

4 Results and discussion

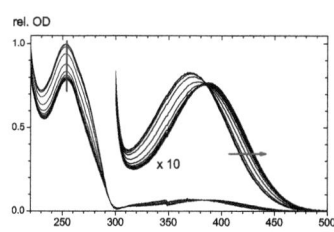

Fig. 4.26: Absorption changes upon hybridisation, when lowering temperature from 85 to 25°C. The well-known hyperchromism of the UV absorbance (blue arrow) is accompanied by a 1190 cm^{-1} red-shift of the HNF absorption band.

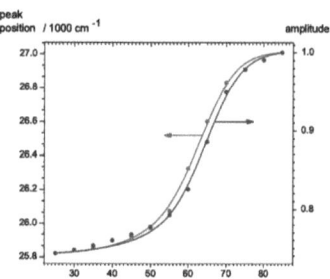

Fig. 4.27: Temperature-dependence of the spectra in Fig. 4.26. The amplitude of the absorption peak around 260 nm (blue points, right scale) yields a melting point of 64°C. The peak position of the HNF absorption band (red points, left scale) shows a melting point which is lower by 3°C.

4.2.2.3 Duplex melting

UV/vis absorption spectra of the DNA-HNF duplex are shown in Fig. 4.26 as a function of temperature. The nucleobases absorb intensely at 260 nm while the HNF chromophor is seen by a weak band around 380 nm, corresponding to the S1→S0 transition. At 85 °C only single strands are present. As the temperature is lowered to 25 °C, the HNF absorption band experiences a red-shift while the nucleobase absorption decreases.

Spectral change with temperature is quantified by "melting curvesäs shown in Fig. 4.27. The relative UV absorption amplitude decreases from 1 in the single strands to 0.74 in the duplex (right scale, blue points). A melting point T_m of 64 °C is found for the total concentration c_T=23.5 mM of the single strands. The HNF peak position is shown as red points (left scale) in Fig. 4.27. With this measure the melting point is located 3 K lower.

4.2.3 Discussion

Orientational flexibility of the HNF residue is indicated by the existence of two equally strong NOE cross-peaks between the H1" of the sugar and the H1 and H3 protons, respectively (cf. Fig. 4.24a). From their integrals a 1:1 population ratio could be estimated. Another possible explanation for the observation of a 1:1 integral ratio for these peaks might be spin diffusion effects. This possibility can be ruled out, as the integral ratio exhibited no dependence on the NOESY mixing time parameter. Calculations on both orientations and subsequent spectrum prediction support this conclusion. Interchange between the two orientations of HNF involves a 180° flip around the C1'-O2-bond which links the HNF to the sugar (cf. Fig. 4.17 and 4.22). This can take place only during transient opening of the formal HNF-abasic site base pair. For natural base pairs such opening motions are observed on a millisecond timescale . Another argument for orientational flip on that timescale is that NOE cross-peak signals would appear as an average of both orientations. This might explain why only one NOE data set is observed with some NOE cross-peaks involving the HNF chromophore being exclusively consistent with either orientation and others which can be used to describe both orientations but only with increased error bounds. The consistency of the RDC data with either orientation can be explained by the orientational degeneracy of the small experimental RDC set measured in a single orientational medium (cf. section 2.3.3). Thus one can conclude that the HNF chromophore undergoes orientational flip on a ms timescale.

Possible causes for the orientational degeneracy might be the introduction of the abasic site. The latter has already been reported in the literature to severely increase the flexibility of the helical structure in the vicinity [Coppel et al., 1997, Lin et al., 1998, Hoehn et al., 2001, Lin and de los Santos, 2001, Smirnov et al., 2002, Chen et al., 2007, 2008]. Another reason might be that the HNF chromophore is a conjugated π-system which is largely devoid of functional groups (with the exception of the hydroxyl and the nitro groups at position 2 and 7, respectively). The charge distribution obtained by the DFT calculations described in section 3.1.3 is roughly symmetrical to a 180°-flip around the long HNF axis. Thus no preference in terms of electronic interaction with

4 Results and discussion

the adjacent base pairs can occur.

In both orientations the HNF chromophore intercalates into the DNA double helix despite its α-glycosidic attachment to the sugar (instead of β for natural nucleotides). This is indicated by the existence of interstrand NOE cross-peaks from the T19, ABA20 and T21 residues to the H5, H6, and H8 protons of HNF as well as the intrastrand ones from the A6 and A8 residues to the H1, H3 and H4 protons. The comparison of the CSD data for 13merHNF, 13merRef and 13merRefGC shows that significant CSDs occur exclusively for the central five base pairs. This suggests that by HNF incorporation the helical structure of the DNA is mainly perturbed two base pairs in each direction while the overall B-DNA conformation remains intact.

Stacking interactions with residues 6, 8, 21 and partly 19 are illustrated in Fig. 4.23. While T21 is fully stacked with the HNF moiety, the T19 residue is turned outward, away from HNF. Thus stacking interactions do not stabilize the T19:A8 base pair as well as the other base pairs. As a consequence the T19 imino proton is shifted up-field and broadened, similar to what is observed with semi-terminal imino protons [Nonin et al., 1995]. This is illustrated by the NMR spectra in Fig. 4.19 where the imino protons of the HNF-substituted duplex (a) are compared to those of the same duplex containing a central A:T (b) or G:C (c) base pair. The T21 imino proton on the other hand, though also shifted upwards significantly, exhibits a very sharp resonance due to the ring current induced by HNF and the stabilizing effect of strong stacking interactions with the latter.

The thermodynamics of duplex formation for 13merHNF has been examined, to probe whether HNF intercalates and how stable the local and global helical arrangement is. When monitored via the temperature-dependent 260 nm absorbance of the nucleobases, a melting point of 64.0 °C is found. A comparison with the melting point of 13merRefGC is instructive. Under the same conditions, it is determined to be T_m=69.35 °C. From a comparison of standard hybridization enthalpies and with the assumption of a common hybridization entropy of -1.2 kJ/mol [Xia et al., 1998], 13merHNF is found to be less stable by 6.9 kJ/mol compared to 13merRefGC. This is equivalent to the lack of enthalpy from hydrogen bonding. When following the red-shift of the absorption band of HNF

4.2 2-Hydroxy-7nitrofluorene

at 380 nm, the melting point is estimated at 61.0 °C. The difference in melting points of 3.5 K can be compared to the 0.5 K which are found for the 13mer2AP duplex (see section 4.1.2.4). The large difference found for 13merHNF indicates considerable premelting around the HNF chromophore. This is in line with the results from the structure calculations and chemical shift analysis, which show orientational exchange and less favorable stacking interactions of HNF with T19.

In the past, base pair mimics devoid of hydrogen bonding have been demonstrated to be incorporated by DNA polymerases with comparable efficiency and even higher selectivity than natural bases due to steric complementarity [Morales and Kool, 1998, Guckian et al., 1998, Matray and Kool, 1999, Guckian et al., 2000]. The pyrene nucleotide, for example, can sterically mimic a WC base pair and is incorporated into DNA duplexes opposite to an abasic site without disruption of structure or decrease in duplex stability [Matray and Kool, 1998, Singh et al., 2002, Smirnov et al., 2002]. But pyrene has a pronounced effect on the local dynamics of adjacent base pairs, indicated by the presence of two interconverting resonances for the thymine imino proton to the 5'-side and broadening of the imino proton of the adjacent G:C base pair [Smirnov et al., 2002]. The fact that for the HNF-containing DNA duplex we do not observe interconverting signals, indicates that the perturbation of local dynamics is weaker than for a pyrene residue. However, substantial broadening of the T19 imino proton and orientational flip of the HNF indicate that local flexibility is also induced.

Summary

Structural and dynamic perturbations in DNA upon incorporation of either fluorophore, 2-Aminopurine (2AP) or 2-Hydroxy-7-nitrofluorene (HNF), are characterized by NMR spectroscopy. For this purpose the NMR solution structures of the modified DNA duplexes with the sequence 5'-GCTGCAXACGTCG-3' are solved. For X=2AP (13mer2AP) the partner base in the complementary strand is T, while for X=HNF (13merHNF) an abasic site is introduced to avoid steric strain.

As a structural isomer of A, the fluorescence properties of 2AP are commonly utilized to monitor stacking-unstacking transitions in molecular biology. By comparing results on 13mer2AP with the corresponding unmodified DNA duplex (13merRef, X=A), any perturbation can be unambiguously assigned to 2AP incorporation. For the NMR solution structure of 13merRef and 13mer2AP small but significant changes in helical parameters are found throughout the helix. Imino proton exchange measurements reveal an extended, distributed effect of 2AP incorporation on the lifetimes of the central seven base pair. This effect is explained by decreased activation enthalpy for base pair opening due to weakened stacking interactions and hydrogen bonding. The latter are indicated by the reduced melting point of 13mer2AP compared to 13merRef. Local melting around the modification site, as sensed by 2AP fluorescence, further supports the results from base pair dynamics. However, the reduced base pair lifetime of 2AP:T cannot fully account for the rapid water exchange observed with saturation transfer experiments in the absence of base catalyst. This indicates enhanced intrinsic catalysis. As a possible catalytic site the T O4 atom opposite 2AP is discussed, which is easily accessible through the major groove and lacks a hydrogen bonding partner within the base pair.

4 Results and discussion

HNF is a fluorene derivative which was designed as a molecular probe for THz vibrational activity in biomolecules. Due to its elongated shape it is introduced opposite to an abasic site, which is known to induce orientational flexibility into DNA helices. The overall NMR solution structure is found to be B-DNA. However the NOE cross-peaks involving the HNF residue can only be accounted for by two different orientations of the HNF inside the DNA helical stack. Their population ratio is estimated to be 1:1. Dynamical perturbation is indicated by the increased linewidth and strong upfield shift of the T residue to the 5'-side of the abasic site. This can be explained with the help of the solution structure, which shows that this T residue cannot stack efficiently with the HNF. Lack of one stacking partner leads to destabilization of the base pair, as is oberved for the helical termini.

The dynamic as well as the structural perturbation due to HNF incorporation is large compared to the perturbations induced upon 2AP incorporation. This is not surprising considering the rather artificial shape of HNF as compared to 2AP. Furthermore, 2AP incorporation does not require introduction of an abasic site, which causes orientational flexibility within the double helix. In conclusion, HNF is ill-suited for application to biological problems. Changes observed with 2AP are much smaller. Structural differences are weak but the extended, distributed effect of 2AP incorporation on base pair opening rates limits its value for biologically significant applications such as monitoring stacking-unstacking transitions.

Future development should concentrate on the design of a fluorophore with the demonstrated spectral properties of HNF [Dallmann et al., 2009] but the more native shape of 2AP. For example, the introduction of an electron withdrawing group at the 6-position of 2AP would be an interesting future goal. While a relevant structure (6-Cyano-2-Aminopurine) has already been designed [Hocek and Holý, 1995], attachment to the sugar and incorporation into a DNA sequence have not been demonstrated yet. The introduction of another functional group to 2AP at the 6-position would circumvent the problem of solvent attack and thus reduce the dynamic as well as thermodynamic destabilization.

Zusammenfassung

Mittels NMR-Spektroskopie werden Störungen in Struktur und Dynamik von DNA untersucht, die durch den Einbau jeweils eines der beiden Fluorophore 2-Aminopurin (2AP) und 2-Hydroxy-7-nitrofluoren (HNF) hervorgerufen werden. Zu diesem Zweck werden die NMR-Strukturen der modifizierten Duplexe mit der Sequenz 5'-GCTGCAXACGTCG-3' berechnet. Im Fall X=2AP (13mer2AP) ist die Partnerbase im Komplementärstrang ein T, während gegenüber X=HNF (13merHNF) eine abasische Stelle eingeführt wird.

Als Strukturisomer von A, werden die Fluoreszenzeigenschaften von 2AP gern genutzt, um stacking-unstacking Übergänge in der Molekularbiologie zu verfolgen. Durch den Vergleich der Ergebnisse zum 13mer2AP mit denjenigen des entsprechenden unmodifizierten DNA Doppelstranges (13merRef, X=A) konnte jegliche Änderung eindeutig dem Einbau von 2AP zugordnet werden. Für die NMR-Strukturen von 13merRef und 13mer2AP können kleine aber signifikante, über die gesamte Helix verteilte Strukturstörungen nachgewiesen werden. Experimente zum Iminoprotonenaustausch mit Wasser ergeben, daß der Einbau von 2AP die Basenpaarlebensdauern der 7 zentralen Basenpaare erniedrigt. Dieser ausgedehnte Effekt kann durch eine verminderte Aktivierungsenthalpie für die Öffnung des Basenpaares erklärt werden. Ursachen dafür können schwächere Stapelwechselwirkungen und Wasserstoffbrückenbindungen sein. Geringere Energiebeiträge aus letzteren Wechselwirkungen können aus dem niedrigeren Schmelzpunkt des 13mer2AP Duplex abgeschätzt werden. Die DNA Schmelzpunktsbestimmung mittels 2AP-Fluoreszenz deutet lokales Schmelzen um die Modifikationsstelle herum an, was die Ergebnisse zur Basenpaardynamik zusätzlich unterstützt. Die kürzere Lebensdauer des 2AP:T Basenpaares kann jedoch nicht den schnellen Wasseraustausch

4 Results and discussion

im Sättigungstransfer-Experiment ohne Zugabe von Basenkatalysator erklären. Als Erklärung für diese Diskrepanz wird eine effizientere intrinsische Katalyse vermutet. Als mögliche, katalytisch aktive Stelle wird das T O4 Atom diskutiert, welches über die große Furche leicht zugänglich ist und das keine Wasserstoffbrückenbindung innerhalb des Basenpaares ausbilden kann.

HNF ist ein Fluorenderivat das als molekulare Sonde für vibratorische Aktivität von Biomolekülen im THz-Bereich entwickelt wurde. Aufgrund seiner langgestreckten Form, wird statt einer Partnerbase eine abasische Stelle eingesetzt. Von letzterer ist jedoch bekannt, daß sie die helikale Struktur lokal flexibilisiert. Die übergeordnete Struktur des 13merHNF ist eine B-Form DNA Helix. Die NOE Kreuzpeaks zu den Protonen im HNF können jedoch nur durch zwei verschiedene Orientierungen des HNFs in der helikalen Anordnung beschrieben werden. Das Verhältnis der beiden Orientierungen untereinander wird als 1:1 abgeschätzt. Störungen in der Basenpaardynamik werden durch die höhere Linienbreite und die starke Hochfeldverschiebung des T auf der 5'-Seite ausgehend von der abasischen Stelle angedeutet. Eine Erklärung hierfür kann mit Hilfe der berechneten Struktur gegeben werden. Diese zeigt, daß das T keine effektiven Basenstapelwechselwirkungen mit HNF eingehen kann. Das Fehlen eines Wechselwirkungspartners führt zur Destabilisierung des Basenpaares, was z.B. auch für Basenpaare an den Helixtermini beobachtet wird.

Sowohl die dynamischen als auch strukturellen Änderungen nach Einbau von HNF sind groß im Vergleich zu den Störungen die durch den Einbau von 2AP verursacht werden. Eine Erklärung ist die stärkere Abweichung von der Struktur natürlicher Nukleobasen von HNF im Vergleich zu 2AP. Außerdem wird beim Einbau von 2AP keine abasische Stelle im Komplementärstrang benötigt, welche die Flexibilität der DNA-Helix lokal erhöhen würde. Das Fazit ist, das HNF für Anwendungen im biologischen Bereich nicht geeignet ist. Die Änderungen beim Einbau von 2AP sind wesentlich kleiner. Die strukturellen Unterschiede sind schwach, aber der ausgedehnte Effekt des 2AP-Einbaus auf die Basenpaarlebensdauern schränkt die Einsatzmöglichkeiten für biologische Fragestellungen wie stacking-unstacking Übergänge ein.

4.2 2-Hydroxy-7nitrofluorene

Die zukünftige Entwicklung neuartiger Fluorophore sollte sich an den bereits demonstrierten spektralen Eigenschaften des HNF [Dallmann et al., 2009] und der deutlich natürlicheren Struktur des 2AP orientieren. Die Einführung einer elektronenziehenden Gruppe in der 6-Position des 2AP stellt ein attraktives Syntheseziel dar. Während eine relevante Struktur (6-Cyano-2-Aminopurine) eines entsprechenden Fluorophores bereits beschrieben wurde [Hocek and Holý, 1995], konnte die glycosidische Verknüpfung und der Einbau in DNA noch nicht realisiert werden. Durch das Einführen einer weiteren funktionellen Gruppe in der 6-Position des 2AP könnte sowohl die dynamische als auch die thermodynamische Destabilisierung verhindert werden.

Literaturverzeichnis

H. M. Al-Hashimi, H. Valafar, M. Terrell, E. R. Zartler, M. K. Eidsness, and J. H. Prestegard. Variation of molecular alignment as a means of resolving orientational ambiguities in protein structures from dipolar couplings. *Journal of Magnetic Resonance*, 143(2):402–406, April 2000. ISSN 1090-7807. URL http://www.sciencedirect.com/science/article/B6WJX-45F4W4J-7B/2/c8035b09e10d580a3a95586b9bdd9518.

B.W. Allan and N.O. Reich. Targeted base stacking disruption by the ecori dna methyltransferase. *Biochemistry*, 35(47):14757–62, 1996.

B.W. Allan, J.M. Beechem, W.M. Lindstrom, and N.O. Reich. Direct real time observation of base flipping by the ecori dna methyltransferase. *J Biol Chem*, 273(4):2368–73, 1998.

B.W. Allan, N.O. Reich, and J.M. Beechem. Measurement of the absolute temporal coupling between dna binding and base flipping. *Biochemistry*, 38(17):5308–14, 1999.

C. Altona and M. Sundaralingam. Conformational analysis of the sugar ring in nucleosides and nucleotides. new description using the concept of pseudorotation. *Journal of the American Chemical Society*, 94(23):8205–8212, 1972. ISSN 0002-7863. URL http://pubs3.acs.org/acs/journals/doilookup?in_doi=10.1021/ja00778a043.

Robert H. Austin, Mark W. Roberson, and Paul Mansky. Far-infrared perturbation of reaction rates in myoglobin at low temperatures. *Phys. Rev. Lett.*, 62(16):1912–1915, Apr 1989. doi: 10.1103/PhysRevLett.62.1912.

R.P. Bandwar and S.S. Patel. Peculiar 2-aminopurine fluorescence monitors the dynamics

Literaturverzeichnis

of open complex formation by bacteriophage t7 rna polymerase. *J Biol Chem*, 276 (17):14075–82, 2001.

Laura G. Barrientos, Caroline Dolan, and Angela M. Gronenborn. Characterization of surfactant liquid crystal phases suitable for molecular alignment and measurement of dipolar couplings. *Journal of Biomolecular NMR*, 16(4):329–337, April 2000. URL http://dx.doi.org/10.1023/A:1008356618658.

Christopher Bauer, Ray Freeman, Tom Frenkiel, James Keeler, and A. J. Shaka. Gaussian pulses. *Journal of Magnetic Resonance (1969)*, 58(3):442–457, July 1984. ISSN 0022-2364. URL http://www.sciencedirect.com/science/article/B7GXD-4CRGG0P-RY/2/621e7b2792d539626697048805db7d1c.

Ad Bax and Nico Tjandra. High-resolution heteronuclear nmr of human ubiquitin in an aqueous liquid crystalline medium. *Journal of Biomolecular NMR*, 10(3):289–292, October 1997. URL http://dx.doi.org/10.1023/A:1018308717741.

Edwin D. Becker, James A. Ferretti, Raj K. Gupta, and George H. Weiss. The choice of optimal parameters for measurement of spin-lattice relaxation times. ii. comparison of saturation recovery, inversion recovery, and fast inversion recovery experiments. *Journal of Magnetic Resonance (1969)*, 37(3):381–394, February 1980. ISSN 0022-2364. URL http://www.sciencedirect.com/science/article/B7GXD-4CRG61F-4R/2/4d6573e03198e4c0b40ca3af11c17c36.

J.M. Beechem, M.R. Otto, L.B. Bloom, R. Eritja, L.J. Reha-Krantz, and M.F. Goodman. Exonuclease-polymerase active site partitioning of primer-template dna strands and equilibrium mg2+ binding properties of bacteriophage t4 dna polymerase. *Biochemistry*, 37(28):10144–55, 1998.

A.S. Bernards, J.K. Miller, K.K. Bao, and I. Wong. Flipping duplex dna inside out: a double base-flipping reaction mechanism by escherichia coli muty adenine glycosylase. *J Biol Chem*, 277(23):20960–4, 2002.

Literaturverzeichnis

C. Bernstein, D. Morgan, H.L. Gensler, S. Schneider, and G.E. Holmes. The dependence of hno2 mutagenesis in phage t4 on ligase and the lack of dependence of 2ap mutagenesis on repair functions. *Mol Gen Genet*, 148(2):213–20, 1976.

Brent H. Besler, Kenneth M. Merz Jr., and Peter A. Kollman. Atomic charges derived from semiempirical methods. *Journal of Computational Chemistry*, 11(4):431–439, 1990. URL http://dx.doi.org/10.1002/jcc.540110404.

Laurie Betts, John A. Josey, James M. Veal, and Steven R. Jordan. A nucleic acid triple helix formed by a peptide nucleic acid-dna complex. *Science*, 270(5243):1838–1841, 1995. doi: 10.1126/science.270.5243.1838. URL http://www.sciencemag.org/cgi/content/abstract/270/5243/1838.

Pratip K. Bhattacharya, Julie Cha, and Jacqueline K. Barton. 1h nmr determination of base-pair lifetimes in oligonucleotides containing single base mismatches. *Nucl. Acids Res.*, 30(21):4740–4750, 2002. URL http://nar.oxfordjournals.org/cgi/content/abstract/30/21/4740.

A. Bierzynski, H. Kozlowska, and K.I. Wierzchowski. Investigations on purine and pyrimidine bases stacking associations in aqueous solutions by the fluorescence quenching method. ii. heteroassociation between 2-aminopurine and thymidine. *Biophys Chem*, 6(3):223–9, 1977a.

A. Bierzynski, H. Kozlowska, and K.L. Wierzchowski. Investigations on purine and pyrimidine bases stacking associations in aqueous solutions by the fluorescence quenching method. i. autoassociation of 2-aminopurine. *Biophys Chem*, 6(3):213–22, 1977b.

A. J. Birchall and A. N. Lane. Anisotropic rotation in nucleic acid fragments: significance for determination of structures from nmr data. *European Biophysics Journal*, 19(2): 73–78, November 1990. URL http://dx.doi.org/10.1007/BF00185089.

Martin Blackledge. Recent progress in the study of biomolecular structure and dynamics in solution from residual dipolar couplings. *Progress in Nucle-*

Literaturverzeichnis

ar *Magnetic Resonance Spectroscopy*, 46(1):23–61, March 2005. ISSN 0079-6565. URL http://www.sciencedirect.com/science/article/B6THC-4FB3X3Y-1/2/2afbc520ee6ba524e6d767fc45649df7.

L.B. Bloom, M.R. Otto, J.M. Beechem, and M.F. Goodman. Influence of 5'-nearest neighbors on the insertion kinetics of the fluorescent nucleotide analog 2-aminopurine by klenow fragment. *Biochemistry*, 32(41):11247–58, 1993.

L.B. Bloom, M.R. Otto, R. Eritja, L.J. Reha-Krantz, M.F. Goodman, and J.M. Beechem. Pre-steady-state kinetic analysis of sequence-dependent nucleotide excision by the 3'-exonuclease activity of bacteriophage t4 dna polymerase. *Biochemistry*, 33(24):7576–86, 1994.

Victor A. Bloomfield, Donald M. Crothers, and Jr. Ignacio Tinoco. *Nucleic Acids: structures, properties and functions*. University Science Books, 2000.

R. Boelens, T. M. G. Koning, and R. Kaptein. Determination of biomolecular structures from proton-proton noe's using a relaxation matrix approach. *Journal of Molecular Structure*, 173:299–311, 1988. URL http://www.sciencedirect.com/science/article/B6TGS-44F6NNT-MD/2/4ea0b2e8985486472a28bca6aca26718.

J. M. Bohlen and G. Bodenhausen. Experimental aspects of chirp nmr spectroscopy. *Journal of Magnetic Resonance, Series A*, 102(3):293–301, May 1993. URL http://www.sciencedirect.com/science/article/B6WJY-45PKP5K-48/2/5fe826bb492454665795d2c3d73c3534.

B. A. Borgias and T. L. James. Mardigras - a procedure for matrix analysis of relaxation for discerning geometry of an aqueous structure. *J. Magn. Res.*, 87:475–487, 1990.

Curt M. Breneman and Kenneth B. Wiberg. Determining atom-centered monopoles from molecular electrostatic potentials. the need for high sampling density in formamide conformational analysis. *Journal of Computational Chemistry*, 11(3):361–373, 1990. URL http://dx.doi.org/10.1002/jcc.540110311.

Literaturverzeichnis

C.A. Brennan, M.D. Van Cleve, and R.I. Gumport. The effects of base analogue substitutions on the cleavage by the ecori restriction endonuclease of octadeoxyribonucleotides containing modified ecori recognition sequences. *J Biol Chem*, 261(16):7270–8, 1986.

A. Broo. A theoretical investigation of the physical reason for the very different luminescence properties of the two isomers adenine and 2-aminopurine. *J Phys Chem A*, 102(3):526–531, 1998. URL http://pubs3.acs.org/acs/journals/doilookup?in_doi=10.1021/jp9713625.

Axel T. Brünger. *X-PLOR (Version 4.0) A System for X-Ray Crystallography and NMR*. Yale University, 1996.

Patrik R. Callis. Polarized fluorescence and estimated lifetimes of the dna bases at room temperature. *Chemical Physics Letters*, 61(3):568–570, March 1979. URL http://www.sciencedirect.com/science/article/B6TFN-44JMRXJ-16/2/ed885054c65d1027130c2f2809df0368.

I.W. Caras, M.A. MacInnes, D.H. Persing, P. Coffino, and D.W. Martin Jr. Mechanism of 2-aminopurine mutagenesis in mouse t-lymphosarcoma cells. *Mol Cell Biol*, 2(9): 1096–103, 1982.

M. H. Caruthers, A. D. Barone, S. L. Beaucage, D. R. Dodds, E. F. Fisher, L. J. McBride, M. Matteucci, Z. Stabinsky, and J. Y. Tang. Chemical synthesis of deoxyoligonucleotides by the phosphoramidite method. *Methods In Enzymology*, 154:287–313, 1987.

Erwin Chargaff. Chemical specificity of nucleic acids and mechanism of their enzymatic degradation. *Experientia*, 6:201–209, 1950.

Congju Chen and Irina M. Russu. Sequence-dependence of the energetics of opening of at basepairs in dna. *Biophysical Journal*, 87(4):2545–2551, October 2004. ISSN 0006-3495. URL http://linkinghub.elsevier.com/retrieve/pii/S000634950473725X.

Jingyang Chen, Francois-Yves Dupradeau, David A. Case, Christopher J. Turner, and JoAnne Stubbe. Nuclear magnetic resonance structural studies and molecular mode-

Literaturverzeichnis

ling of duplex dna containing normal and 4'-oxidized abasic sites;. *Biochemistry*, 46 (11):3096–3107, 2007. URL http://pubs.acs.org/doi/abs/10.1021/bi6024269.

Jingyang Chen, Francois-Yves Dupradeau, David A. Case, Christopher J. Turner, and JoAnne Stubbe. Dna oligonucleotides with a, t, g or c opposite an abasic site: structure and dynamics. *Nucl. Acids Res.*, 36(1):253–262, 2008. URL http://nar.oxfordjournals.org/cgi/content/abstract/36/1/253.

Lisa Emily Chirlian and Michelle Miller Francl. Atomic charges derived from electrostatic potentials: A detailed study. *Journal of Computational Chemistry*, 8(6):894–905, 1987. URL http://dx.doi.org/10.1002/jcc.540080616.

James Chou, Sander Gaemers, Bernard Howder, John Louis, and Ad Bax. A simple apparatus for generating stretched polyacrylamide gels, yielding uniform alignment of proteins and detergent micelles*. *Journal of Biomolecular NMR*, 21(4):377–382, December 2001. URL http://dx.doi.org/10.1023/A:1013336502594.

L.K. Clayton, M.F. Goodman, E.W. Branscomb, and D.J. Galas. Error induction and correction by mutant and wild type t4 dna polymerases. kinetic error discrimination mechanisms. *J Biol Chem*, 254(6):1902–12, 1979.

G. Marius Clore and Angela M. Gronenborn. A nuclear-overhauser-enhancement study of the solution structure of a double-stranded dna undecamer comprising a portion of the specific target site for the cyclic-amp-receptor protein in the <i>gal</i> operon. *European Journal of Biochemistry*, 141(1):119–129, 1984. URL http://dx.doi.org/10.1111/j.1432-1033.1984.tb08166.x.

G.Marius Clore and Angela M. Gronenborn. Probing the three-dimensional structures of dna and rna oligonucleotides in solution by nuclear overhauser enhancement measurements. *FEBS Letters*, 179(2):187–198, January 1985. ISSN 0014-5793. URL http://www.sciencedirect.com/science/article/B6T36-44G8DJN-C2/2/639200ce81acaa6864e59ff25de1d5c1.

Literaturverzeichnis

Simona Cocco and Remi Monasson. Theoretical study of collective modes in dna at ambient temperature. *The Journal of Chemical Physics*, 112(22):10017–10033, 2000. doi: 10.1063/1.481646. URL http://link.aip.org/link/?JCP/112/10017/1.

James G. Colson, Peter T. Lansbury, and Franklin D. Saeva. 7,12-dihydropleiadenes. vii. application of nuclear overhauser effects to stereochemical problems. *Journal of the American Chemical Society*, 89(19):4987–4990, September 1967. ISSN 0002-7863. URL http://dx.doi.org/10.1021/ja00995a028.

Daniel Coman and Irina M. Russu. A nuclear magnetic resonance investigation of the energetics of basepair opening pathways in dna. *Biophysical Journal*, 89(5):3285–3292, November 2005. ISSN 0006-3495. URL http://linkinghub.elsevier.com/retrieve/pii/S0006349505729702.

Benjamin N. Conner, Tsunehiro Takano, Shoji Tanaka, Keiichi Itakura, and Richard E. Dickerson. The molecular structure of d(icpcpgpg), a fragment of right-handed double helical a-dna. *Nature*, 295(5847):294–299, January 1982. URL http://dx.doi.org/10.1038/295294a0.

Benjamin N. Conner, Chun Yoon, Joyce L. Dickerson, and Richard E. Dickerson. Helix geometry and hydration in an a-dna tetramer: Ic-c-g-g. *Journal of Molecular Biology*, 174(4):663–695, April 1984. ISSN 0022-2836. URL http://www.sciencedirect.com/science/article/B6WK7-4FNGD09-70/2/96b986d21306a89683b1a0c9c999e3b9.

M. L. Connolly. Analytical molecular surface calculation. *Journal of Applied Crystallography*, 16(5):548–558, Oct 1983. doi: 10.1107/S0021889883010985. URL http://dx.doi.org/10.1107/S0021889883010985.

Yannick Coppel, Nathalie Berthet, Christian Coulombeau, Christiane Coulombeau, Julian Garcia, and Jean Lhomme. Solution conformation of an abasic dna undecamer duplex d(cgcacxcacgc)·d(gcgtgtgtgcg): the unpaired thymine stacks inside the helix;. *Biochemistry*, 36(16):4817–4830, 1997. URL http://pubs.acs.org/doi/abs/10.1021/bi962677y.

111

Literaturverzeichnis

B. H. Dahl, J. Nielsen, and O. Dahl. Mechanistic studies on the phosphoramidite coupling reaction in oligonucleotide synthesis .1. evidence for nucleophilic catalysis by tetrazole and rate variations with the phosphorus substituents. *Nucleic Acids Research*, 15(4): 1729–1743, 1987.

Andre Dallmann, Matthias Pfaffe, Clemens Muegge, Rainer Mahrwald, Sergey A. Kovalenko, and Nikolaus P. Ernsting. Local thz time domain spectroscopy of duplex dna via fluorescence of an embedded probe. *The Journal of Physical Chemistry B*, pages –, September 2009. ISSN 1520-6106. URL http://dx.doi.org/10.1021/jp906037g.

D. Daujotyte, S. Serva, G. Vilkaitis, E. Merkiene, C. Venclovas, and S. Klimasauskas. Hhai dna methyltransferase uses the protruding gln237 for active flipping of its target cytosine. *Structure (Camb)*, 12(6):1047–55, 2004.

David B. Davies, Valery I. Pahomov, and Alexei N. Veselkov. Nmr determination of the conformational and drug binding properties of the dna heptamer d(gpcpgpapapgpc) in aqueous solution. *Nucleic Acids Research*, 25(22):4523–4531, 1997.

Eva de Alba and Nico Tjandra. Nmr dipolar couplings for the structure determination of biopolymers in solution. *Progress in Nuclear Magnetic Resonance Spectroscopy*, 40(2):175–197, February 2002. ISSN 0079-6565. URL http://www.sciencedirect.com/science/article/B6THC-44J3W0T-1/2/a9c21795b70e6c8a9f52dafd370fbd33.

RE Dickerson, HR Drew, BN Conner, RM Wing, AV Fratini, and ML Kopka. The anatomy of a-, b-, and z-dna. *Science*, 216(4545):475–485, 1982. doi: 10.1126/science.7071593. URL http://www.sciencemag.org/cgi/content/abstract/216/4545/475.

Richard E. Dickerson and Horace R. Drew. Structure of a b-dna dodecamer: Influence of base sequence on helix structure. *J. Mol. Biol.*, 149:761–786, 1981a.

Richard E. Dickerson and Horace R. Drew. Structure of a b-dna dodecamer: Geometry of hydration. *J. Mol. Biol.*, 151:535–556, 1981b.

Literaturverzeichnis

Utz Dornberger, Mikael Leijon, and Hartmut Fritzsche. High base pair opening rates in tracts of gc base pairs. *J. Biol. Chem.*, 274(11):6957–6962, 1999. URL http://www.jbc.org/cgi/content/abstract/274/11/6957.

Horace Drew, Tsunehiro Takano, Shoji Tanaka, Keiichi Itakura, and Richard E. Dickerson. High-salt d(cpgpcpg), a left-handed z[prime] dna double helix. *Nature*, 286(5773): 567–573, August 1980. URL http://dx.doi.org/10.1038/286567a0.

Horace R. Drew and Richard E. Dickerson. Conformation and dynamics in a z-dna tetramer. *Journal of Molecular Biology*, 152(4):723–736, November 1981. ISSN 0022-2836. URL http://www.sciencedirect.com/science/article/B6WK7-4DNGVR9-CH/2/f9b8606c35043516b099d914c179118f.

Horace R. Drew, Richard M. Wing, Tsunehiro Takano, Christopher Broka, Shoji Tanaka, Keiichi Itakura, and Richard E. Dickerson. Structure of a b-dna dodecamer: Conformation and dynamics. *Proc Natl Acad Sci U S A*, 78(4):2179–2183, April 1981.

M. Eigen. Proton transfer, acid-base catalysis, and enzymatic hydrolysis. part i: Elementary processes. *Angewandte Chemie International Edition in English*, 3(1):1–19, 1964. URL http://dx.doi.org/10.1002/anie.196400011.

Lyndon Emsley and Geoffrey Bodenhausen. Gaussian pulse cascades: New analytical functions for rectangular selective inversion and in-phase excitation in nmr. *Chemical Physics Letters*, 165(6):469–476, February 1990. ISSN 0009-2614. URL http://www.sciencedirect.com/science/article/B6TFN-44XDTGM-6V/2/c0ec5d5b82d70c27204b6a7ef63bf4e9.

Lyndon Emsley and Geoffrey Bodenhausen. Optimization of shaped selective pulses for nmr using a quaternion description of their overall propagators. *Journal of Magnetic Resonance (1969)*, 97(1):135–148, March 1992. ISSN 0022-2364. URL http://www.sciencedirect.com/science/article/B7GXD-4CRGCPM-B/2/2e076ec69e76ed752a2a40d9502e0430.

Literaturverzeichnis

K. Cecilia Engman, Peter Sandin, Sadie Osborne, Tom Brown, Martin Billeter, Per Lincoln, Bengt Norden, Bo Albinsson, and L. Marcus Wilhelmsson. Dna adopts normal b-form upon incorporation of highly fluorescent dna base analogue tc: Nmr structure and uv-vis spectroscopy characterization. *Nucleic Acids Research*, 34(17):5087–5095, 2004.

R. Eritja, B.E. Kaplan, D. Mhaskar, L.C. Sowers, J. Petruska, and M.F. Goodman. Synthesis and properties of defined dna oligomers containing base mispairs involving 2-aminopurine. *Nucleic Acids Res*, 14(14):5869–84, 1986.

F. E. Evans and R. A. Levine. Nmr-study of stacking interactions and conformational adjustments in the dinucleotide carcinogen adduct 2'-deoxycytidylyl-(3-]5)-2'-deoxy-8-(n-fluoren-2-ylacetamido)guanosine. *Biochemistry*, 27(8):3046–3055, 1988.

Kervin Evans, Daguang Xu, Younsik Kim, and Thomas M. Nordlund. 2-aminopurine optical spectra: Solvent, pentose ring, and dna helix melting dependence. *Journal of Fluorescence*, 2(4):209–216, December 1992. URL http://dx.doi.org/10.1007/BF00865278.

Brent Ewing, Steffen J. Glaser, and Gary P. Drobny. Development and optimization of shaped nmr pulses for the study of coupled spin systems. *Chemical Physics*, 147 (1):121–129, October 1990. ISSN 0301-0104. URL http://www.sciencedirect.com/science/article/B6TFM-44WK50F-5J/2/182bba5c16baae497e669d014593ab7d.

P.A. Fagan, C. Fabrega, R. Eritja, M.F. Goodman, and D.E. Wemmer. Nmr study of the conformation of the 2-aminopurine:cytosine mismatch in dna. *Biochemistry*, 35 (13):4026–33, 1996.

G.V. Fazakerley, L.C. Sowers, R. Eritja, B.E. Kaplan, and M.F. Goodman. Nmr studies on an oligodeoxynucleotide containing 2-aminopurine opposite adenine. *Biochemistry*, 26(18):5641–6, 1987.

Agostino Fede, Martin Billeter, Werner Leupin, and Kurt Wuethrich. Determination of the nmr solution structure of the hoechst 33258-d(gtggaattccac)2 complex and comparison with the x-ray crystal structure. *Structure*, 1(3):177–186, November 1993. ISSN 0969-2126. URL http://www.sciencedirect.com/science/article/B6VSR-4CT6089-M/2/75132ea71234b6e40638586b45c91232.

Juli Feigon, Andrew H.-J. Wang, Gijs A. van der Marel, Jacques H. Van Boom, and Alexander Rich. A one- and two-dimensional nmr study of the b to z transition of (m5dc-dg)3 in methanolic solution. *Nucl. Acids Res.*, 12(2):1243–1263, 1984. doi: 10.1093/nar/12.2.1243. URL http://nar.oxfordjournals.org/cgi/content/abstract/12/2/1243.

E. Fidalgo da Silva, S.S. Mandal, and L.J. Reha-Krantz. Using 2-aminopurine fluorescence to measure incorporation of incorrect nucleotides by wild type and mutant bacteriophage t4 dna polymerases. *J Biol Chem*, 277(43):40640–9, 2002.

T. Fiebig, C. Wan, and A.H. Zewail. Femtosecond charge transfer dynamics of a modified dna base: 2-aminopurine in complexes with nucleotides. *Chemphyschem*, 3(9):781–8, 2002.

Peter F. Flynn, Ramona J. Bieber Urbauer, Hui Zhang, Andrew L. Lee, and A. Joshua Wand. Main chain and side chain dynamics of a heme protein: 15n and 2h nmr relaxation studies of r. capsulatus ferrocytochrome c2. *Biochemistry*, 40(22):6559–6569, June 2001. ISSN 0006-2960. URL http://dx.doi.org/10.1021/bi0102252.

E. Folta-Stogniew and I. M. Russu. Sequence dependence of base-pair opening in a dna dodecamer containing the caca/gtgt sequence motif. *Biochemistry*, 33(36):11016–11024, Sep 1994. doi: http://pubs.acs.org/doi/pdf/10.1021/bi00202a022.

E. Folta-Stogniew and I. M. Russu. Base-catalysis of imino proton exchange in dna: effects of catalyst upon dna structure and dynamics. *Biochemistry*, 35(25):8439–8449, Jun 1996. doi: 10.1021/bi952932z. URL http://dx.doi.org/10.1021/bi952932z.

R. Franklin and R.G. Gosling. Molecular configuration in sodium thymonucleate. *Nature*, 171:740–741, 1953.

Literaturverzeichnis

Ray Freeman. High resolution nmr using selective excitation. *Journal of Molecular Structure*, 266:39–51, March 1992. ISSN 0022-2860. URL http://www.sciencedirect.com/science/article/B6TGS-44WD0JG-JD/2/d19884f57c3baafd3e1ddbe991b1fa87.

E. Freese. Specific mutagenic effect of base analogues on phage-t4. *J Mol Biol*, 1(2):87–105, 1959.

Riqiang Fu and Geoffrey Bodenhausen. Broadband decoupling in nmr with frequency-modulated [']chirp' pulses. *Chemical Physics Letters*, 245(4-5):415–420, November 1995. ISSN 0009-2614. URL http://www.sciencedirect.com/science/article/B6TFN-3YF4B83-G/2/2b1f880c76fc5ba29a3ee2d982ce9e41.

J. Fujimoto, Z. Nuesca, M. Mazurek, and L.C. Sowers. Synthesis and hydrolysis of oligodeoxyribonucleotides containing 2-aminopurine. *Nucleic Acids Res*, 24(4):754–9, 1996.

J. Gajewska, A. Bierzynski, K. Bolewska, K.L. Wierzchowski, A.I. Petrov, and B.I. Sukhorukov. Fluorescence quenching and spin label electron spin resonance studies of stacking self-association in aqueous solutions of 2-aminopurine riboside and its 5'-mono- and -diphosphate. *Biophys Chem*, 15(3):191–204, 1982.

Michael Garwood and Lance DelaBarre. The return of the frequency sweep: Designing adiabatic pulses for contemporary nmr. *Journal of Magnetic Resonance*, 153(2):155–177, December 2001. URL http://www.sciencedirect.com/science/article/B6WJX-457VF60-J/2/e21bc243e1e29bc5c76aedc31a3169e7.

M.F. Goodman and R.L. Ratliff. Evidence of 2-aminopurine-cytosine base mispairs involving two hydrogen bonds. *J Biol Chem*, 258(21):12842–6, 1983.

M.F. Goodman, R. Hopkins, and W.C. Gore. 2-aminopurine-induced mutagenesis in t4 bacteriophage: a model relating mutation frequency to 2-aminopurine incorporation in dna. *Proc Natl Acad Sci U S A*, 74(11):4806–10, 1977.

Literaturverzeichnis

H. Gowher and A. Jeltsch. Molecular enzymology of the ecorv dna-(adenine-n (6))-methyltransferase: kinetics of dna binding and bending, kinetic mechanism and linear diffusion of the enzyme on dna. *J Mol Biol*, 303(1):93–110, 2000.

D. Grossberger and W. Clough. Incorporation into dna of the base analog 2-aminopurine by the epstein-barr virus-induced dna polymerase in vivo and in vitro. *Proc Natl Acad Sci U S A*, 78(12):7271–5, 1981.

K. M. Guckian, T. R. Krugh, and E. T. Kool. Solution structure of a nonpolar, non-hydrogen-bonded base pair surrogate in dna. *Journal Of The American Chemical Society*, 122(29):6841–6847, July 2000. URL http://pubs.acs.org/doi/abs/10.1021/ja994164v.

Kevin M. Guckian, Thomas R. Krugh, and Eric T. Kool. Solution structure of a dna duplex containing a replicable difluorotoluene-adenine. *Nat Struct Mol Biol*, 5(11): 954–959, November 1998. ISSN 1072-8368. URL http://dx.doi.org/10.1038/2930.

Maurice Gueron, Michel Kochoyan, and Jean-Louis Leroy. A single mode of dna base-pair opening drives imino proton exchange. *Nature*, 328(6125):89–92, July 1987. URL http://dx.doi.org/10.1038/328089a0.

C.R. Guest, R.A. Hochstrasser, L.C. Sowers, and D.P. Millar. Dynamics of mismatched base pairs in dna. *Biochemistry*, 30(13):3271–9, 1991.

P. J. Hajduk, D. A. Horita, and L. E. Lerner. Theoretical-analysis of relaxation during shaped pulses .1. the effects of short t(1) and t(2). *Journal Of Magnetic Resonance Series A*, 103(1):40–52, June 1993.

Mark R. Hansen, Luciano Mueller, and Arthur Pardi. Tunable alignment of macromolecules by filamentous phage yields dipolar coupling interactions. *Nat Struct Mol Biol*, 5(12):1065–1074, December 1998. ISSN 1072-8368. URL http://dx.doi.org/10.1038/4176.

Literaturverzeichnis

S. J. O. Hardman and K. C. Thompson. The fluorescence transition of 2-aminopurine in double- and single-stranded dna. *Int J Quant Chem*, 107(11):2092–2099, September 2007.

S.J.O. Hardman and K.C. Thompson. Influence of base stacking and hydrogen bonding on the fluorescence of 2-aminopurine and pyrrolocytosine in nucleic acids. *Biochemistry*, 45(30):9145–55, 2006.

D. R. Hare, D. E. Wemmer, S. H. Chou, G. Drobny, and B. R. Reid. Assignment of the non-exchangeable proton resonances of d(c-g-c-g-a-a-t-t-c-g-c-g) using two-dimensional nuclear magnetic resonance methods. *J Mol Biol*, 171(3):319–336, Dec 1983.

C. Hariharan and L.J. Reha-Krantz. Using 2-aminopurine fluorescence to detect bacteriophage t4 dna polymerase-dna complexes that are important for primer extension and proofreading reactions. *Biochemistry*, 44(48):15674–84, 2005.

C. Hariharan, L.B. Bloom, S.A. Helquist, E.T. Kool, and L.J. Reha-Krantz. Dynamics of nucleotide incorporation: snapshots revealed by 2-aminopurine fluorescence studies. *Biochemistry*, 45(9):2836–44, 2006.

Udo Heinemann and Claudia Alings. Crystallographic study of one turn of g/c-rich b-dna. *Journal of Molecular Biology*, 210(2):369–381, November 1989. ISSN 0022-2836. URL http://www.sciencedirect.com/science/article/B6WK7-4FNGB0M-22/2/10981508f8a706013859d2f8d819f847.

C. W. Hilbers and D. J. Patel. Proton nuclear magnetic resonance investigations of the nucleation and propagation reactions associated with the helix-coil transition of d-aptpgpcpapt in h2o solution. *Biochemistry*, 14(12):2656–2660, Jun 1975.

Michal Hocek and Antonín Holý. A facile synthesis of 6-cyanopurine bases. *Collection of Czechoslovak Chemical Communications*, 60(8):1386–1389, 1995. URL http://dlib.lib.cas.cz/835/.

Literaturverzeichnis

R.A. Hochstrasser, T.E. Carver, L.C. Sowers, and D.P. Millar. Melting of a dna helix terminus within the active site of a dna polymerase. *Biochemistry*, 33(39):11971–9, 1994.

Silvia T. Hoehn, Christopher J. Turner, and JoAnne Stubbe. Solution structure of an oligonucleotide containing an abasic site: evidence for an unusual deoxyribose conformation. *Nucl. Acids Res.*, 29(16):3413–3423, 2001. URL http://nar.oxfordjournals.org/cgi/content/abstract/29/16/3413.

B. Holz, S. Klimasauskas, S. Serva, and E. Weinhold. 2-aminopurine as a fluorescent probe for dna base flipping by methyltransferases. *Nucleic Acids Res*, 26(4):1076–83, 1998.

M. L. Horng, J. A. Gardecki, A. Papazyan, and M. Maroncelli. Subpicosecond measurements of polar solvation dynamics: Coumarin 153 revisited. *The Journal of Physical Chemistry*, 99(48):17311–17337, November 1995. ISSN 0022-3654. URL http://dx.doi.org/10.1021/j100048a004.

R.V. Hosur, M. Ravikumar, K.V.R. Chary, Anu Sheth, Girjesh Govil, Tan Zu-Kun, and H.Todd Miles. Solution structure of d-gaattcgaattc by 2d nmr: A new approach to determination of sugar geometries in dna segments. *FEBS Letters*, 205(1):71–76, September 1986. ISSN 0014-5793. URL http://www.sciencedirect.com/science/article/B6T36-449T565-PY/2/caed07596396740205be9e7c513283ea.

S Jain and M Sundaralingam. Effect of crystal packing environment on conformation of the dna duplex. molecular structure of the a-dna octamer d(g-t-g-t-a-c-a-c) in two crystal forms. *Journal of Biological Chemistry*, 264(22):12780–12784, 1989. URL http://www.jbc.org/content/264/22/12780.abstract.

J.M. Jean and K.B. Hall. 2-aminopurine electronic structure and fluorescence properties in dna. *Biochemistry*, 41(44):13152–61, 2002.

J.M. Jean and K.B. Hall. Stacking-unstacking dynamics of oligodeoxynucleotide trimers. *Biochemistry*, 43(31):10277–84, 2004.

Literaturverzeichnis

J.M. Jean and B.P. Krueger. Structural fluctuations and excitation transfer between adenine and 2-aminopurine in single-stranded deoxytrinucleotides. *J Phys Chem B Condens Matter Mater Surf Interfaces Biophys*, 110(6):2899–909, 2006.

J. Jeener, B. H. Meier, P. Bachmann, and R. R. Ernst. Investigation of exchange processes by two-dimensional nmr spectroscopy. *The Journal of Chemical Physics*, 71(11): 4546–4553, 1979. doi: 10.1063/1.438208. URL http://link.aip.org/link/?JCP/71/4546/1.

Y. Jia, A. Kumar, and S.S. Patel. Equilibrium and stopped-flow kinetic studies of interaction between t7 rna polymerase and its promoters measured by protein and 2-aminopurine fluorescence changes. *J Biol Chem*, 271(48):30451–8, 1996.

Eric Johansson, Gary Parkinson, and Stephen Neidle. A new crystal form for the dodecamer c-g-c-g-a-a-t-t-c-g-c-g: symmetry effects on sequence-dependent dna structure. *Journal of Molecular Biology*, 300(3):551–561, July 2000. ISSN 0022-2836. URL http://www.sciencedirect.com/science/article/B6WK7-45F51CK-CG/2/2620a69460950d3e9ccf8c644d0c746c.

Neil P. Johnson, Walter A. Baase, and Peter H. von Hippel. Investigating local conformations of double-stranded dna by low-energy circular dichroism of pyrrolo-cytosine. *Proceedings of the National Academy of Sciences of the United States of America*, 102(20): 7169–7173, 2005a. URL http://www.pnas.org/content/102/20/7169.abstract.

N.P. Johnson, W.A. Baase, and P.H. Von Hippel. Low-energy circular dichroism of 2-aminopurine dinucleotide as a probe of local conformation of dna and rna. *Proc Natl Acad Sci U S A*, 101(10):3426–31, 2004.

N.P. Johnson, W.A. Baase, and P.H. von Hippel. Low energy cd of rna hairpin unveils a loop conformation required for lambdan antitermination activity. *J Biol Chem*, 280 (37):32177–83, 2005b.

M. W. Kalnik, M. Kouchakdjian, B. F. L. Li, P. F. Swann, and D. J. Patel. Base pair

Literaturverzeichnis

mismatches and carcinogen-modified bases in dna - an nmr-study of a.c and a.o4met pairing in dodecanucleotide duplexes. *Biochemistry*, 27(1):100–108, 1988.

Venugopal Karunakaran, Matthias Pfaffe, Ilya Ioffe, Tamara Senyushkina, Sergey A. Kovalenko, Rainer Mahrwald, Vadim Fartzdinov, Heinz Sklenar, and Nikolaus P. Ernsting. Solvation oscillations and excited-state dynamics of 2-amino- and 2-hydroxy-7-nitrofluorene and its 2"-deoxyriboside. *The Journal of Physical Chemistry A*, 112 (18):4294–4307, May 2008. ISSN 1089-5639. URL http://dx.doi.org/10.1021/jp712176m.

Rochus Keller. *The Computer Aided Resonance Assignment Tutorial*. CANTINA Verlag, 1st edition, 2004. ISBN 3-85600-112-3.

S.O. Kelley and J.K. Barton. Electron transfer between bases in double helical dna. *Science*, 283(5400):375–81, 1999.

S.R. Kirk, N.W. Luedtke, and Y. Tor. 2-aminopurine as a real-time probe of enzymatic cleavage and inhibition of hammerhead ribozymes. *Bioorg Med Chem*, 9(9):2295–301, 2001.

Douglas A. Klewer, Aaron Hoskins, Peiming Zhang, V. J. Davisson, Donald E. Bergstrom, and Andy C. LiWang. Nmr structure of a dna duplex containing nucleoside analog 1-(2'-deoxy-beta-d-ribofuranosyl)-3-nitropyrrole and the structure of the unmodified control. *Nucl. Acids Res.*, 28(22):4514–4522, 2000. URL http://nar.oxfordjournals.org/cgi/content/abstract/28/22/4514.

Gerard J. Kleywegt and T. Alwyn Jones. Databases in protein crystallography. *Acta Crystallographica Section D*, 54(6 Part 1):1119–1131, 1998. doi: doi:10.1107/S0907444998007100. URL http://dx.doi.org/10.1107/S0907444998007100.

Dennis D. Klug, Marek Z. Zgierski, John S. Tse, Zhenxian Liu, James R. Kincaid, Kazimierz Czarnecki, and Russell J. Hemley. Doming modes and dynamics of model heme compounds. *Proceedings of the National Academy of Sciences of the United*

States of America, 99(20):12526–12530, 2002. doi: 10.1073/pnas.152464699. URL http://www.pnas.org/content/99/20/12526.abstract.

Michel Kochoyan, Jean Louis Leroy, and Maurice Guéron. Proton exchange and base-pair lifetimes in a deoxy-duplex containing a purine-pyrimidine step and in the duplex of inverse sequence. *Journal of Molecular Biology*, 196(3):599–609, August 1987. ISSN 0022-2836. URL http://www.sciencedirect.com/science/article/B6WK7-4DPBVX9-12/2/958221adb458bdd0677519cea2525f9b.

Michel Kochoyan, Gerard Lancelot, and Jean Louis Leroy. Study of structure, basepair opening kinetics and proton exchange mechanism of the d-(aattgcaatt) self-complementary oligodeoxynucleotide in solution. *Nucl. Acids Res.*, 16(15):7685–7702, 1988. URL http://nar.oxfordjournals.org/cgi/content/abstract/16/15/7685.

Olaf Koehler, Dilip Venkatrao Jarikote, and Oliver Seitz. Forced intercalation probes (fit probes): Thiazole orange as a fluorescent base in peptide nucleic acids for homogeneous single-nucleotide-polymorphism detection. *ChemBioChem*, 6(1):69–77, 2005. URL http://dx.doi.org/10.1002/cbic.200400260.

Mary L. Kopka, Albert V. Fratini, Horace R. Drew, and Richard E. Dickerson. Ordered water structure around a b-dna dodecamer : A quantitative study. *Journal of Molecular Biology*, 163(1):129–146, January 1983. ISSN 0022-2836. URL http://www.sciencedirect.com/science/article/B6WK7-4DM18GH-96/2/27ce7f6bfc9f0cc34b8614e94b943692.

Jozef Kowalewski, George C. Levy, LeRoy F. Johnson, and Loren Palmer. A three-parameter non-linear procedure for fitting inversion-recovery measurements of spin-lattice relaxation times. *Journal of Magnetic Resonance (1969)*, 26(3):533–536, June 1977. ISSN 0022-2364. URL http://www.sciencedirect.com/science/article/B7GXD-4CRG986-18M/2/b26b5f9bdc9f961d9cb09ed993aa089e.

T. R. Krugh, D. E. Graves, and M. P. Stone. 2-dimensional nmr-studies on the anthramycin-d(atgcat)2 adduct. *Biochemistry*, 28(26):9988–9994, 1989.

Literaturverzeichnis

Anil Kumar, R. R. Ernst, and K. Wuthrich. A two-dimensional nuclear overhauser enhancement (2d noe) experiment for the elucidation of complete proton-proton cross-relaxation networks in biological macromolecules. *Biochemical and Biophysical Research Communications*, 95(1):1–6, July 1980a. URL http://www.sciencedirect.com/science/article/B6WBK-4DN95M1-XV/2/4c770b0bfd98d66737ad3a98cbc511f8.

Anil Kumar, G. Wagner, R. R. Ernst, and K. Wüthrich. Studies of j-connectivities and selective 1h-1h overhauser effects in h2o solutions of biological macromolecules by two-dimensional nmr experiments. *Biochemical and Biophysical Research Communications*, 96(3):1156–1163, October 1980b. ISSN 0006-291X. URL http://www.sciencedirect.com/science/article/B6WBK-4DYN48C-V0/2/a62dbf3327a9a22a93bda0b39d7c3a29.

Anil Kumar, Gerhard Wagner, Richard R. Ernst, and Kurt Wuethrich. Buildup rates of the nuclear overhauser effect measured by two-dimensional proton magnetic resonance spectroscopy: implications for studies of protein conformation. *Journal of the American Chemical Society*, 103(13):3654–3658, July 1981. ISSN 0002-7863. URL http://dx.doi.org/10.1021/ja00403a008.

E. Kupce and R. Freeman. Adiabatic pulses for wideband inversion and broadband decoupling. *Journal of Magnetic Resonance, Series A*, 115(2):273–276, August 1995. URL http://www.sciencedirect.com/science/article/B6WJY-45S997T-5G/2/a15a876f98d5100f76f133e83243b583.

Eriks Kupce and Ray Freeman. Compensation for spin-spin coupling effects during adiabatic pulses. *Journal of Magnetic Resonance*, 127(1):36–48, July 1997. ISSN 1090-7807. URL http://www.sciencedirect.com/science/article/B6WJX-45M2WY9-4/2/0491a258f92b12c8141d4608e7116eae.

Eriks Kupce and Ray Freeman. Compensated adiabatic inversion pulses: Broadband inept and hsqc. *Journal of Magnetic Resonance*, 187(2):258–265, August

Literaturverzeichnis

2007. ISSN 1090-7807. URL http://www.sciencedirect.com/science/article/ B6WJX-4NRT3K5-1/2/3ff1262346456f2eff2cc3480673fee8.

Andrew N. Lane. Nmr studies of dynamics in nucleic acids. *Progress in Nuclear Magnetic Resonance Spectroscopy*, 25(5):481–505, 1993. ISSN 0079-6565. URL http://www.sciencedirect.com/science/article/B6THC-44JMY5R-K/ 2/6fee0f21a4d31f54325514538ae96713.

Andrew N. Lane. Influence of conformational averaging on ^{1}h-^{1}h noes and structure determination in dna. *Magnetic Resonance in Chemistry*, 34(13):S3–S10, 1996. URL http://dx.doi.org/10.1002/(SICI)1097-458X(199612)34:13<S3:: AID-OMR15>3.0.CO;2-Q.

Filip Lankas, Thomas E. Cheatham III, Nad'a Spackova, Pavel Hobza, Jorg Langowski, and Jiri Sponer. Critical effect of the n2 amino group on structure, dynamics, and elasticity of dna polypurine tracts. *Biophys. J.*, 82(5):2592–2609, 2002. URL http: //www.biophysj.org/cgi/content/abstract/82/5/2592.

R. Lavery, M. Moakher, J. H. Maddocks, D. Petkeviciute, and K. Zakrzewska. Conformational analysis of nucleic acids revisited: Curves+. *Nucl. Acids Res.*, pages 1–13, 2009. doi: 10.1093/nar/gkp608. URL http://nar.oxfordjournals.org/cgi/ content/abstract/gkp608v1.

S.M. Law, R. Eritja, M.F. Goodman, and K.J. Breslauer. Spectroscopic and calorimetric characterizations of dna duplexes containing 2-aminopurine. *Biochemistry*, 35(38): 12329–37, 1996.

M. Leijon and A. Graslund. Effects of sequence and length on imino proton exchange and base pair opening kinetics in dna oligonucleotide duplexes. *Nucl. Acids Res.*, 20(20): 5339–5343, 1992. doi: 10.1093/nar/20.20.5339. URL http://nar.oxfordjournals. org/cgi/content/abstract/20/20/5339.

Mikael Leijon. Proton exchange rates measured by saturation transfer using delayed randomization of the solvent magnetization. *Journal of Magnetic*

Resonance, Series B, 112(2):181–185, August 1996. ISSN 1064-1866. URL http://www.sciencedirect.com/science/article/B6WK0-45MG470-2F/2/ d6b3a8528bb65b6c986956de49af39ac.

T. Lenz, E. Bonnist, G. Pljevaljcic, R. Neely, D. Dryden, A. Scheidig, A. Jones, and E. Weinhold. 2-aminopurine flipped into the active site of the adenine-specific dna methyltransferase m.taqi: Crystal structures and time-resolved fluorescence. *J Am Chem Soc*, 129:6240–6248, 2007.

Jean-Louis Leroy, Daniel Broseta, and Maurice Gueron. Proton exchange and base-pair kinetics of poly(ra) poly(ru) and poly(ri) poly(rc). *Journal of Molecular Biology*, 184 (1):165–178, July 1985. URL http://www.sciencedirect.com/science/article/ B6WK7-4DN8WBP-HG/2/0869b2833b0e626e983b0f6cab91324a.

Jean Louis Leroy, Eric Charretier, Michel Kochoyan, and Maurice Gueron. Evidence from base-pair kinetics for two types of adenine tract structures in solution: their relation to dna curvature. *Biochemistry*, 27(25):8894–8898, 1988. URL http://pubs3.acs. org/acs/journals/doilookup?in_doi=10.1021/bi00425a004.

Jean Louis Leroy, Kalle Gehring, Abdelali Kettani, and Maurice Gueron. Acid multimers of oligodeoxycytidine strands: Stoichiometry, base-pair characterization, and proton exchange properties. *Biochemistry*, 32(23):6019–6031, 1993. doi: 10.1021/bi00074a013. URL http://pubs.acs.org/doi/abs/10.1021/bi00074a013.

George C. Levy and Ian R. Peat. The experimental approach to accurate carbon-13 spin-lattice relaxation measurements. *Journal of Magnetic Resonance (1969)*, 18(3):500–521, June 1975. ISSN 0022-2364. URL http://www.sciencedirect.com/science/ article/B7GXD-4CRG84W-WH/2/c162996b23e515c58deab58541d5e76d.

B. F. Li, P. F. Swann, M. Kalnik, and D. J. Patel. Synthesis and structural studies by nuclear magnetic resonance of dodecadeoxynucleotides containing o6-methylguanine, o6-ethylguanine and o4-methylthymine. *IARC Sci Publ*, (84):44–8, 1987.

Literaturverzeichnis

David MJ Lilley and Timothy J Wilson. Fluorescence resonance energy transfer as a structural tool for nucleic acids. *Current Opinion in Chemical Biology*, 4(5):507–517, October 2000. ISSN 1367-5931. URL http://www.sciencedirect.com/science/article/B6VRX-417MKGV-7/2/dce71d76a8ca0711e4a597bb621f9920.

Z Lin, KN Hung, AP Grollman, and C de los Santos. Solution structure of duplex dna containing an extrahelical abasic site analog determined by nmr spectroscopy and molecular dynamics. *Nucl. Acids Res.*, 26(10):2385–2391, 1998. doi: 10.1093/nar/26.10.2385. URL http://nar.oxfordjournals.org/cgi/content/abstract/26/10/2385.

Zhen Lin and Carlos de los Santos. Nmr characterization of clustered bistrand abasic site lesions: effect of orientation on their solution structure. *Journal of Molecular Biology*, 308(2):341–352, April 2001. URL http://www.sciencedirect.com/science/article/B6WK7-457D196-5C/2/dca2668c222f82f12b7471a711796d8b.

Giovanni Lipari and Attila Szabo. Model-free approach to the interpretation of nuclear magnetic resonance relaxation in macromolecules. 1. theory and range of validity. *Journal of the American Chemical Society*, 104(17):4546–4559, August 1982a. ISSN 0002-7863. URL http://dx.doi.org/10.1021/ja00381a009.

Giovanni Lipari and Attila Szabo. Model-free approach to the interpretation of nuclear magnetic resonance relaxation in macromolecules. 2. analysis of experimental results. *Journal of the American Chemical Society*, 104(17):4559–4570, August 1982b. ISSN 0002-7863. URL http://dx.doi.org/10.1021/ja00381a010.

Rebecca S. Lipsitz and Nico Tjandra. Residual dipolar couplings in nmr structure analysis. *Annual Review of Biophysics and Biomolecular Structure*, 33(1):387–413, 2004. URL http://arjournals.annualreviews.org/doi/abs/10.1146/annurev.biophys.33.110502.140306.

C. Liu and C.T. Martin. Promoter clearance by t7 rna polymerase. initial bubble collapse and transcript dissociation monitored by base analog fluorescence. *J Biol Chem*, 277(4):2725–31, 2002.

Literaturverzeichnis

J. W. Longworth, R. O. Rahn, and R. G. Shulman. Luminescence of pyrimidines, purines, nucleosides, and nucleotides at 77[degree]k. the effect of ionization and tautomerization. *The Journal of Chemical Physics*, 45(8):2930–2939, 1966. doi: 10.1063/1.1728048. URL http://link.aip.org/link/?JCP/45/2930/1.

F.G. Loontiens, L.W. McLaughlin, S. Diekmann, and R.M. Clegg. Binding of hoechst 33258 and 4',6'-diamidino-2-phenylindole to self-complementary decadeoxynucleotides with modified exocyclic base substituents. *Biochemistry*, 30:182–189, 1991.

Judit A. Losonczi, Michael Andrec, Mark W. F. Fischer, and James H. Prestegard. Order matrix analysis of residual dipolar couplings using singular value decomposition. *Journal of Magnetic Resonance*, 138(2):334–342, June 1999. ISSN 1090-7807. URL http://www.sciencedirect.com/science/article/B6WJX-45GMX90-N/2/6b49d710012edda18ce3e7f172aec495.

Xiang-Jun Lu and Wilma K. Olson. 3dna: a software package for the analysis, rebuilding and visualization of three-dimensional nucleic acid structures. *Nucl. Acids Res.*, 31 (17):5108–5121, 2003. doi: 10.1093/nar/gkg680. URL http://nar.oxfordjournals.org/cgi/content/abstract/31/17/5108.

J. Luis Perez Lustres, Sergey A. Kovalenko, Manuel Mosquera, Tamara Senyushkina, Wolfgang Flasche, and Nikolaus P. Ernsting. Ultrafast solvation of <i>n</i>-methyl-6-quinolone probes local ir spectrum13. *Angewandte Chemie International Edition*, 44(35):5635–5639, 2005. URL http://dx.doi.org/10.1002/anie.200501397.

P.O. Lycksell, A. Graeslund, F. Claesens, L.W. McLaughlin, U. Larsson, and R. Rigler. Base pair opening dynamics of a 2-aminopurine substituted eco ri restriction sequence and its unsubstituted counterpart in oligonucleotides. *Nucleic Acids Res*, 15(21):9011–25, 1987.

Junhe Ma, Gregory I. Goldberg, and Nico Tjandra. Weak alignment of biomacromolecules in collagen gels: An alternative way to yield residual dipolar couplings for

Literaturverzeichnis

nmr measurements. *Journal of the American Chemical Society*, 130(48):16148–16149, December 2008. ISSN 0002-7863. URL http://dx.doi.org/10.1021/ja807064k.

Douglas MacDonald and Ponzy Lu. Residual dipolar couplings in nucleic acid structure determination. *Current Opinion in Structural Biology*, 12(3):337–343, June 2002. ISSN 0959-440X. URL http://www.sciencedirect.com/science/article/B6VS6-469GK4C-C/2/efb6d16a758d77f4ae1b97310152141c.

Douglas MacDonald, Kristina Herbert, Xiaolin Zhang, Thomas Polgruto, and Ponzy Lu. Solution structure of an a-tract dna bend. *Journal of Molecular Biology*, 306(5): 1081–1098, March 2001. URL http://www.sciencedirect.com/science/article/B6WK7-45KNCGY-31/1/ecf95f30009f8a7b86a6a6ff3e4f23b3.

Alexander D. MacKerell, Nilesh Banavali, and Nicolas Foloppe. Development and current status of the charmm force field for nucleic acids. *Biopolymers*, 56(4):257–265, 2000. URL http://dx.doi.org/10.1002/1097-0282(2000)56:4<257::AID-BIP10029>3.0.CO;2-W.

S. Macura and R. R. Ernst. Elucidation of cross relaxation in liquids by two-dimensional n.m.r. spectroscopy. *Molecular Physics: An International Journal at the Interface Between Chemistry and Physics*, 41(1):95–117, 1980. URL http://www.informaworld.com/10.1080/00268978000102601.

EG Malygin, VV Zinoviev, NA Petrov, AA Evdokimov, L Jen-Jacobson, VG Kossykh, and S Hattman. Effect of base analog substitutions in the specific gatc site on binding and methylation of oligonucleotide duplexes by the bacteriophage t4 dam dna-[n6-adenine] methyltransferase. *Nucl. Acids Res.*, 27(4):1135–1144, 1999. URL http://nar.oxfordjournals.org/cgi/content/abstract/27/4/1135.

S.S. Mandal, E. Fidalgo da Silva, and L.J. Reha-Krantz. Using 2-aminopurine fluorescence to detect base unstacking in the template strand during nucleotide incorporation by the bacteriophage t4 dna polymerase. *Biochemistry*, 41(13):4399–406, 2002.

Literaturverzeichnis

L.A. Marquez and L.J. Reha-Krantz. Using 2-aminopurine fluorescence and mutational analysis to demonstrate an active role of bacteriophage t4 dna polymerase in strand separation required for 3' –> 5'-exonuclease activity. *J Biol Chem*, 271(46):28903–11, 1996.

Tracy J. Matray and Eric T. Kool. Selective and stable dna base pairing without hydrogen bonds. *Journal of the American Chemical Society*, 120(24):6191–6192, 1998. URL http://pubs.acs.org/doi/abs/10.1021/ja9803310.

Tracy J. Matray and Eric T. Kool. A specific partner for abasic damage in dna. *Nature*, 399(6737):704–708, June 1999. ISSN 0028-0836. URL http://dx.doi.org/10.1038/21453.

Olivier Mauffret, Georges Tevanian, and Serge Fermandjian. Residual dipolar coupling constants and structure determination of large dna duplexes. *Journal of Biomolecular NMR*, 24(4):317–328, December 2002. URL http://dx.doi.org/10.1023/A:1021645131882.

Sylvia McDonald and Warren S. Warren. Uses of shaped pulses in nmr: A primer. *Concepts in Magnetic Resonance*, 3(2):55–81, 1991. URL http://dx.doi.org/10.1002/cmr.1820030202.

L.W. McLaughlin, T. Leong, F. Benseler, and N. Piel. A new approach to the synthesis of a protected 2-aminopurine derivative and its incorporation into oligodeoxynucleotides containing the eco ri and bam hi recognition sites. *Nucleic Acids Res*, 16(12):5631–44, 1988.

M. Menger, T. Tuschl, F. Eckstein, and D. Porschke. Mg(2+)-dependent conformational changes in the hammerhead ribozyme. *Biochemistry*, 35(47):14710–6, 1996.

M. Menger, F. Eckstein, and D. Porschke. Multiple conformational states of the hammerhead ribozyme, broad time range of relaxation and topology of dynamics. *Nucleic Acids Res*, 28(22):4428–34, 2000.

Literaturverzeichnis

Mihaela-Rita Mihailescu and Irina M. Russu. A signature of the t to r transition in human hemoglobin. *Proceedings of the National Academy of Sciences of the United States of America*, 98(7):3773–3777, 2001. URL http://www.pnas.org/content/98/7/3773.abstract.

S.K. Mishra, M.K. Shukla, and P.C. Mishra. Electronic spectra of adenine and 2-aminopurine: an ab initio study of energy level diagrams of different tautomers in gas phase and aqueous solution. *Spectrochim Acta A Mol Biomol Spectrosc*, 56A(7): 1355–84, 2000.

J. G. Moe, E. Folta-Stogniew, and I. M. Russu. Energetics of base pair opening in a dna dodecamer containing an a3t3 tract. *Nucleic Acids Res*, 23(11):1984–1989, Jun 1995. doi: http://nar.oxfordjournals.org/cgi/reprint/23/11/1984.

James G. Moe and Irina M. Russu. Proton exchange and base-pair opening kinetics in 5'-d(cgcgaattcgcg)-3' and related dodecamers. *Nucl. Acids Res.*, 18(4):821–827, 1990. URL http://nar.oxfordjournals.org/cgi/content/abstract/18/4/821.

James G. Moe and Irina M. Russu. Kinetics and energetics of base-pair opening in 5'-d(cgcgaattcgcg)-3' and a substituted dodecamer containing g.cntdot.t mismatches. *Biochemistry*, 31(36):8421–8428, 1992. doi: 10.1021/bi00151a005. URL http://pubs.acs.org/doi/abs/10.1021/bi00151a005.

Juan C. Morales and Eric T. Kool. Efficient replication between non-hydrogen-bonded nucleoside shape analogs. *Nat Struct Mol Biol*, 5(11):950–954, November 1998. ISSN 1072-8368. URL http://dx.doi.org/10.1038/2925.

R.K. Neely, D. Daujotyte, S. Grazulis, S.W. Magennis, D.T.F. Dryden, S. Klimasauskas, and A.C. Jones. Time-resolved fluorescence of 2-aminopurine as a probe of base flipping in m.hhai-dna complexes. *Nucleic Acids Res*, 33(22):6953–6960, 2005.

Jakob T. Nielsen, Khalil Arar, and Michael Petersen. Solution structure of a locked nucleic acid modified quadruplex: Introducing the v4 folding topology13. *Ange-*

wandte Chemie, 121(17):3145–3149, 2009. URL http://dx.doi.org/10.1002/ange.200806244.

Sylvie Nonin, Jean-Louis Leroy, and Maurice Gueron. Terminal base pairs of oligodeoxynucleotides: Imino proton exchange and fraying. Biochemistry, 34:10652–10659, 1995.

T.M. Nordlund, S. Andersson, L. Nilsson, R. Rigler, A. Graeslund, and L.W. McLaughlin. Structure and dynamics of a fluorescent dna oligomer containing the ecori recognition sequence: fluorescence, molecular dynamics, and nmr studies. Biochemistry, 28(23):9095–103, 1989.

M.A. O'Neill and J.K. Barton. 2-aminopurine: a probe of structural dynamics and charge transfer in dna and dna:rna hybrids. J Am Chem Soc, 124(44):13053–66, 2002a.

M.A. O'Neill and J.K. Barton. Effects of strand and directional asymmetry on base-base coupling and charge transfer in double-helical dna. Proc Natl Acad Sci U S A, 99(26):16543–50, 2002b.

M.A. O'Neill and J.K. Barton. Dna charge transport: conformationally gated hopping through stacked domains. J Am Chem Soc, 126(37):11471–83, 2004.

Marcel Ottiger and Ad Bax. Bicelle-based liquid crystals for nmr-measurement of dipolar couplings at acidic and basic ph values. Journal of Biomolecular NMR, 13(2):187–191, February 1999a. URL http://dx.doi.org/10.1023/A:1008395916985.

Marcel Ottiger and Ad Bax. How tetrahedral are methyl groups in proteins? a liquid crystal nmr study. Journal of the American Chemical Society, 121(19):4690–4695, May 1999b. ISSN 0002-7863. URL http://dx.doi.org/10.1021/ja984484z.

Albert W. Overhauser. Polarization of nuclei in metals. Phys. Rev., 92(2):411–, October 1953. URL http://link.aps.org/abstract/PR/v92/p411.

D. J. Patel and C. W. Hilbers. Proton nuclear magnetic resonance investigations of

Literaturverzeichnis

fraying in double-stranded d-aptpgpcpapt in h2o solution. *Biochemistry*, 14(12):2651–2656, Jun 1975.

N. Patel, H. Berglund, L. Nilsson, R. Rigler, L.W. McLaughlin, and A. Graeslund. Thermodynamics of interaction of a fluorescent dna oligomer with the anti-tumour drug netropsin. *Eur J Biochem*, 203(3):361–6, 1992.

O.V. Petrauskene, S. Schmidt, A.S. Karyagina, I.I. Nikolskaya, E.S. Gromova, and D. Cech. The interaction of dna duplexes containing 2-aminopurine with restriction endonucleases ecorii and ssoii. *Nucleic Acids Res*, 23(12):2192–7, 1995.

J. M. L. Pieters, E. Devroom, G. A. van der Marel, J. H. van Boom, and C. ALTONA. Conformational consequences of the incorporation of arabinofuranosylcytidine in dna - an nmr-study of the dna fragments d(cgctagcg) and d(cgactagcg) in solution. *European Journal of Biochemistry*, 184(2):415–425, 1989.

R.C. Pless and M.J. Bessman. Influence of local nucleotide sequence on substitution of 2-aminopurine for adenine during deoxyribonucleic acid synthesis in vitro. *Biochemistry*, 22(21):4905–15, 1983.

Jean-Luc Pons, Therese E. Malliavin, and Marc A. Delsuc. Gifa v4: A complete package for nmr data-set processing. *J Biomol NMR*, 8:445–452, 1996.

J. H. Prestegard, H. M. Al-Hashimi, and J. R. Tolman. Nmr structures of biomolecules using field oriented media and residual dipolar couplings. *Quarterly Reviews of Biophysics*, 33(04):371–424, 2000.

E. W. Prohofsky, K. C. Lu, L. L. Van Zandt, and B. F. Putnam. Breathing modes and induced resonant melting of the double helix. *Physics Letters A*, 70(5-6):492 – 494, 1979. ISSN 0375-9601. doi: DOI:10.1016/0375-9601(79)90376-1. URL http://www.sciencedirect.com/science/article/B6TVM-46SPKGW-MX/2/fdc50d5fdce51e2ecd5d0c100043c4c8.

Literaturverzeichnis

Y.V. Reddy and D.N. Rao. Binding of ecop15i dna methyltransferase to dna reveals a large structural distortion within the recognition sequence. *J Mol Biol*, 298(4): 597–610, 2000.

Brian R. Reid, Kevin Banks, Peter Flynn, and Willy Nerdal. Nmr distance measurements in dna duplexes: Sugars and bases have the same correlation times. *Biochemistry*, 28: 10001–10007, 1989.

A Rich, A Nordheim, and A H J Wang. The chemistry and biology of left-handed z-dna. *Annu. Rev. Biochem.*, 53(1):791–, July 1984. URL http://dx.doi.org/10.1146/annurev.bi.53.070184.004043.

G. C. K. Roberts, editor. *NMR of Macromolecules: A Practical Approach*. Oxford University Press, 1993.

E.G. Rogan and M.J. Bessman. Studies on the pathway of incorporation of 2-aminopurine into the deoxyribonucleic acid of escherichia coli. *J Bacteriol*, 103(3): 622–33, 1970.

A. Ronen. 2-aminopurine. *Mut. Res./Reviews in Genetic Toxicology*, 75(1):1–47, January 1979.

V. A. Roongta, C. R. Jones, and D. G. Gorenstein. Effect of distortions in the deoxyribose phosphate backbone conformation of duplex oligodeoxyribonucleotide dodecamers containing gt, gg, ga, ac, and gu base-pair mismatches on p-31 nmr-spectra. *Biochemistry*, 29(22):5245–5258, 1990.

Albrecht Roscher, Lyndon Emsley, and Claude Roby. The effect of imperfect saturation in saturation-recoveryt1measurements. *Journal of Magnetic Resonance, Series A*, 118 (1):108–112, January 1996. ISSN 1064-1858. URL http://www.sciencedirect.com/science/article/B6WJY-45MG42N-71/2/518e80259c8dd1b41792f7ce9649c5d8.

Markus Ruckert and Gottfried Otting. Alignment of biological macromolecules in novel nonionic liquid crystalline media for nmr experiments. *Journal of the Ame-*

Literaturverzeichnis

rican Chemical Society, 122(32):7793–7797, August 2000. ISSN 0002-7863. URL http://dx.doi.org/10.1021/ja001068h.

J. Ruthmann, S. A. Kovalenko, N. P. Ernsting, and D. Ouw. Femtosecond relaxation of 2-amino-7-nitrofluorene in acetonitrile: Observation of the oscillatory contribution to the solvent response. *The Journal of Chemical Physics*, 109(13):5466–5468, 1998. doi: 10.1063/1.477164. URL http://link.aip.org/link/?JCP/109/5466/1.

M. Sass and D. Ziessow. Error analysis for optimized inversion recovery spin-lattice relaxation measurements. *Journal of Magnetic Resonance (1969)*, 25(2):263–276, February 1977. ISSN 0022-2364. URL http://www.sciencedirect.com/science/article/B7GXD-4CRG70F-GK/2/793c66bf441c7e295cb2135d932d5f7e.

A. Saupe and G. Englert. High-resolution nuclear magnetic resonance spectra of orientated molecules. *Phys. Rev. Lett.*, 11(10):462–, November 1963. URL http://link.aps.org/abstract/PRL/v11/p462.

Roger E. Schirmer and Joseph H. Noggle. Quantitative application of the nuclear overhauser effect to the determination of molecular structure. *Journal of the American Chemical Society*, 94(9):2947–2952, May 1972. ISSN 0002-7863. URL http://dx.doi.org/10.1021/ja00764a009.

Roger E. Schirmer, Joseph H. Noggle, Jeffrey Paul. Davis, and Phillip A. Hart. Determination of molecular geometry by quantitative application of the nuclear overhauser effect. *Journal of the American Chemical Society*, 92(11):3266–3273, June 1970. ISSN 0002-7863. URL http://dx.doi.org/10.1021/ja00714a005.

J. C. Schulhof, D. Molko, and R. Teoule. The final deprotection step in oligonucleotide synthesis is reduced to a mild and rapid ammonia treatment by using labile base-protecting groups. *Nucleic Acids Research*, 15(2):397–416, 1987.

J. L. Schwartz, J. S. Rice, B. A. Luxon, J. M. Sayer, G. Xie, H. J. C. Yeh, X. Liu, D. M. Jerina, and D. G. Gorenstein. Solution structure of the minor conformer of a dna duplex containing a dg mismatch opposite a benzo[a]pyrene diol epoxide/da adduct: Glycosidic rotation from syn to anti at the modified deoxyadenosine. *Biochemistry*, 36 (37):11069–11076, September 1997. ISSN 0006-2960. URL http://dx.doi.org/10.1021/bi971306u.

Charles D. Schwieters, John J. Kuszewski, Nico Tjandra, and G. Marius Clore. The xplor-nih nmr molecular structure determination package. *J. Magn. Reson.*, 160:65–73, 2003.

L. Serrano-Andres, M. Merchan, and A.C. Borin. Adenine and 2-aminopurine: Paradigms of modern theoretical photochemistry. *Proc Natl Acad Sci U S A*, 103(23):86918696, 2006.

Zippora Shakked, D. Rabinovich, W. B. T. Cruse, E. Egert, Olga Kennard, Graciela Sala, S. A. Salisbury, and M. A. Viswamitra. Crystalline a-dna: The x-ray analysis of the fragment d(g-g-t-a-t-a-c-c). *Proceedings of the Royal Society of London. Series B. Biological Sciences*, 213(1193):479–487, 1981. doi: 10.1098/rspb.1981.0076. URL http://rspb.royalsocietypublishing.org/content/213/1193/479.abstract.

Zippora Shakked, Gali Guerstein-Guzikevich, Miriam Eisenstein, Felix Frolow, and Dov Rabinovich. The conformation of the dna double helix in the crystal is dependent on its environment. *Nature*, 342(6248):456–460, November 1989. URL http://dx.doi.org/10.1038/342456a0.

Edward C. Sherer and Christopher J. Cramer. Quantum chemical characterization of the cytosine: 2-aminopurine base pair. *J Comp Chem*, 22(11):1167–1179, 2001. URL http://dx.doi.org/10.1002/jcc.1075.

M. S. Silver, R. I. Joseph, and D. I. Hoult. Highly selective [pi]/2 and [pi] pulse generation. *Journal of Magnetic Resonance (1969)*, 59(2):347–351, September 1984. URL http://www.sciencedirect.com/science/article/B7GXD-4CRGG62-TW/2/595590a803c9b083099a6587fb51bb91.

Literaturverzeichnis

M. S. Silver, R. I. Joseph, and D. I. Hoult. Selective spin inversion in nuclear magnetic resonance and coherent optics through an exact solution of the bloch-riccati equation. *Phys. Rev. A*, 31(4):2753–2755, April 1985. URL http://link.aps.org/abstract/PRA/v31/p2753.

Ishwar Singh, Walburga Hecker, Ashok K. Prasad, Virinder S. Parmar, and Oliver Seitz. Local disruption of dna-base stacking by bulky base surrogates. *Chemical Communications*, (5):500–501, 2002. URL http://dx.doi.org/10.1039/b110842e.

U. Chandra Singh and Peter A. Kollman. An approach to computing electrostatic charges for molecules. *Journal of Computational Chemistry*, 5(2):129–145, 1984. URL http://dx.doi.org/10.1002/jcc.540050204.

N. D. Sinha, J. Biernat, J. McManus, and H. Koester. Polymer support oligonucleotide synthesis .18. use of beta-cyanoethyl-n,n-dialkylamino-/n-morpholino phosphoramidite of deoxynucleosides for the synthesis of dna fragments simplifying deprotection and isolation of the final product. *Nucleic Acids Research*, 12(11):4539–4557, 1984.

Serge Smirnov, Tracy J. Matray, Eric T. Kool, and Carlos de los Santos. Integrity of duplex structures without hydrogen bonding: Dna with pyrene paired at abasic sites. *Nucl. Acids Res.*, 30(24):5561–5569, 2002. URL http://nar.oxfordjournals.org/cgi/content/abstract/30/24/5561.

K. Snoussi and J.-L. Leroy. Imino proton exchange and base-pair kinetics in rna duplexes. *Biochemistry*, 40(30):8898–8904, 2001. URL http://pubs.acs.org/doi/abs/10.1021/bi010385d.

L.C. Sowers, G.V. Fazakerley, R. Eritja, B.E. Kaplan, and M.F. Goodman. Base pairing and mutagenesis: observation of a protonated base pair between 2-aminopurine and cytosine in an oligonucleotide by proton nmr. *Proc Natl Acad Sci U S A*, 83(15): 5434–8, 1986.

L.C. Sowers, R. Eritja, F.M. Chen, T. Khwaja, B.E. Kaplan, M.F. Goodman, and G.V. Fazakerley. Characterization of the high ph wobble structure of the 2-

aminopurine.cytosine mismatch by n-15 nmr spectroscopy. *Biochem Biophys Res Commun*, 165(1):89–92, 1989.

L.C. Sowers, Y. Boulard, and G.V. Fazakerley. Multiple structures for the 2-aminopurine-cytosine mispair. *Biochemistry*, 39(25):7613–20, 2000.

H. Peter Spielmann, David E. Wemmer, and Jens Peter Jacobsen. Solution structure of a dna complex with the fluorescent bis-intercalator toto determined by nmr spectroscopy. *Biochemistry*, 34:8543–8553, 1995.

Richard Stefl, Haihong Wu, Sapna Ravindranathan, Vladimir Sklenar, and Juli Feigon. Dna a-tract bending in three dimensions: Solving the da4t4 vs. dt4a4 conundrum. *Proceedings of the National Academy of Sciences*, 101(5):1177–1182, February 2004. URL http://www.pnas.org/cgi/content/abstract/101/5/1177.

J.T. Stivers, K.W. Pankiewicz, and K.A. Watanabe. Kinetic mechanism of damage site recognition and uracil flipping by escherichia coli uracil dna glycosylase. *Biochemistry*, 38(3):952–63, 1999.

Lucjan Strekowski, Rebecca A. Watson, and W.David Wilson. Selective catalysis of a-t base pair proton exchange in dna complexes: imino proton nmr analysis. *Nucl. Acids Res.*, 15(20):8511–8519, 1987. URL http://nar.oxfordjournals.org/cgi/content/abstract/15/20/8511.

T. Su, B.A. Connolly, C. Darlington, R. Mallin, and D.T.F. Dryden. Unusual 2-aminopurine fluorescence from a complex of dna and the ecoki methyltransferase. *Nucleic Acids Res*, 32(7):2223–30, 2004.

Gopalakrishnan Subramaniam, Manuel M. Paz, Gopinatha Suresh Kumar, Arunangshu Das, Yolanda Palom, Cristina C. Clement, Dinshaw J. Patel, and Maria Tomasz. Solution structure of a guanine-n7-linked complex of the mitomycin c metabolite 2,7-diaminomitosene and dna. basis of sequence selectivity. *Biochemistry*, 40(35): 10473–10484, September 2001. ISSN 0006-2960. URL http://dx.doi.org/10.1021/bi010965a.

Literaturverzeichnis

Weihong Tan, Kemim Wang, and Timothy J Drake. Molecular beacons. *Current Opinion in Chemical Biology*, 8(5):547–553, October 2004. ISSN 1367-5931. URL http://www.sciencedirect.com/science/article/B6VRX-4D4XP0H-1/2/5ab55a3cba40fa0beaf488eed013e01c.

Alberto Tannús and Michael Garwood. Adiabatic pulses. *NMR in Biomedicine*, 10(8): 423–434, 1997. URL http://dx.doi.org/10.1002/(SICI)1099-1492(199712)10: 8<423::AID-NBM488>3.0.CO;2-X.

H. Teitelbaum and S. W. Englander. Open states in native polynucleotides. ii. hydrogen-exchange study of cytosine-containing double helices. *J Mol Biol*, 92(1):79–92, Feb 1975a.

H. Teitelbaum and S. W. Englander. Open states in native polynucleotides. i. hydrogen-exchange study of adenine-containing double helices. *J Mol Biol*, 92(1):55–78, Feb 1975b.

Nico Tjandra and Ad Bax. Direct measurement of distances and angles in biomolecules by nmr in a dilute liquid crystalline medium. *Science*, 278(5340):1111–1114, 1997. URL http://www.sciencemag.org/cgi/content/abstract/278/5340/1111.

Nico Tjandra, Stephan Grzesiek, and Ad Bax. Magnetic field dependence of nitrogen15 proton j splittings in 15n-enriched human ubiquitin resulting from relaxation interference and residual dipolar coupling. *Journal of the American Chemical Society*, 118 (26):6264–6272, January 1996. ISSN 0002-7863. URL http://dx.doi.org/10.1021/ja960106n.

Nico Tjandra, James G. Omichinski, Angela M. Gronenborn, G. Marius Clore, and Ad Bax. Use of dipolar 1h-15n and 1h-13c couplings in the structure determination of magnetically oriented macromolecules in solution. *Nat Struct Mol Biol*, 4(9):732–738, September 1997. URL http://dx.doi.org/10.1038/nsb0997-732.

Nico Tjandra, Shin-ichi Tate, Akira Ono, Masatsune Kainosho, and Ad Bax. The nmr structure of a dna dodecamer in an aqueous dilute liquid crystalline phase. *Journal of* the *American Chemical Society*, 122(26):6190–6200, July 2000. ISSN 0002-7863. URL http://dx.doi.org/10.1021/ja000324n.

D. Tleugabulova and L. Reha-Krantz. Probing dna polymerase-dna interactions: Examining the template strand in exonuclease complexes using 2-aminopurine fluorescence and acrylamide quenching. *Biochemistry*, 46:6559–6569, 2007.

J R Tolman, J M Flanagan, M A Kennedy, and J H Prestegard. Nuclear magnetic dipole interactions in field-oriented proteins: information for structure determination in solution. *Proceedings of the National Academy of Sciences of the United States of America*, 92(20):9279–9283, 1995. URL http://www.pnas.org/content/92/20/9279.abstract.

M. Tonelli and T.L. James. Insights into the dynamic nature of dna duplex structure via analysis of nuclear overhauser effect intensities. *Biochemistry*, 37(33):11478–11487, 1998. URL http://pubs3.acs.org/acs/journals/doilookup?in_doi=10.1021/bi980905d.

A. Ujvari and C.T. Martin. Thermodynamic and kinetic measurements of promoter binding by t7 rna polymerase. *Biochemistry*, 35(46):14574–82, 1996.

Annaleen Vermeulen, Hongjun Zhou, and Arthur Pardi. Determining dna global structure and dna bending by application of nmr residual dipolar couplings. *Journal of the American Chemical Society*, 122(40):9638–9647, 2000. URL http://pubs.acs.org/doi/abs/10.1021/ja0019191.

Hans-Achim Wagenknecht. Fluorescent dna base modifications and substitutes: Multiple fluorophore labeling and the deteq concept. *Annals of the New York Academy of Sciences*, 1130(Fluorescence Methods and Applications: Spectroscopy, Imaging, and Probes):122–130, 2008. URL http://dx.doi.org/10.1196/annals.1430.001.

Gerhard Wagner, Anil Kumar, and Kurt Wuethrich. Systematic application of two-dimensional ^{1}h nuclear-magnetic-resonance techniques for studies of

Literaturverzeichnis

proteins. *European Journal of Biochemistry*, 114(2):375–384, 1981. URL http://dx.doi.org/10.1111/j.1432-1033.1981.tb05157.x.

C. Wan, T. Fiebig, O. Schiemann, J.K. Barton, and A.H. Zewail. Femtosecond direct observation of charge transfer between bases in dna. *Proc Natl Acad Sci U S A*, 97 (26):14052–5, 2000.

AJ Wang, GJ Quigley, FJ Kolpak, G van der Marel, JH van Boom, and A Rich. Left-handed double helical dna: variations in the backbone conformation. *Science*, 211 (4478):171–176, 1981. doi: 10.1126/science.7444458. URL http://www.sciencemag.org/cgi/content/abstract/211/4478/171.

Q Wang, RW Schoenlein, LA Peteanu, RA Mathies, and CV Shank. Vibrationally coherent photochemistry in the femtosecond primary event of vision. *Science*, 266 (5184):422–424, 1994. doi: 10.1126/science.7939680. URL http://www.sciencemag.org/cgi/content/abstract/266/5184/422.

D.C. Ward, E. Reich, and L. Stryer. Fluorescence studies of nucleotides and polynucleotides. i. formycin, 2-aminopurine riboside, 2,6-diaminopurine riboside, and their derivatives. *J Biol Chem*, 244(5):1228–37, 1969.

W. S. Warren, S. L. Hammes, and J. L. Bates. Dynamics of radiation damping in nuclear magnetic resonance. *J. Chem. Phys.*, 91(10):5895–5904, November 1989. URL http://link.aip.org/link/?JCP/91/5895/1.

Warren S. Warren. Effects of arbitrary laser or nmr pulse shapes on population inversion and coherence. *J. Chem. Phys.*, 81(12):5437–5448, December 1984. URL http://link.aip.org/link/?JCP/81/5437/1.

S.M. Watanabe and M.F. Goodman. Kinetic measurement of 2-aminopurine x cytosine and 2-aminopurine x thymine base pairs as a test of dna polymerase fidelity mechanisms. *Proc Natl Acad Sci U S A*, 79(21):6429–33, 1982.

Literaturverzeichnis

Susan M. Watanabe and Myron F. Goodman. On the molecular basis of transition mutations: Frequencies of forming 2-aminopurinemiddle dot cytosine and adeninemiddle dot cytosine base mispairs in vitro. *Proc Natl Acad Sci U S A*, 78(5):2864–2868, 1981. URL http://www.pnas.org/cgi/content/abstract/78/5/2864.

J. Watson and F. Crick. Molecular structure of nucleic acids. *Nature*, 177:737–738, 1953.

Scott J. Weiner, Peter A. Kollman, David A. Case, U. Chandra Singh, Caterina Ghio, Guliano Alagona, Salvatore Profeta, and Paul Weiner. A new force field for molecular mechanical simulation of nucleic acids and proteins. *Journal of the American Chemical Society*, 106(3):765–784, 1984. ISSN 0002-7863. URL http://pubs3.acs.org/acs/journals/doilookup?in_doi=10.1021/ja00315a051.

George H. Weiss, Raj K. Guptaj, James A. Ferretti, and Edwin D. Becker. The choice of optimal parameters for measurement of spin-lattice relaxation times. i. mathematical formulation. *Journal of Magnetic Resonance (1969)*, 37(3):369–379, February 1980. ISSN 0022-2364. URL http://www.sciencedirect.com/science/article/B7GXD-4CRG61F-4P/2/2a40aa3863b723e792ea64fb155081f8.

Sybren S. Wijmenga, Martijn Kruithof, and Cees W. Hilbers. Analysis of 1h chemical shifts in dna: Assessment of the reliability of 1h chemical shift calculations for use in structure refinement. *Journal of Biomolecular NMR*, 10(4):337–350, December 1997. URL http://dx.doi.org/10.1023/A:1018348123074.

Michael P. Williamson, Timothy F. Havel, and Kurt Wuthrich. Solution conformation of proteinase inhibitor iia from bull seminal plasma by 1h nuclear magnetic resonance and distance geometry. *Journal of Molecular Biology*, 182(2):295–315, March 1985. URL http://www.sciencedirect.com/science/article/B6WK7-4DMP5KW-C/2/c8e51df47bac997838c4f97cf08f480c.

Richard Wing, Horace Drew, Tsunehiro Takano, Chris Broka, Shoji Tanaka, Keiichi Itakura, and Richard E. Dickerson. Crystal structure analysis of a complete turn of b-

Literaturverzeichnis

dna. *Nature*, 287(5784):755–758, October 1980. URL http://dx.doi.org/10.1038/287755a0.

Jens Wohnert, Katherine J. Franz, Mark Nitz, Barbara Imperiali, and Harald Schwalbe. Protein alignment by a coexpressed lanthanide-binding tag for the measurement of residual dipolar couplings. *Journal of the American Chemical Society*, 125(44): 13338–13339, November 2003. ISSN 0002-7863. URL http://dx.doi.org/10.1021/ja036022d.

Christian Wojczewski, Karen Stolze, and Joachim W. Engels. Fluorescent oligonucleotides - versatile tools as probes and primers for dna and rna analysis. *Synlett*, 1999 (10):1667–1678, 1999.

Marion C. Woods, Hung-Che. Chiang, Yasuhiro. Nakadaira, and Koji. Nakanishi. Nuclear overhauser effect, a unique method of defining the relative stereochemistry and conformation of taxane derivatives. *Journal of the American Chemical Society*, 90 (2):522–523, January 1968. ISSN 0002-7863. URL http://dx.doi.org/10.1021/ja01004a074.

Bin Wu, Frederic Girard, Bernd van Buuren, Jurgen Schleucher, Marco Tessari, and Sybren Wijmenga. Global structure of a dna three-way junction by solution nmr: towards prediction of 3h fold. *Nucl. Acids Res.*, 32(10):3228–3239, 2004. URL http://nar.oxfordjournals.org/cgi/content/abstract/32/10/3228.

Kurt Wuthrich, Gerhard Wider, Gerhard Wagner, and Werner Braun. Sequential resonance assignments as a basis for determination of spatial protein structures by high resolution proton nuclear magnetic resonance. *Journal of Molecular Biology*, 155 (3):311–319, March 1982. URL http://www.sciencedirect.com/science/article/B6WK7-4DNGVST-CY/2/fced6c4fdc2a4f852e3dcf56d8c0e35c.

Tianbing Xia, Jr. John SantaLucia, Mark E. Burkard, Ryszard Kierzek, Susan J. Schroeder, Xiaoqi Jiao, Christopher Cox, and Douglas H. Turner. Thermodynamic parameters for an expanded nearest-neighbor model for formation of rna duplexes with watson-crick base pairs. *Biochemistry*, 37:14719–14735, 1998.

Aihua Xie, Alexander F. G. van der Meer, and Robert H. Austin. Excited-state lifetimes of far-infrared collective modes in proteins. *Phys. Rev. Lett.*, 88(1):018102, Dec 2001. doi: 10.1103/PhysRevLett.88.018102.

D. Xu, K.O. Evans, and T.M. Nordlund. Melting and premelting transitions of an oligomer measured by dna base fluorescence and absorption. *Biochemistry*, 33(32): 9592–9, 1994.

A. H. Zewail. Femtochemistry: Atomic-scale dynamics of the chemical bond. *Journal of Physical Chemistry A*, 104(24):5660–5694, June 2000.

Lijuan Zhao, J. Luis Perez Lustres, Vadim Farztdinov, and Nikolaus P. Ernsting. Femtosecond fluorescence spectroscopy by upconversion with tilted gate pulses. *Physical Chemistry Chemical Physics*, 7(8):1716–1725, 2005. URL http://dx.doi.org/10.1039/b500108k.

Lukasz Zidek, Haihong Wu, Juli Feigon, and Vladimir Sklenar. Measurement of small scalar and dipolar couplings in purine and pyrimidine bases*. *Journal of Biomolecular NMR*, 21(2):153–160, October 2001. URL http://dx.doi.org/10.1023/A:1012435106858.

E R Zuiderweg, R Kaptein, and K Wuethrich. Secondary structure of the lac repressor dna-binding domain by two-dimensional 1h nuclear magnetic resonance in solution. *Proceedings of the National Academy of Sciences of the United States of America*, 80 (19):5837–5841, 1983. URL http://www.pnas.org/content/80/19/5837.abstract.

Erik R.P. Zuiderweg, Ruud M. Scheek, Rolf Boelens, Wilfred F. van Gunsteren, and Robert Kaptein. Determination of protein structures from nuclear magnetic resonance data using a restrained molecular dynamics approach: The lac repressor dna binding domain. *Biochimie*, 67(7-8):707–715, July 1985. ISSN 0300-

Literaturverzeichnis

9084. URL http://www.sciencedirect.com/science/article/B6VRJ-4MSJRKT-M/2/e1d77b2363c9c7bd51bba545fa4262dd.

M. Zweckstetter and A. Bax. Prediction of sterically induced alignment in a dilute liquid crystalline phase: Aid to protein structure determination by nmr. *J. Am. Chem. Soc.*, 122(15):3791–3792, 2000. URL http://pubs3.acs.org/acs/journals/doilookup?in_doi=10.1021/ja0000908.

Markus Zweckstetter. Nmr: prediction of molecular alignment from structure using the pales software. *Nat. Protocols*, 3(4):679–690, March 2008. ISSN 1754-2189. URL http://dx.doi.org/10.1038/nprot.2008.36.

Appendices

Script code

1 Force field parameter and topology files

In this section the force field parameter and topology files, respectively, that were used throughout the calculations are presented. Changes or additions that were introduced by the author are marked as "mod by anda".

```
language
!RNA PARAMETER FILE 'FRAMEOWRK' FROM PARALLHDG.DNA AND ATOM NAMES
! AND HEAVY ATOM PARAMETERS FROM DNA-RNA.PARAM
!INCLUDES ALL NONEXCHANGEABLE HYDROGEN TERMS FOR BOND, ANGLE, AND
!IMPROPERS WITH ENERGY CONSTANT VARIABLES: $kchbond, $kchangle, AND $kchimpr.
!BOND, ANGLE, AND IMPROPERS WERE ESTIMATED FROM VALUES FROM THE STANDARD
!NUCLEOTIDES OF INSIGHTII 95.0 (BIOSYM/MOLECULAR SIMULATIONS).
!CREATED 2/24/96-- JASON P. RIFE AND PETER B. MOORE
! DNA-RNA-ALLATOM.PARAM

set echo=off message=off end


! checkversion 1.0


evaluate ($kchbond = 2000)
evaluate ($kchangle = 1000)
evaluate ($kchimpr = 1000)


!****************** mod by anda - HNF******************

BOND PX1  OX2    1489.209   1.485 ! Nobs =    1
BOND PX1  OX3    1489.209   1.485 ! Nobs =    1
BOND PX1  OX4    3350.720   1.593 ! Nobs =    1
BOND OX4  CX5    1709.551   1.427 ! Nobs =    1
BOND CX5  HX6    $kchbond   1.090 ! Nobs =    1
BOND CX5  HX7    $kchbond   1.090 ! Nobs =    1
BOND CX5  CX8    5235.500   1.511 ! Nobs =    1
BOND CX8  HX9    $kchbond   1.090 ! Nobs =    1
BOND CX8  OX10   2769.190   1.446 ! Nobs =    1
BOND CX8  CX12   3350.720   1.528 ! Nobs =    1
BOND OX10 CX11   1982.674   1.420 ! Nobs =    1
```

Script code

```
BOND CX11 CX14    1709.551    1.521 ! Nobs =    1
BOND CX11 OX18    1982.674    1.420 ! mod by anda, taken from C1'-O4' distance
BOND CX11 HX43    $kchbond    1.090 ! Nobs =    1
BOND CX12 HX13    $kchbond    1.090 ! Nobs =    1
BOND CX12 CX14    3350.720    1.518 ! Nobs =    1
BOND CX12 OX17    1982.674    1.431 ! Nobs =    1
BOND CX14 HX15    $kchbond    1.090 ! Nobs =    1
BOND CX14 HX16    $kchbond    1.090 ! Nobs =    1
BOND OX18 CX19    2769.190    1.446 ! mod by anda, taken from C4'-O4' distance
BOND CX19 CX20    1500.000    1.391 ! mod by anda, extrapolated from AGCT C5-C6 values, mean
      distance from calc
BOND CX19 CX24    1500.000    1.391 ! mod by anda, extrapolated from AGCT C5-C6 values, mean
      distance from calc
BOND CX20 CX21    1500.000    1.391 ! mod by anda, extrapolated from AGCT C5-C6 values, mean
      distance from calc
BOND CX20 HX36    $kchbond    1.090 ! Nobs =    1
BOND CX21 CX22    1500.000    1.391 ! mod by anda, extrapolated from AGCT C5-C6 values, mean
      distance from calc
BOND CX21 HX35    $kchbond    1.090 ! Nobs =    1
BOND CX22 CX23    1500.000    1.391 ! mod by anda, extrapolated from AGCT C5-C6 values, mean
      distance from calc
BOND CX22 CX25    1500.000    1.451 ! mod by anda, extrapolated from AGCT C5-C6 values
BOND CX23 CX24    1500.000    1.391 ! mod by anda, extrapolated from AGCT C5-C6 values, mean
      distance from calc
BOND CX23 CX27    1000.000    1.509 ! mod by anda, extrapolated from AGCT C5-C6 values
BOND CX24 HX40    $kchbond    1.090 ! Nobs =    1
BOND CX25 CX26    1500.000    1.391 ! mod by anda, extrapolated from AGCT C5-C6 values, mean
      distance from calc
BOND CX25 CX28    1500.000    1.391 ! mod by anda, extrapolated from AGCT C5-C6 values, mean
      distance from calc
BOND CX26 CX27    1500.000    1.509 ! mod by anda, extrapolated from AGCT C5-C6 values
BOND CX26 CX31    1500.000    1.391 ! mod by anda, extrapolated from AGCT C5-C6 values, mean
      distance from calc
BOND CX27 HX41    $kchbond    1.090 ! Nobs =    1
BOND CX27 HX42    $kchbond    1.090 ! Nobs =    1
BOND CX28 CX29    1500.000    1.391 ! mod by anda, extrapolated from AGCT C5-C6 values, mean
      distance from calc
BOND CX28 HX37    $kchbond    1.090 ! Nobs =    1
BOND CX29 CX30    1500.000    1.391 ! mod by anda, extrapolated from AGCT C5-C6 values, mean
      distance from calc
BOND CX29 HX38    $kchbond    1.090 ! Nobs =    1
BOND CX30 CX31    1500.000    1.391 ! mod by anda, extrapolated from AGCT C5-C6 values, mean
      distance from calc
BOND CX30 NX32    1370.370    1.471 ! mod by anda, k taken from C C4-N4, length from calc
BOND CX31 HX39    $kchbond    1.090 ! Nobs =    1
BOND NX32 OX33    1734.375    1.227 ! mod by anda, taken from T C2 ON
BOND NX32 OX34    1734.375    1.227 ! mod by anda, taken from T C2 ON

!******************* end mod by anda - HNF *****************

!the generic bonds were taken from param11.dna with 3*kq
BOND  C5R  OH      876.000    1.4300    ! 5' end
BOND  C5D  OH      876.000    1.4300    ! 5' end
BOND  C3R  OH      876.000    1.4300    ! 3' end
BOND  C3D  OH      876.000    1.4300    ! 3' end
BOND  O2R  HO     1350.000    0.9572
```

```
!Phos.  - combined RNA/DNA statistics used
!                      kq         x_eq       sigma
BOND  P    O1P     1489.209      1.485     ! 0.015  Phos
BOND  P    O2P     1489.209      1.485     ! 0.015  P
BOND  P    O5R     3350.720      1.593     ! 0.010  P
BOND  P    OH      3350.720      1.593     ! 0.010  P    ! For 5pho patch
BOND  P    O3R     2326.889      1.607     ! 0.012  P
BOND  P    OX17    2326.889      1.607     ! 0.012  P    ! mod by anda
BOND  PX1  O3R     2326.889      1.607     ! 0.012  P    ! mod by anda


!Sugars
!RNA statistics
BOND  O5R  C5R     1709.551      1.425     ! 0.014  Sugar
BOND  C5R  C4R     1982.674      1.510     ! 0.013  S
BOND  C4R  C3R     2769.190      1.524     ! 0.011  S
BOND  C3R  C2R     2769.190      1.525     ! 0.011  S
BOND  C2R  C1R     3350.720      1.528     ! 0.010  S
BOND  O4R  C1R     2326.888      1.414     ! 0.012  S
BOND  O4R  C4R     2326.888      1.453     ! 0.012  S
BOND  O3R  C3R     1982.674      1.423     ! 0.013  S
BOND  C2R  O2R     1982.674      1.413     ! 0.013  S

!DNA statistics
BOND  O5R  C5D     1709.551      1.427     ! 0.014  Sugar
BOND  C5D  C4D     5235.500      1.511     ! 0.008  S
BOND  C4D  C3D     3350.720      1.528     ! 0.010  S
BOND  C3D  C2D     3350.720      1.518     ! 0.010  S
BOND  C2D  C1D     1709.551      1.521     ! 0.014  S
BOND  O4D  C1D     1982.674      1.420     ! 0.013  S
BOND  O4D  C4D     2769.190      1.446     ! 0.011  S
BOND  O3R  C3D     1982.674      1.431     ! 0.013  S

!hydrogen/carbon
BOND  C4R  H       $kchbond      1.09
BOND  C3R  H       $kchbond      1.09
BOND  C2R  H       $kchbond      1.09
BOND  C1R  H       $kchbond      1.09
BOND  C5R  H       $kchbond      1.09

BOND  C4D  H       $kchbond      1.09
BOND  C3D  H       $kchbond      1.09
BOND  C2D  H       $kchbond      1.09
BOND  C1D  H       $kchbond      1.09
BOND  C5D  H       $kchbond      1.09

!Bases
!base specific bonds taken from param11.dna , 3*kq
BOND  O2U  HO      1350.000      0.957     ! UR
BOND  HN   NNA     1416.000      1.010     ! URA
BOND  HN   N1T     1416.000      1.010     ! Infer.
BOND  HN   N1C     1416.000      1.010
BOND  HN   N9G     1416.000      1.010
BOND  HN   N9A     1416.000      1.010
BOND  HN   N9P     1416.000      1.010
```

Script code

```
BOND   HN    N3U      1416.000    1.010
BOND   HN    N3T      1416.000    1.010
BOND   H2    N2       1416.000    1.010
BOND   H2    N4C      1416.000    1.010
BOND   H2    N2G      1416.000    1.010
BOND   H2    N6A      1416.000    1.010

BOND   HO    OH       1350.000    0.960      ! PARAM7 (IR stretch 3400 cm-1)

!Base sugar joint bonds (scale from sugar)
!                       kq          x_eq       sigma
BOND   C1R   N1T      1709.551    1.473      ! 0.014   Base
BOND   C1R   N1U      4136.691    1.469      ! 0.009   B
BOND   C1R   N1C      2326.889    1.470      ! 0.012   B
BOND   C1R   N9G      4136.691    1.459      ! 0.009   B
BOND   C1R   N9A      3350.720    1.462      ! 0.010   B
BOND   C1R   N9P      3350.720    1.462      ! 0.010   B

BOND   C1D   N1T      1709.551    1.473      ! 0.014   B    !DNA
BOND   C1D   N1U      4136.691    1.469      ! 0.009   B
BOND   C1D   N1C      2326.889    1.470      ! 0.012   B
BOND   C1D   N9G      4136.691    1.459      ! 0.009   B
BOND   C1D   N9A      3350.720    1.462      ! 0.010   B
BOND   C1D   N9P      3350.720    1.462      ! 0.010   B

!cytosine                           kq         x_eq   sigma
BOND         C2C   ON             1370.370   1.240  !0.009  B
BOND         C4C   N4C            1370.370   1.335  !0.009  B
BOND         N1C   C2C            1110.000   1.397  !0.010  B
BOND         N1C   C6C            3083.333   1.367  !0.006  B
BOND         C2C   NC             1734.375   1.353  !0.008  B
BOND         NC    C4C            2265.306   1.335  !0.007  B
BOND         C4C   C5C            1734.375   1.425  !0.008  B
BOND         C5C   C6C            1734.375   1.339  !0.008  B
BOND         C5C   H              $kchbond   1.09
BOND         C6C   H              $kchbond   1.09

!thymine
BOND         N1T   C2T            1734.375   1.376  !0.008  B
BOND         C2T   N3T            1734.375   1.373  !0.008  B
BOND         N3T   C4T            1734.375   1.382  !0.008  B
BOND         C4T   C5T            1370.370   1.445  !0.009  B
BOND         C5T   C6T            2265.306   1.339  !0.007  B
BOND         C6T   N1T            2265.306   1.378  !0.007  B
BOND         C2T   ON             1734.375   1.220  !0.008  B
BOND         C4T   ON             1370.370   1.228  !0.008  B
BOND         C5T   CC3E           3083.333   1.496  !0.006  B
BOND         C6T   H              $kchbond   1.09
BOND         CC3E  H              $kchbond   1.09

!adenine
BOND         NC    C2A            1370.370   1.339  !0.009  B
BOND         C2A   N3A            1370.370   1.331  !0.009  B
BOND         N3A   C4A            3083.333   1.344  !0.006  B
BOND         C4A   C5A            2265.306   1.383  !0.007  B
BOND         C5A   C6A            1370.370   1.406  !0.009  B
```

1 Force field parameter and topology files

```
BOND    C6A    NC      2265.306    1.351  !0.007  B
BOND    C5A    N7A     3083.333    1.388  !0.006  B
BOND    N7A    C8A     2265.306    1.311  !0.007  B
BOND    C8A    N9A     1734.375    1.373  !0.008  B
BOND    N9A    C4A     3083.333    1.374  !0.006  B
BOND    C6A    N6A     1734.375    1.335  !0.008  B
BOND    C8A    H       $kchbond    1.08
BOND    C2A    H       $kchbond    1.09

!purine
BOND    NC     C2P     1370.370    1.339  !0.009  B
BOND    C2P    N3P     1370.370    1.331  !0.009  B
BOND    N3P    C4P     3083.333    1.344  !0.006  B
BOND    C4P    C5P     2265.306    1.383  !0.007  B
BOND    C5P    C6P     1370.370    1.406  !0.009  B
BOND    C6P    NC      2265.306    1.351  !0.007  B
BOND    C5P    N7P     3083.333    1.388  !0.006  B
BOND    N7P    C8P     2265.306    1.311  !0.007  B
BOND    C8P    N9P     1734.375    1.373  !0.008  B
BOND    N9P    C4P     3083.333    1.374  !0.006  B
BOND    C6P    H       $kchbond    1.09   !0.008  B
BOND    C8P    H       $kchbond    1.08
BOND    C2P    H       $kchbond    1.09

!guanine
BOND    NNA    C2G     1734.375    1.373  !0.008  B
BOND    C2G    N3G     1734.375    1.323  !0.008  B
BOND    N3G    C4G     2265.306    1.350  !0.007  B
BOND    C4G    C5G     2265.306    1.379  !0.007  B
BOND    C5G    C6G     1110.000    1.419  !0.010  B
BOND    C6G    NNA     2265.306    1.391  !0.007  B
BOND    C5G    N7G     3083.333    1.388  !0.006  B
BOND    N7G    C8G     3083.333    1.305  !0.006  B
BOND    C8G    N9G     2265.306    1.374  !0.007  B
BOND    N9G    C4G     1734.375    1.375  !0.008  B
BOND    C2G    N2G     1110.000    1.341  !0.010  B
BOND    C6G    O6G     1370.370    1.237  !0.009  B
BOND    C8G    H       $kchbond    1.08

!uracil
BOND    C2U    ON      1370.370    1.219  !0.009  B
BOND    C4U    ON      1734.375    1.232  !0.008  B
BOND    N1U    C2U     1370.370    1.381  !0.009  B
BOND    N1U    C6U     1370.370    1.375  !0.009  B
BOND    C2U    N3U     2265.306    1.373  !0.007  B
BOND    N3U    C4U     1370.370    1.380  !0.009  B
BOND    C4U    C5U     1370.370    1.431  !0.009  B
BOND    C5U    C6U     1370.370    1.337  !0.009  B
BOND    C5U    H       $kchbond    1.09
BOND    C6U    H       $kchbond    1.09
BOND    C2D    NX29    2265.306    1.479  !check param, added for pyr

!****************** mod by anda - HNF ******************

ANGLe  OX2  PX1  OX3    1337.074   119.600  ! Nobs =    1
ANGLe  OX2  PX1  OX4     357.719   108.100  ! Nobs =    1
```

Script code

```
ANGLe OX3   PX1  OX4      412.677   108.300 ! Nobs =   1
ANGLe PX1   OX4  CX5     1175.163   120.900 ! Nobs =   1
ANGLe OX4   CX5  HX6     $kchangle  109.70  ! Nobs =   1
ANGLe OX4   CX5  HX7     $kchangle  109.70  ! Nobs =   1
ANGLe OX4   CX5  CX8     1534.906  110.200  ! Nobs =   1
ANGLe HX6   CX5  HX7     $kchangle  109.17  ! Nobs =   1
ANGLe HX6   CX5  CX8     $kchangle  109.17  ! Nobs =   1
ANGLe HX7   CX5  CX8     $kchangle  109.17  ! Nobs =   1
ANGLe CX5   CX8  HX9     $kchangle  107.78  ! Nobs =   1
ANGLe CX5   CX8  OX10     429.678  109.400  ! Nobs =   1
ANGLe CX5   CX8  CX12     488.878  114.700  ! Nobs =   1
ANGLe HX9   CX8  OX10    $kchangle  112.98  ! Nobs =   1
ANGLe HX9   CX8  CX12    $kchangle  106.91  ! Nobs =   1
ANGLe OX10  CX8  CX12    1099.976  105.600  ! Nobs =   1
ANGLe CX8   OX10 CX11     650.874  109.700  ! Nobs =   1
ANGLe OX10  CX11 CX14     909.071  106.100  ! Nobs =   1
ANGLe OX10  CX11 OX18     357.719  107.016  ! Nobs =   1
ANGLe OX10  CX11 HX43    $kchangle  107.95  ! Nobs =   1
ANGLe CX14  CX11 OX18     488.878  111.73   ! mod by anda, gaussian value, kforce appr. from
       G,C,A,T
ANGLe CX14  CX11 HX43    $kchangle  112.29  ! Nobs =   1
ANGLe OX18  CX11 HX43    $kchangle  108.56  ! mod by anda, gaussian value
ANGLe CX8   CX12 HX13    $kchangle  111.16  ! Nobs =   1
ANGLe CX8   CX12 CX14    1099.976  103.200  ! Nobs =   1
ANGLe CX8   CX12 OX17     621.574  110.300  ! Nobs =   1
ANGLe HX13  CX12 CX14    $kchangle  111.98  ! Nobs =   1
ANGLe HX13  CX12 OX17    $kchangle  109.34  ! Nobs =   1
ANGLe CX14  CX12 OX17     412.677  110.600  ! Nobs =   1
ANGLe CX11  CX14 CX12     650.874  102.700  ! Nobs =   1
ANGLe CX11  CX14 HX15    $kchangle  112.29  ! Nobs =   1
ANGLe CX11  CX14 HX16    $kchangle  112.29  ! Nobs =   1
ANGLe CX12  CX14 HX15    $kchangle  111.36  ! Nobs =   1
ANGLe CX12  CX14 HX16    $kchangle  111.36  ! Nobs =   1
ANGLe HX15  CX14 HX16    $kchangle  107.52  ! Nobs =   1
ANGLe CX11  OX18 CX19     650.874  120.224  ! mod by anda, gaussian value, kforce c1-O4-C4
ANGLe OX18  CX19 CX20     621.574  124.456  ! mod by anda, gaussian value, kforce C4-C3-O3
ANGLe OX18  CX19 CX24     621.574  115.284  ! mod by anda, gaussian value, kforce C4-C3-O3
ANGLe CX20  CX19 CX24    1457.566  120.259  ! mod by anda, mean gaussian value, kforce C,A,G
       C4-C5-C6
ANGLe CX19  CX20 CX21    1457.566  120.215  ! mod by anda, mean gaussian value, kforce C,A,G
       C4-C5-C6
ANGLe CX19  CX20 HX36    $kchangle  120.775 ! mod by anda, gaussian value
ANGLe CX21  CX20 HX36    $kchangle  119.007 ! mod by anda, gaussian value
ANGLe CX20  CX21 CX22    1457.566  119.689  ! mod by anda, mean gaussian value, kforce C,A,G
       C4-C5-C6
ANGLe CX20  CX21 HX35    $kchangle  119.297 ! mod by anda, gaussian value
ANGLe CX22  CX21 HX35    $kchangle  121.012 ! mod by anda, gaussian value
ANGLe CX21  CX22 CX23    1457.566  119.780  ! mod by anda, mean gaussian value, kforce C,A,G
       C4-C5-C6
ANGLe CX21  CX22 CX25    1457.566  131.505  ! mod by anda, mean gaussian value, kforce C,A,G
       C4-C5-C6
ANGLe CX23  CX22 CX25    1457.566  108.714  ! mod by anda, mean gaussian value, kforce C,A,G
       C4-C5-C6
ANGLe CX22  CX23 CX24    1457.566  120.959  ! mod by anda, mean gaussian value, kforce C,A,G
       C4-C5-C6
ANGLe CX22  CX23 CX27    1457.566  109.983  ! mod by anda, mean gaussian value, kforce C,A,G
```

1 Force field parameter and topology files

```
             C4–C5–C6
ANGLe CX24 CX23 CX27    1457.566    129.058 ! mod by anda, mean gaussian value, kforce C,A,G
             C4–C5–C6
ANGLe CX19 CX24 CX23    1457.566    119.094 ! mod by anda, mean gaussian value, kforce C,A,G
             C4–C5–C6
ANGLe CX19 CX24 HX40    $kchangle   118.578 ! mod by anda, gaussian value
ANGLe CX23 CX24 HX40    $kchangle   122.328 ! mod by anda, gaussian value
ANGLe CX22 CX25 CX26    1457.566    108.593 ! mod by anda, mean gaussian value, kforce C,A,G
             C4–C5–C6
ANGLe CX22 CX25 CX28    1457.566    131.102 ! mod by anda, mean gaussian value, kforce C,A,G
             C4–C5–C6
ANGLe CX26 CX25 CX28    1457.566    120.305 ! mod by anda, mean gaussian value, kforce C,A,G
             C4–C5–C6
ANGLe CX25 CX26 CX27    1457.566    110.022 ! mod by anda, mean gaussian value, kforce C,A,G
             C4–C5–C6
ANGLe CX25 CX26 CX31    1457.566    120.807 ! mod by anda, mean gaussian value, kforce C,A,G
             C4–C5–C6
ANGLe CX27 CX26 CX31    1457.566    129.171 ! mod by anda, mean gaussian value, kforce C,A,G
             C4–C5–C6
ANGLe CX23 CX27 CX26    1457.566    102.689 ! mod by anda, mean gaussian value, kforce C,A,G
             C4–C5–C6
ANGLe CX23 CX27 HX41    $kchangle   111.842 ! mod by anda, gaussian value
ANGLe CX23 CX27 HX42    $kchangle   111.916 ! mod by anda, gaussian value
ANGLe CX26 CX27 HX41    $kchangle   111.832 ! mod by anda, gaussian value
ANGLe CX26 CX27 HX42    $kchangle   111.824 ! mod by anda, gaussian value
ANGLe HX41 CX27 HX42    $kchangle   106.844 ! mod by anda, gaussian value
ANGLe CX25 CX28 CX29    1457.566    119.272 ! mod by anda, mean gaussian value, kforce C,A,G
             C4–C5–C6
ANGLe CX25 CX28 HX37    $kchangle   121.017 ! mod by anda, gaussian value
ANGLe CX29 CX28 HX37    $kchangle   119.711 ! mod by anda, gaussian value
ANGLe CX28 CX29 CX30    1457.566    119.411 ! mod by anda, mean gaussian value, kforce C,A,G
             C4–C5–C6
ANGLe CX28 CX29 HX38    $kchangle   121.419 ! mod by anda, gaussian value
ANGLe CX30 CX29 HX38    $kchangle   119.171 ! mod by anda, gaussian value
ANGLe CX29 CX30 CX31    1457.566    122.394 ! mod by anda, mean gaussian value, kforce C,A,G
             C4–C5–C6
ANGLe CX29 CX30 NX32    569.362     118.868 ! mod by anda, mean gaussian value, kforce A
             C5–C6–N6
ANGLe CX31 CX30 NX32    569.362     118.738 ! mod by anda, mean gaussian value, kforce A
             C5–C6–N6
ANGLe CX26 CX31 CX30    1457.566    117.812 ! mod by anda, mean gaussian value, kforce C,A,G
             C4–C5–C6
ANGLe CX26 CX31 HX39    $kchangle   122.587 ! mod by anda, gaussian value
ANGLe CX30 CX31 HX39    $kchangle   119.602 ! mod by anda, gaussian value
ANGLe CX30 NX32 OX33    210.000     117.818 ! mod by anda, mean gaussian value, kforce A C–N–H
             double
ANGLe CX30 NX32 OX34    210.000     117.978 ! mod by anda, mean gaussian value, kforce A C–N–H
             double
ANGLe OX33 NX32 OX34    1337.074    124.204 ! mod by anda, mean gaussian value, kforce O–P–O

!****************** end of mod by anda – HNF *************

!Phos.
!the ANGLes were taken from param11.dna with 3*kq
ANGLe        HO     OH     C5R      139.500   107.300
ANGLe        HO     O5R    C5R      139.500   107.300
```

Script code

				kq	x_eq	sigma	
ANGLe	HO	OH	C5D	139.500	107.300		
ANGLe	HO	O5R	C5D	139.500	107.300		
ANGLe	HO	O3R	P	139.500	107.300		
ANGLe	HO	OH	P	139.500	107.300	! For 5pho patch	
ANGLe	HO	O2R	C2R	139.500	107.300		
ANGLe	OH	P	O3R	144.300	102.600	!	
ANGLe	OH	P	O5R	144.300	102.600	!	
ANGLe	OH	P	O1P	296.700	108.230	!	
ANGLe	OH	P	O2P	296.700	108.230	!	
ANGLe	OH	C5R	C4R	210.000	112.000	!	
ANGLe	OH	C5D	C4D	210.000	112.000	!	
ANGLe	C4D	C3D	OH	139.500	111.000	!	
ANGLe	C4R	C3R	OH	139.500	111.000	!	
ANGLe	C2D	C3D	OH	139.500	111.000	!	
ANGLe	C2R	C3R	OH	139.500	111.000	!	
ANGLe	C3R	OH	HO	139.500	107.300	!	
ANGLe	C3D	OH	HO	139.500	107.300	!	

!Phos. — combined RNA/DNA statistics used
!

				kq	x_eq	sigma	
ANGLe	O1P	P	O2P	1337.074	119.600	!1.5 P	
ANGLe	O5R	P	O1P	357.719	108.100	!2.9 P	
ANGLe	O5R	P	O2P	412.677	108.300	!2.7 P	
ANGLe	O3R	P	O5R	833.356	104.000	!1.9 P	
ANGLe	OX17	P	O5R	833.356	104.000	!1.9 P	!mod by anda
ANGLe	O3R	PX1	OX4	833.356	104.000	!1.9 P	!mod by anda
ANGLe	O2P	P	O3R	293.791	108.300	!3.2 P	
ANGLe	O1P	P	O3R	293.791	107.400	!3.2 P	
ANGLe	OX3	PX1	O3R	293.791	108.300	!3.2 P	!mod by anda
ANGLe	OX2	PX1	O3R	293.791	107.400	!3.2 P	!mod by anda
ANGLe	O2P	P	OX17	293.791	108.300	!3.2 P	!mod by anda
ANGLe	O1P	P	OX17	293.791	107.400	!3.2 P	!mod by anda
ANGLe	O5R	C5R	C4R	1534.906	110.200	!1.4 P	
ANGLe	P	O5R	C5R	1175.163	120.900	!1.6 P	
ANGLe	P	O3R	C3R	2089.178	119.700	!1.2 P	
ANGLe	O5R	C5D	C4D	1534.906	110.200	!1.4 P	!DNA
ANGLe	P	O5R	C5D	1175.163	120.900	!1.6 P	
ANGLe	P	O3R	C3D	2089.178	119.700	!1.2 P	
ANGLe	PX1	O3R	C3D	2089.178	119.700	!1.2 P	!mod by anda
ANGLe	P	OX17	CX12	2089.178	119.700	!1.2 P	!mod by anda

!Sugars
!RNA statistics
!

				kq	x_eq	sigma	
ANGLe	O4R	C4R	C3R	561.212	105.500	!1.4 S	
ANGLe	C5R	C4R	C3R	488.878	115.500	!1.5 S	
ANGLe	C5R	C4R	O4R	561.212	109.200	!1.4 S	
ANGLe	C1R	O4R	C4R	1357.996	109.600	!0.9 S	
ANGLe	C4R	C3R	C2R	1099.976	102.700	!1.0 S	
ANGLe	C3R	C2R	C1R	1357.996	101.500	!0.9 S	
ANGLe	O4R	C1R	C2R	561.212	106.400	!1.4 S	
ANGLe	N1T	C1R	C2R	429.678	113.400	!1.6 S	

1 Force field parameter and topology files

```
ANGLe    N1C    C1R    C2R     429.678   113.400  !1.6  S
ANGLe    N1U    C1R    C2R     429.678   113.400  !1.6  S
ANGLe    N9G    C1R    C2R     429.678   113.400  !1.6  S
ANGLe    N9A    C1R    C2R     429.678   113.400  !1.6  S
ANGLe    N9P    C1R    C2R     429.678   113.400  !1.6  S
ANGLe    O4R    C1R    N1T    1099.976   108.200  !1.0  S
ANGLe    O4R    C1R    N1C    1099.976   108.200  !1.0  S
ANGLe    O4R    C1R    N1U    1099.976   108.200  !1.0  S
ANGLe    O4R    C1R    N9A    1099.976   108.200  !1.0  S
ANGLe    O4R    C1R    N9P    1099.976   108.200  !1.0  S
ANGLe    O4R    C1R    N9G    1099.976   108.200  !1.0  S
ANGLe    C1R    C2R    O2R     357.719   110.600  !2.9  S scale from phos.
ANGLe    C3R    C2R    O2R     357.719   113.300  !2.9  S scale from phos.
ANGLe    C4R    C3R    O3R     445.032   110.500  !2.6  S scale from phos.
ANGLe    C2R    C3R    O3R     383.726   111.000  !2.8  S scale from phos.

!DNA statistics
ANGLe    O4D    C4D    C3D    1099.976   105.600  !1.0  S
ANGLe    C5D    C4D    C3D     488.878   114.700  !1.5  S
ANGLe    C5D    C4D    O4D     429.678   109.400  !1.6  S
ANGLe    C1D    O4D    C4D     650.874   109.700  !1.3  S
ANGLe    C4D    C3D    C2D    1099.976   103.200  !1.0  S
ANGLe    C3D    C2D    C1D     650.874   102.700  !1.3  S
ANGLe    O4D    C1D    C2D     909.071   106.100  !1.1  S
ANGLe    N1T    C1D    C2D     488.878   114.200  !1.5  S
ANGLe    N1C    C1D    C2D     488.878   114.200  !1.5  S
ANGLe    N1U    C1D    C2D     488.878   114.200  !1.5  S
ANGLe    N9G    C1D    C2D     488.878   114.200  !1.5  S
ANGLe    N9A    C1D    C2D     488.878   114.200  !1.5  S
ANGLe    N9P    C1D    C2D     488.878   114.200  !1.5  S
ANGLe    O4D    C1D    N1T    1357.996   107.800  !0.9  S
ANGLe    O4D    C1D    N1C    1357.996   107.800  !0.9  S
ANGLe    O4D    C1D    N1U    1357.996   107.800  !0.9  S
ANGLe    O4D    C1D    N9A    1357.996   107.800  !0.9  S
ANGLe    O4D    C1D    N9P    1357.996   107.800  !0.9  S
ANGLe    O4D    C1D    N9G    1357.996   107.800  !0.9  S
ANGLe    C4D    C3D    O3R     621.574   110.300  !2.2  S scale from phos.
ANGLe    C2D    C3D    O3R     412.677   110.600  !2.7  S scale from phos.

!Ribose terms involving non-exchageables
ANGLe    OH     C5R    H      $kchangle  109.83
ANGLe    O5R    C5R    H      $kchangle  109.83
ANGLe    H      C5R    H      $kchangle  109.11
ANGLe    H      C5R    H      $kchangle  109.11
ANGLe    C4R    C5R    H      $kchangle  107.93
ANGLe    H      C4R    C3R    $kchangle  107.13
ANGLe    H      C4R    O4R    $kchangle  113.74
ANGLe    H      C3R    C4R    $kchangle  111.35
ANGLe    H      C3R    O3R    $kchangle  105.87
ANGLe    H      C3R    OH     $kchangle  105.87
ANGLe    H      C3R    C2R    $kchangle  112.27
ANGLe    H      C2R    C3R    $kchangle  111.41
ANGLe    H      C2R    O2R    $kchangle  113.07
ANGLe    H      C2R    C1R    $kchangle  112.38
ANGLe    H      C1R    C2R    $kchangle  111.95
ANGLe    H      C1R    N1C    $kchangle  107.70
```

Script code

```
ANGLe       H       C1R     N1U     $kchangle       107.70
ANGLe       H       C1R     N1T     $kchangle       107.70
ANGLe       H       C1R     N9A     $kchangle       107.70
ANGLe       H       C1R     N9P     $kchangle       107.70
ANGLe       H       C1R     N9G     $kchangle       107.70
ANGLe       H       C1R     O4R     $kchangle       106.86

!Deoxyribose terms involving non-exchageables
!
ANGLe       OH      C5D     H       $kchangle       109.70
ANGLe       O5R     C5D     H       $kchangle       109.70
ANGLe       H       C5D     H       $kchangle       109.17
ANGLe       C4D     C5D     H       $kchangle       109.17
ANGLe       C5D     C4D     H       $kchangle       107.78
ANGLe       H       C4D     C3D     $kchangle       106.91
ANGLe       H       C4D     O4D     $kchangle       112.98
ANGLe       H       C3D     C4D     $kchangle       111.16
ANGLe       H       C3D     O3R     $kchangle       109.34
ANGLe       H       C3D     OH      $kchangle       109.34
ANGLe       H       C3D     C2D     $kchangle       111.98
ANGLe       H       C2D     C3D     $kchangle       111.36
ANGLe       H       C2D     H       $kchangle       107.52
ANGLe       H       C1D     H       $kchangle       107.52  !mod by anda
ANGLe       H       C2D     C1D     $kchangle       112.29
ANGLe       H       C1D     C2D     $kchangle       110.94
ANGLe       H       C1D     N1C     $kchangle       108.25
ANGLe       H       C1D     N1U     $kchangle       108.25
ANGLe       H       C1D     N1T     $kchangle       108.25
ANGLe       H       C1D     N9A     $kchangle       108.25
ANGLe       H       C1D     N9P     $kchangle       108.25
ANGLe       H       C1D     N9G     $kchangle       108.25
ANGLe       H       C1D     O4D     $kchangle       107.95

!Bases
!cytosine                           kq         x_eq    sigma
ANGLe       C6C     N1C     C2C     2277.447   120.300 !0.40 B
ANGLe       N1C     C2C     NC      743.656    119.200 !0.70 B
ANGLe       C2C     NC      C4C     1457.566   119.900 !0.50 B
ANGLe       NC      C4C     C5C     2277.447   121.900 !0.40 B
ANGLe       C4C     C5C     C6C     1457.566   117.400 !0.50 B
ANGLe       C5C     C6C     N1C     1457.566   121.000 !0.50 B
ANGLe       N1C     C2C     ON      1012.199   118.900 !0.60 B
ANGLe       NC      C2C     ON      743.656    121.900 !0.70 B
ANGLe       NC      C4C     N4C     743.656    118.000 !0.70 B
ANGLe       C5C     C4C     N4C     743.656    120.200 !0.70 B
ANGLe       C6C     N1C     C1R     763.873    120.800 !1.20 B scale from sugar
ANGLe       C2C     N1C     C1R     909.071    118.800 !1.10 B scale from sugar
ANGLe       C6C     N1C     C1D     763.873    120.800 !1.20 B !DNA
ANGLe       C2C     N1C     C1D     909.071    118.800 !1.10 B
ANGLe       C4C     N4C     H2      105.000    120.000 !from param11.dna, 3*keq
ANGLe       H2      N4C     H2      105.000    120.000
ANGLe       N1C     C6C     H       $kchangle       119.63
ANGLe       C5C     C6C     H       $kchangle       119.36
ANGLe       C4C     C5C     H       $kchangle       121.54
ANGLe       C6C     C5C     H       $kchangle       121.54
```

1 Force field parameter and topology files

```
!thymine                          kq         x_eq      sigma
ANGLe     C6T    N1T    C2T       1457.566   121.300   !0.50 B
ANGLe     N1T    C2T    N3T       1012.199   114.600   !0.60 B
ANGLe     C2T    N3T    C4T       1012.199   127.200   !0.60 B
ANGLe     N3T    C4T    C5T       1012.199   115.200   !0.60 B
ANGLe     C4T    C5T    C6T       1012.199   118.000   !0.60 B
ANGLe     C5T    C6T    N1T       1012.199   123.700   !0.60 B
ANGLe     N1T    C2T    ON         569.362   123.100   !0.80 B
ANGLe     N3T    C2T    ON        1012.199   122.300   !0.60 B
ANGLe     N3T    C4T    ON        1012.199   119.900   !0.60 B
ANGLe     C5T    C4T    ON         743.656   124.900   !0.70 B
ANGLe     C4T    C5T    CC3E      1012.199   119.000   !0.60 B
ANGLe     C6T    C5T    CC3E      1012.199   122.900   !0.60 B
ANGLe     C6T    N1T    C1R        488.878   120.400   !1.50 B  scale from sugar
ANGLe     C2T    N1T    C1R        429.678   118.200   !1.60 B  scale from sugar
ANGLe     C6T    N1T    C1D        488.878   120.400   !1.50 B  !DNA
ANGLe     C2T    N1T    C1D        429.678   118.200   !1.60 B
ANGLe     C2T    N3T    HN         105.000   116.500   !from param11.dna, 3*keq
ANGLe     C4T    N3T    HN         105.000   116.500
ANGLe     C5T    CC3E   H         $kchange   109.50
ANGLe     H      CC3E   H         $kchange   109.44
ANGLe     N1T    C6T    H         $kchange   119.52
ANGLe     C5T    C6T    H         $kchange   119.52

!adenine                          kq         x_eq      sigma
ANGLe     C6A    NC     C2A       1012.199   118.600   !0.60 B
ANGLe     NC     C2A    N3A       1457.566   129.300   !0.50 B
ANGLe     C2A    N3A    C4A       1457.566   110.600   !0.50 B
ANGLe     N3A    C4A    C5A        743.656   126.800   !0.70 B
ANGLe     C4A    C5A    C6A       1457.566   117.000   !0.50 B
ANGLe     C5A    C6A    NC        1457.566   117.700   !0.50 B
ANGLe     C4A    C5A    N7A       1457.566   110.700   !0.50 B
ANGLe     C5A    N7A    C8A       1457.566   103.900   !0.50 B
ANGLe     N7A    C8A    N9A       1457.566   113.800   !0.50 B
ANGLe     C8A    N9A    C4A       2277.447   105.800   !0.40 B
ANGLe     N9A    C4A    C5A       2277.447   105.800   !0.40 B
ANGLe     N3A    C4A    N9A        569.362   127.400   !0.80 B
ANGLe     C6A    C5A    N7A        743.656   132.300   !0.70 B
ANGLe     NC     C6A    N6A       1012.199   118.600   !0.60 B
ANGLe     C5A    C6A    N6A        569.362   123.700   !0.80 B
ANGLe     C8A    N9A    C1R        339.499   127.700   !1.80 B  scale from sugar
ANGLe     C4A    N9A    C1R        339.499   126.300   !1.80 B  scale from sugar
ANGLe     C8A    N9A    C1D        339.499   127.700   !1.80 B  !DNA
ANGLe     C4A    N9A    C1D        339.499   126.300   !1.80 B
ANGLe     C6A    N6A    H2         105.000   120.000   !from param11.dna, 3*keq
ANGLe     H2     N6A    H2         105.000   120.000
ANGLe     N7A    C8A    H         $kchange   123.16
ANGLe     N9A    C8A    H         $kchange   123.16
ANGLe     NC     C2A    H         $kchange   115.54
ANGLe     N3A    C2A    H         $kchange   115.54

!purine                           kq         x_eq      sigma
ANGLe     C6P    NC     C2P       1012.199   118.600   !0.60 B
ANGLe     NC     C2P    N3P       1457.566   129.300   !0.50 B
ANGLe     C2P    N3P    C4P       1457.566   110.600   !0.50 B
ANGLe     N3P    C4P    C5P        743.656   126.800   !0.70 B
```

Script code

```
ANGLe      C4P     C5P     C6P     1457.566    117.000   !0.50 B
ANGLe      C5P     C6P     NC      1457.566    117.700   !0.50 B
ANGLe      C4P     C5P     N7P     1457.566    110.700   !0.50 B
ANGLe      C5P     N7P     C8P     1457.566    103.900   !0.50 B
ANGLe      N7P     C8P     N9P     1457.566    113.800   !0.50 B
ANGLe      C8P     N9P     C4P     2277.447    105.800   !0.40 B
ANGLe      N9P     C4P     C5P     2277.447    105.800   !0.40 B
ANGLe      N3P     C4P     N9P      569.362    127.400   !0.80 B
ANGLe      C6P     C5P     N7P      743.656    132.300   !0.70 B
ANGLe      NC      C6P     H       $kchangle   120.164   !0.60 B !modified by anda
ANGLe      C5P     C6P     H       $kchangle   120.164   !0.80 B !modified by anda
ANGLe      C8P     N9P     C1R      339.499    127.700   !1.80 B  scale from sugar
ANGLe      C4P     N9P     C1R      339.499    126.300   !1.80 B  scale from sugar
ANGLe      C8P     N9P     C1D      339.499    127.700   !1.80 B !DNA
ANGLe      C4P     N9P     C1D      339.499    126.300   !1.80 B
ANGLe      N7P     C8P     H       $kchangle   123.16
ANGLe      N9P     C8P     H       $kchangle   123.16
ANGLe      NC      C2P     H       $kchangle   115.54
ANGLe      N3P     C2P     H       $kchangle   115.54

!guanine                             kq         x_eq     sigma
ANGLe      O6G     NNA     C2G     1012.199    125.100   !0.60 B
ANGLe      NNA     C2G     N3G     1012.199    123.900   !0.60 B
ANGLe      C2G     N3G     C4G     1457.566    111.900   !0.50 B
ANGLe      N3G     C4G     C5G     1457.566    128.600   !0.50 B
ANGLe      C4G     C5G     C6G     1012.199    118.800   !0.60 B
ANGLe      C5G     C6G     NNA     1457.566    111.500   !0.50 B
ANGLe      C4G     C5G     N7G     2277.447    110.800   !0.40 B
ANGLe      C5G     N7G     C8G     1457.566    104.300   !0.50 B
ANGLe      N7G     C8G     N9G     1457.566    113.100   !0.50 B
ANGLe      C8G     N9G     C4G     2277.447    106.400   !0.40 B
ANGLe      N9G     C4G     C5G     2277.447    105.400   !0.40 B
ANGLe      N3G     C4G     N9G     1012.199    126.000   !0.60 B
ANGLe      O6G     C5G     N7G     1012.199    130.400   !0.60 B
ANGLe      NNA     C2G     N2G      449.866    116.20    !0.90 B
ANGLe      N3G     C2G     N2G      743.656    119.900   !0.70 B
ANGLe      NNA     C6G     O6G     1012.199    119.900   !0.60 B
ANGLe      C5G     C6G     O6G     1012.199    128.600   !0.60 B
ANGLe      C8G     N9G     C1R      650.874    127.000   !1.30 B  scale from sugar
ANGLe      C4G     N9G     C1R      650.874    126.500   !1.30 B  scale from sugar
ANGLe      C8G     N9G     C1D      650.874    127.000   !1.30 B !DNA
ANGLe      C4G     N9G     C1D      650.874    126.500   !1.30 B
ANGLe      C2G     N2G     H2       105.000    120.000            !from param11.dna, 3*keq
ANGLe      H2      N2G     H2       105.000    120.000
ANGLe      C2G     NNA     HN       105.000    119.300
ANGLe      C6G     NNA     HN       105.000    119.300
ANGLe      N7G     C8G     H       $kchangle   122.91
ANGLe      N9G     C8G     H       $kchangle   122.91

!uracile                             kq         x_eq     sigma
ANGLe      C6U     N1U     C2U     1012.199    121.000   !0.60 B
ANGLe      N1U     C2U     N3U     1012.199    114.900   !0.60 B
ANGLe      C2U     N3U     C4U     1012.199    127.000   !0.60 B
ANGLe      N3U     C4U     C5U     1012.199    114.600   !0.60 B
ANGLe      C4U     C5U     C6U     1012.199    119.700   !0.60 B
ANGLe      C5U     C6U     N1U     1457.566    122.700   !0.50 B
```

1 Force field parameter and topology files

```
ANGLe    N1U   C2U   ON              743.656   122.800  !0.70 B
ANGLe    N3U   C2U   ON              743.656   122.200  !0.70 B
ANGLe    N3U   C4U   ON              743.656   119.400  !0.70 B
ANGLe    C5U   C4U   ON             1012.199   125.900  !0.60 B
ANGLe    C6U   N1U   C1R             561.212   121.200  !1.40 B
ANGLe    C2U   N1U   C1R             763.872   117.700  !1.20 B
ANGLe    C6U   N1U   C1D             561.212   121.200  !1.40 B !DNA
ANGLe    C2U   N1U   C1D             763.872   117.700  !1.20 B
ANGLe    C4U   ON    HO              105.000   120.000  !from param11.dna, 3*keq
ANGLe    C2U   N3U   HN              105.000   116.500
ANGLe    C4U   N3U   HN              105.000   116.500
ANGLe    N1U   C6U   H              $kchangle  119.38
ANGLe    C5U   C6U   H              $kchangle  119.38
ANGLe    C4U   C5U   H              $kchangle  119.56
ANGLe    C6U   C5U   H              $kchangle  119.56
ANGLe    C3D   C2D   NX29           1457.566   110.00   !check param, added fpr pyr
ANGLe    H     C2D   NX29            500.00    109.51   !check param, added fpr pyr

{
!******************* mod by anda - HNF *******************

DIHEdral OX3   PX1   OX4   CX5      750.0 0    180.00  ! Nobs =    1 ... Value =    172.75
DIHEdral PX1   OX4   CX5   HX6      750.0 0    -60.00  ! Nobs =    1 ... Value =    -63.48
DIHEdral PX1   OX4   CX5   HX7      750.0 0     60.00  ! Nobs =    1 ... Value =     55.20
DIHEdral PX1   OX4   CX5   CX8      750.0 0    180.00  ! Nobs =    1 ... Value =    176.08
DIHEdral OX4   CX5   CX8   HX9      750.0 0    180.00  ! Nobs =    1 ... Value =    172.35
DIHEdral OX4   CX5   CX8   OX10     750.0 0    -60.00  ! Nobs =    1 ... Value =    -69.24
DIHEdral OX4   CX5   CX8   CX12     750.0 0     60.00  ! Nobs =    1 ... Value =     52.07
DIHEdral HX6   CX5   CX8   HX9      750.0 0     60.00  ! Nobs =    1 ... Value =     51.91
DIHEdral HX6   CX5   CX8   OX10     750.0 0    180.00  ! Nobs =    1 ... Value =    170.31
DIHEdral HX6   CX5   CX8   CX12     750.0 0    -60.00  ! Nobs =    1 ... Value =    -68.38
DIHEdral HX7   CX5   CX8   HX9      750.0 0    -60.00  ! Nobs =    1 ... Value =    -66.18
DIHEdral HX7   CX5   CX8   OX10     750.0 0     60.00  ! Nobs =    1 ... Value =     52.23
DIHEdral HX7   CX5   CX8   CX12     750.0 0    180.00  ! Nobs =    1 ... Value =    173.54
DIHEdral CX5   CX8   CX12  CX14     750.0 0   -120.00  ! Nobs =    1 ... Value =   -129.37
DIHEdral CX5   CX8   CX12  OX17     750.0 0    120.00  ! Nobs =    1 ... Value =    112.32
DIHEdral HX9   CX8   CX12  CX14     750.0 0    120.00  ! Nobs =    1 ... Value =    110.44
DIHEdral HX9   CX8   CX12  OX17     750.0 0      0.00  ! Nobs =    1 ... Value =     -7.87
DIHEdral OX10  CX8   CX12  HX13     750.0 0    120.00  ! Nobs =    1 ... Value =    111.59
DIHEdral OX10  CX8   CX12  CX14     750.0 0      0.00  ! Nobs =    1 ... Value =     -5.92
DIHEdral OX10  CX8   CX12  OX17     750.0 0   -120.00  ! Nobs =    1 ... Value =   -124.23
DIHEdral OX18  CX11  CX14  HX15     750.0 0     90.00  ! Nobs =    1 ... Value =     80.69
DIHEdral OX10  CX11  OX18  CX19     750.0 0    180.00  ! Nobs =    1 ... Value =    177.32
DIHEdral CX14  CX11  OX18  CX19     750.0 0    -60.00  ! Nobs =    1 ... Value =    -67.60
DIHEdral HX43  CX11  OX18  CX19     750.0 0     60.00  ! Nobs =    1 ... Value =     56.28
DIHEdral CX8   CX12  CX14  HX16     750.0 0    -90.00  ! Nobs =    1 ... Value =    -88.75
DIHEdral HX13  CX12  CX14  CX11     750.0 0    -90.00  ! Nobs =    1 ... Value =    -90.38
DIHEdral OX17  CX12  CX14  HX15     750.0 0    -90.00  ! Nobs =    1 ... Value =    -93.06}
DIHEdral OX18  CX19  CX20  CX21    24000.0 0   180.00  ! Nobs =    1 ... Value =   -179.61
DIHEdral OX18  CX19  CX20  HX36    24000.0 0     0.00  ! Nobs =    1 ... Value =     -0.12
DIHEdral CX24  CX19  CX20  CX21     2400.0 0     0.00  ! Nobs =    1 ... Value =     -0.46
DIHEdral CX24  CX19  CX20  HX36     2400.0 0   180.00  ! Nobs =    1 ... Value =    179.03
DIHEdral OX18  CX19  CX24  CX23    24000.0 0   180.00  ! Nobs =    1 ... Value =    179.62
DIHEdral OX18  CX19  CX24  HX40    24000.0 0     0.00  ! Nobs =    1 ... Value =     -0.53
DIHEdral CX20  CX19  CX24  CX23     2400.0 0     0.00  ! Nobs =    1 ... Value =      0.42
```

Script code

```
DIHEdral CX20 CX19 CX24 HX40    2400.0 0    180.00 ! Nobs =   1 ... Value =  -179.73
DIHEdral CX19 CX20 CX21 CX22    2400.0 0      0.00 ! Nobs =   1 ... Value =     0.26
DIHEdral CX19 CX20 CX21 HX35    2400.0 0    180.00 ! Nobs =   1 ... Value =  -179.84
DIHEdral HX36 CX20 CX21 CX22    2400.0 0    180.00 ! Nobs =   1 ... Value =  -179.24
DIHEdral HX36 CX20 CX21 HX35    2400.0 0      0.00 ! Nobs =   1 ... Value =     0.67
DIHEdral CX20 CX21 CX22 CX23    2400.0 0      0.00 ! Nobs =   1 ... Value =    -0.02
DIHEdral CX20 CX21 CX22 CX25    2400.0 0    180.00 ! Nobs =   1 ... Value =   179.94
DIHEdral HX35 CX21 CX22 CX23    2400.0 0    180.00 ! Nobs =   1 ... Value =  -179.92
DIHEdral HX35 CX21 CX22 CX25    2400.0 0      0.00 ! Nobs =   1 ... Value =     0.03
DIHEdral CX21 CX22 CX23 CX24    2400.0 0      0.00 ! Nobs =   1 ... Value =    -0.02
DIHEdral CX21 CX22 CX23 CX27    2400.0 0    180.00 ! Nobs =   1 ... Value =   179.89
DIHEdral CX25 CX22 CX23 CX24    2400.0 0    180.00 ! Nobs =   1 ... Value =  -179.98
DIHEdral CX25 CX22 CX23 CX27    2400.0 0      0.00 ! Nobs =   1 ... Value =    -0.07
DIHEdral CX21 CX22 CX25 CX26    2400.0 0    180.00 ! Nobs =   1 ... Value =  -179.90
DIHEdral CX21 CX22 CX25 CX28    2400.0 0      0.00 ! Nobs =   1 ... Value =     0.02
DIHEdral CX23 CX22 CX25 CX26    2400.0 0      0.00 ! Nobs =   1 ... Value =     0.06
DIHEdral CX23 CX22 CX25 CX28    2400.0 0    180.00 ! Nobs =   1 ... Value =   179.98
DIHEdral CX22 CX23 CX24 CX19    2400.0 0      0.00 ! Nobs =   1 ... Value =    -0.18
DIHEdral CX22 CX23 CX24 HX40    2400.0 0    180.00 ! Nobs =   1 ... Value =   179.97
DIHEdral CX27 CX23 CX24 CX19    2400.0 0    180.00 ! Nobs =   1 ... Value =   179.93
DIHEdral CX27 CX23 CX24 HX40    2400.0 0      0.00 ! Nobs =   1 ... Value =     0.08
DIHEdral CX22 CX23 CX27 CX26    2400.0 0      0.00 ! Nobs =   1 ... Value =     0.05
DIHEdral CX22 CX23 CX27 HX41    2400.0 0    120.00 ! Nobs =   1 ... Value =   117.87
DIHEdral CX22 CX23 CX27 HX42    2400.0 0   -120.00 ! Nobs =   1 ... Value =  -117.81
DIHEdral CX24 CX23 CX27 CX26    2400.0 0    180.00 ! Nobs =   1 ... Value =   179.95
DIHEdral CX24 CX23 CX27 HX41    2400.0 0    -60.00 ! Nobs =   1 ... Value =   -62.23
DIHEdral CX24 CX23 CX27 HX42    2400.0 0     60.00 ! Nobs =   1 ... Value =    62.09
DIHEdral CX22 CX25 CX26 CX27    2400.0 0      0.00 ! Nobs =   1 ... Value =    -0.03
DIHEdral CX22 CX25 CX26 CX31    2400.0 0    180.00 ! Nobs =   1 ... Value =   179.97
DIHEdral CX28 CX25 CX26 CX27    2400.0 0    180.00 ! Nobs =   1 ... Value =  -179.96
DIHEdral CX28 CX25 CX26 CX31    2400.0 0      0.00 ! Nobs =   1 ... Value =     0.04
DIHEdral CX22 CX25 CX28 CX29    2400.0 0    180.00 ! Nobs =   1 ... Value =  -179.92
DIHEdral CX22 CX25 CX28 HX37    2400.0 0      0.00 ! Nobs =   1 ... Value =     0.06
DIHEdral CX26 CX25 CX28 CX29    2400.0 0      0.00 ! Nobs =   1 ... Value =    -0.01
DIHEdral CX26 CX25 CX28 HX37    2400.0 0    180.00 ! Nobs =   1 ... Value =   179.98
DIHEdral CX25 CX26 CX27 CX23    2400.0 0      0.00 ! Nobs =   1 ... Value =    -0.01
DIHEdral CX25 CX26 CX27 HX41    2400.0 0   -120.00 ! Nobs =   1 ... Value =  -117.79
DIHEdral CX25 CX26 CX27 HX42    2400.0 0    120.00 ! Nobs =   1 ... Value =   117.79
DIHEdral CX31 CX26 CX27 CX23    2400.0 0    180.00 ! Nobs =   1 ... Value =   179.99
DIHEdral CX31 CX26 CX27 HX41    2400.0 0     60.00 ! Nobs =   1 ... Value =    62.22
DIHEdral CX31 CX26 CX27 HX42    2400.0 0    -60.00 ! Nobs =   1 ... Value =   -62.20
DIHEdral CX25 CX26 CX31 CX30    2400.0 0      0.00 ! Nobs =   1 ... Value =    -0.06
DIHEdral CX25 CX26 CX31 HX39    2400.0 0    180.00 ! Nobs =   1 ... Value =  -179.96
DIHEdral CX27 CX26 CX31 CX30    2400.0 0    180.00 ! Nobs =   1 ... Value =   179.94
DIHEdral CX27 CX26 CX31 HX39    2400.0 0      0.00 ! Nobs =   1 ... Value =     0.04
DIHEdral CX25 CX28 CX29 CX30    2400.0 0      0.00 ! Nobs =   1 ... Value =     0.01
DIHEdral CX25 CX28 CX29 HX38    2400.0 0    180.00 ! Nobs =   1 ... Value =  -179.95
DIHEdral HX37 CX28 CX29 CX30    2400.0 0    180.00 ! Nobs =   1 ... Value =  -179.97
DIHEdral HX37 CX28 CX29 HX38    2400.0 0      0.00 ! Nobs =   1 ... Value =     0.07
DIHEdral CX28 CX29 CX30 CX31    2400.0 0      0.00 ! Nobs =   1 ... Value =    -0.03
DIHEdral CX28 CX29 CX30 NX32    2400.0 0    180.00 ! Nobs =   1 ... Value =  -179.98
DIHEdral HX38 CX29 CX30 CX31    2400.0 0    180.00 ! Nobs =   1 ... Value =   179.93
DIHEdral HX38 CX29 CX30 NX32    2400.0 0      0.00 ! Nobs =   1 ... Value =    -0.02
DIHEdral CX29 CX30 CX31 CX26    2400.0 0      0.00 ! Nobs =   1 ... Value =     0.06
DIHEdral CX29 CX30 CX31 HX39    2400.0 0    180.00 ! Nobs =   1 ... Value =   179.95
DIHEdral NX32 CX30 CX31 CX26    2400.0 0    180.00 ! Nobs =   1 ... Value =  -179.99
```

1 Force field parameter and topology files

```
DIHEdral NX32 CX30 CX31 HX39    2400.0 0      0.00 ! Nobs =    1 ... Value =    -0.09
DIHEdral CX29 CX30 NX32 OX33    2400.0 0      0.00 ! Nobs =    1 ... Value =     0.03
DIHEdral CX29 CX30 NX32 OX34    2400.0 0    180.00 ! Nobs =    1 ... Value =   179.92
DIHEdral CX31 CX30 NX32 OX33    2400.0 0    180.00 ! Nobs =    1 ... Value =  -179.93
DIHEdral CX31 CX30 NX32 OX34    2400.0 0      0.00 ! Nobs =    1 ... Value =    -0.04

!****************** end of mod by anda - HNF *************


{
!Dihedrals from param11.dna (included for terminal residues)

!DIHEdral    X    C2R   C3R   X     4.50 3    0.000
!DIHEdral    X    C4R   C3R   X     4.50 3    0.000
!DIHEdral    X    C2R   C1R   X     4.50 3    0.000
!DIHEdral    X    C5R   O5R   X     1.50 3    0.000
!DIHEdral    X    C3R   O3R   X     1.50 3    0.000
 DIHEdral    X    C3R   OH    X     1.50 3    0.000
 DIHEdral    X    C5R   OH    X     1.50 3    0.000
!DIHEdral    X    C2R   O2R   X     1.50 3    0.000
!DIHEdral    X    O5R   P     X     2.25 3    0.000
 DIHEdral    X    OH    P     X     2.25 3    0.000
!DIHEdral    OH   C5R   C4R   O4R   4.50 3    0.000
!DIHEdral    OH   C5R   C4R   C3R   4.50 3    0.000 ! gamma
!DIHEdral    C3R  O3R   P     OH    2.25 3    0.000 ! added by infer
!DIHEdral    C3R  O3R   P     OH    2.25 2    0.000 ! ATB, 7-SEP-84
 DIHEdral    C5R  O5R   P     OH    2.25 3    0.000 ! added by infer
!DIHEdral    C5R  O5R   P     OH    2.25 2    0.000 ! ATB, 7-SEP-84

!DIHEdral    X    C2D   C3D   X     4.50 3    0.000
!DIHEdral    X    C4D   C3D   X     4.50 3    0.000 !DNA
!DIHEdral    X    C2D   C1D   X     4.50 3    0.000
!DIHEdral    X    C5D   O5R   X     1.50 3    0.000
!DIHEdral    X    C3D   O3R   X     1.50 3    0.000
 DIHEdral    X    C3D   OH    X     1.50 3    0.000
 DIHEdral    X    C5D   OH    X     1.50 3    0.000
!DIHEdral    OH   C5D   C4D   O4D   4.50 3    0.000
!DIHEdral    OH   C5D   C4D   C3D   4.50 3    0.000
!DIHEdral    C3D  O3R   P     OH    2.25 3    0.000
!DIHEdral    C3D  O3R   P     OH    2.25 2    0.000
 DIHEdral    C5D  O5R   P     OH    2.25 3    0.000
!DIHEdral    C5D  O5R   P     OH    2.25 2    0.000
}{
!Base hydrogen DIHEdrals taken from param11.dna
 DIHEdral    X    C2G   N2G   X    18.0  2  180.000
 DIHEdral    X    C6A   N6A   X    18.0  2  180.000
!DIHEdral    X    C6A   N4C   X    18.00 2  180.000
 DIHEdral    X    C4C   N4C   X    18.00 2  180.000
}

!****************** mod by anda - HNF ********************

!!! IMPRoper PX1  OX2   OX3   OX4   750.0 0  -35.000 ! Nobs =    1 ... Value =  -36.223
    IMPRoper CX5  OX4   HX6   HX7   $kchimpr   32.868 ! mod by anda, gaussian value
    IMPRoper CX8  CX5   HX9   OX10   94.5 0  -36.228 ! mod by anda, gaussian value, k C5D  X    X
```

Script code

```
            C2D
IMProper  CX11  OX10  CX14  OX18    94.5    0   -35.563  ! mod by anda, gaussian value
IMProper  CX12  CX8   HX13  CX14    94.5    0    41.651  ! mod by anda, gaussian value
IMProper  CX14  CX11  CX12  HX15    94.5    0    28.144  ! mod by anda, gaussian value
IMProper  CX19  OX18  CX20  CX24  2400.0    0     0.000  ! mod by anda, gaussian value
IMProper  CX20  CX19  CX21  HX36  2400.0    0     0.000  ! mod by anda, gaussian value
IMProper  CX21  CX20  CX22  HX35  2400.0    0     0.000  ! mod by anda, gaussian value
IMProper  CX22  CX21  CX23  CX25  2400.0    0     0.000  ! mod by anda, gaussian value
IMProper  CX23  CX22  CX24  CX27  2400.0    0     0.000  ! mod by anda, gaussian value
IMProper  CX24  CX19  CX23  HX40  2400.0    0     0.000  ! mod by anda, gaussian value
IMProper  CX25  CX22  CX26  CX28  2400.0    0     0.000  ! mod by anda, gaussian value
IMProper  CX26  CX25  CX27  CX31  2400.0    0     0.000  ! mod by anda, gaussian value
IMProper  CX27  CX23  CX26  HX41    94.5    0    28.808  ! mod by anda, gaussian value
IMProper  CX28  CX25  CX29  HX37  2400.0    0     0.000  ! mod by anda, gaussian value
IMProper  CX29  CX28  CX30  HX38  2400.0    0     0.000  ! mod by anda, gaussian value
IMProper  CX30  CX29  CX31  NX32  2400.0    0     0.000  ! mod by anda, gaussian value
IMProper  CX31  CX26  CX30  HX39  2400.0    0     0.000  ! mod by anda, gaussian value
IMProper  NX32  CX30  OX33  OX34  2400.0    0     0.000  ! mod by anda, gaussian value
IMProper  CX29  CX31  CX24  CX20  2400.0    0     0.000  ! mod by anda, gaussian value
IMProper  CX28  CX26  CX23  CX21  2400.0    0     0.000  ! mod by anda, gaussian value
IMProper  OX33  NX32  CX30  CX29  2400.0    0     0.000  ! mod by anda, gaussian value
IMProper  OX34  NX32  CX30  CX31  2400.0    0     0.000  ! mod by anda, gaussian value
IMProper  HX38  HX39  HX40  HX36  2400.0    0     0.000  ! mod by anda, gaussian value
IMProper  CX31  CX25  CX23  CX20  2400.0    0   180.000  ! mod by anda, planarity
IMProper  CX29  CX26  CX22  CX19  2400.0    0   180.000  ! mod by anda, planarity
IMProper  CX26  CX25  CX22  CX23  2400.0    0     0.000  ! mod by anda, planarity
IMProper  CX28  CX27  CX22  CX19  2400.0    0   180.000  ! mod by anda, planarity
IMProper  CX31  CX25  CX27  CX21  2400.0    0   180.000  ! mod by anda, planarity
IMProper  NX32  CX25  CX27  CX21  2400.0    0   180.000  ! mod by anda, planarity
IMProper  CX30  CX25  CX27  CX21  2400.0    0   180.000  ! mod by anda, planarity
IMProper  CX31  CX25  CX24  CX20  2400.0    0   180.000  ! mod by anda, planarity
IMProper  HX38  CX26  CX21  HX40  2400.0    0   180.000  ! mod by anda, planarity
IMProper  HX39  CX28  CX23  HX36  2400.0    0   180.000  ! mod by anda, planarity
IMProper  CX31  CX28  CX23  CX20  2400.0    0   180.000  ! mod by anda, planarity

!******************* end of mod by anda - HNF *************

!IMPropers to keep the two purine rings parallel:
!guanine
IMProper     C8G   C4G   C5G   NNA   250.0   2   180.000
IMProper     C8G   C5G   C4G   C2G   250.0   2   180.000
IMProper     N3G   C4G   C5G   N7G   250.0   2   180.000
IMProper     C6G   C5G   C4G   N9G   250.0   2   180.000
!adenine
IMProper     C8A   C4A   C5A   N9A   250.0   2   180.000  ! WYE AND PATCHED RESIDUES
IMProper     C8A   C5A   C4A   C2A   250.0   2   180.000
IMProper     C8A   C4A   C5A   NC    250.0   2   180.000
IMProper     N3A   C4A   C5A   N7A   250.0   2   180.000
IMProper     C6A   C5A   C4A   N9A   250.0   2   180.000
!purine
IMProper     C8P   C4P   C5P   N9P   250.0   2   180.000  ! WYE AND PATCHED RESIDUES
IMProper     C8P   C5P   C4P   C2P   250.0   2   180.000
IMProper     C8P   C4P   C5P   NC    250.0   2   180.000
IMProper     N3P   C4P   C5P   N7P   250.0   2   180.000
IMProper     C6P   C5P   C4P   N9P   250.0   2   180.000
```

1 Force field parameter and topology files

```
!other base specific non-exch hydrogen IMPRopers
IMPRoper    H    C4C   C6C   C5C   $kchimpr   0    0.000
IMPRoper    H    N1C   C5C   C6C   $kchimpr   0    0.000
IMPRoper    H    C4U   C6U   C5U   $kchimpr   0    0.000
IMPRoper    H    N1U   C5U   C6U   $kchimpr   0    0.000
IMPRoper    H    N1T   C5T   C6T   $kchimpr   0    0.000
IMPRoper    H    N7A   N9A   C8A   $kchimpr   0    0.000
IMPRoper    H    NC    N3A   C2A   $kchimpr   0    0.000
IMPRoper    H    N7P   N9P   C8P   $kchimpr   0    0.000
IMPRoper    H    NC    N3P   C2P   $kchimpr   0    0.000
IMPRoper    H    NC    C5P   C6P   $kchimpr   0    0.000
IMPRoper    H    N7G   N9G   C8G   $kchimpr   0    0.000

!Impropers for ribose chirality
IMPRoper    H    C2R   O4R   N9A   $kchimpr   0   -65.000!C1R
IMPRoper    H    C2R   O4R   N9P   $kchimpr   0   -65.000!C1R
IMPRoper    H    C2R   O4R   N9G   $kchimpr   0   -65.000!C1R
IMPRoper    H    C2R   O4R   N1C   $kchimpr   0   -65.000!C1R
IMPRoper    H    C2R   O4R   N1U   $kchimpr   0   -65.000!C1R
IMPRoper    H    C2R   O4R   N1T   $kchimpr   0   -65.000!C1R

IMPRoper    H    C3R   C1R   O2R   $kchimpr   0    65.000!C2R
IMPRoper    H    C4R   C2R   O3R   $kchimpr   0    60.300!C3R
IMPRoper    H    C4R   C2R   OH    $kchimpr   0    60.300!C3R; TERMINAL RES
IMPRoper    H    C5R   C3R   O4R   $kchimpr   0    70.300!C4R
IMPRoper    H    O5R   H     C4R   $kchimpr   0    72.000!C5R;
IMPRoper    H    OH    H     C4R   $kchimpr   0    72.000!C5R; TERMINAL RES
!Impropers for deoxyribose chirality
IMPRoper    H    C2D   O4D   N9A   $kchimpr   0   -65.280!C1D
IMPRoper    H    C2D   O4D   N9P   $kchimpr   0   -65.280!C1D
IMPRoper    H    C2D   O4D   N9G   $kchimpr   0   -65.280!C1D
IMPRoper    H    C2D   O4D   N1C   $kchimpr   0   -65.280!C1D
IMPRoper    H    C2D   O4D   N1T   $kchimpr   0   -65.280!C1D
IMPRoper    H    C2D   O4D   N1U   $kchimpr   0   -65.280!C1D
IMPRoper    H    C2D   O4D   H     $kchimpr   0   -65.280!C1D !mod by anda (ABA)

IMPRoper    H    C3D   H     C1D   $kchimpr   0   -73.500!C2D
IMPRoper    H    C4D   C2D   O3R   $kchimpr   0    62.660!C3D
IMPRoper    H    C4D   C2D   OH    $kchimpr   0    62.660!C3D; TERMINAL RES
IMPRoper    H    C5D   C3D   O4D   $kchimpr   0    70.220!C4D
IMPRoper    H    O5R   H     C4D   $kchimpr   0    71.430!C5D;
IMPRoper    H    OH    H     C4D   $kchimpr   0    71.430!C5D; TERMINAL RES

{
!Phos. - periodical potentials from combined RNA/DNA statistics
!                                  kq      x_eq    (sigma in parenthesis)
DIHEdral    O3R   P     O5R   C5R   1.41  3   24     ! alpha  !P (20.3)
DIHEdral    P     O5R   C5R   C4R   3.45  0   178    ! beta   !P (13.0)
DIHEdral    O5R   C5R   C4R   C3R  12.24  3   18     ! gamma  !S (6.9)
DIHEdral    O5R   C5R   C4R   O4R  24.28  3   14.1   !        !S (4.9)
DIHEdral    C4R   C3R   O3R   P     7.88  0  -153    ! eps    !P (8.6)
DIHEdral    C3R   O3R   P     O5R   1.75  3   33     ! zeta   !P (18.3)

DIHEdral    O3R   P     O5R   C5D   1.41  3    6.0  !DNA
DIHEdral    P     O5R   C5D   C4D   3.45  0  183.5
```

Script code

```
DIHEdral    O5R   C5D   C4D   C3D    12.42  3    18.3
DIHEdral    O5R   C5D   C4D   O4D    24.28  3    14.1
DIHEdral    C4D   C3D   O3R   P       7.88  0   214.0
DIHEdral    C3D   O3R   P     O5R     1.75  3     0.3

!Phos. - discrete values from combined RNA/DNA statistics
!                                     kq      x_eq      (sigma in parenthesis)
!DIHEdral   O3R   P     O5R   C5R     6.07 0  285.3   ! (9.8)  alpha1           !P
!DIHEdral   O3R   P     O5R   C5R     3.98 0   81.1   ! (12.1) alpha2;alpha3=180.
!DIHEdral   P     O5R   C5R   C4R     3.44 0  183.5   ! (13.0) beta             !P
!DIHEdral   O5R   C5R   C4R   C3R    17.94 0   52.5   ! (5.7)  gamma1           !S
!DIHEdral   O5R   C5R   C4R   C3R    14.23 0  179.4   ! (6.4)  gamma2           !S
!DIHEdral   O5R   C5R   C4R   C3R     3.85 0  292.9   ! (12.3) gamma3           !S
!DIHEdral   C4R   C3R   O3R   P       7.88 0  214.0   ! (8.6)  eps              !P
!DIHEdral   C3R   O3R   P     O5R    25.30 0  289.2   ! (4.8)  zeta1            !P
!DIHEdral   C3R   O3R   P     O5R     2.85 0   80.7   ! (14.3) zeta2;zeta3=180.
}{
!Sugars
! c3'-endo conformation as the default for for RNA, c2'-endo for DNA,
!RNA statistics , C3'-endo
DIHEdral    C5R   C4R   C3R   O3R    30.12 0   81.1   ! delta ! c3'-endo  S  (4.4)
DIHEdral    O4R   C4R   C3R   O3R    33.10 0  201.8   !  4.2  ! c3'-endo  S
DIHEdral    O4R   C1R   C2R   C3R    24.28 0  335.4   !  4.9  ! c3'-endo  S
DIHEdral    C1R   C2R   C3R   C4R    74.36 0   35.9   !  2.8  ! c3'-endo  S
DIHEdral    C2R   C3R   C4R   O4R    60.67 0  324.7   !  3.1  ! c3'-endo  S
DIHEdral    C3R   C4R   O4R   C1R    22.42 0   20.5   !  5.1  ! c3'-endo  S
DIHEdral    C4R   O4R   C1R   C2R    15.67 0    2.8   !  6.1  ! c3'-endo  S
DIHEdral    C5R   C4R   C3R   C2R    60.67 0  204.0   !  3.1  ! c3'-endo  S
DIHEdral    O3R   C3R   C2R   O2R    28.79 0   44.3   !  4.5  ! c3'-endo  S
DIHEdral    C4R   O4R   C1R   N1T    13.80 0  241.4   !  6.5  ! c3'-endo  S
DIHEdral    C4R   O4R   C1R   N1C    13.80 0  241.4   !  6.5  ! c3'-endo  S
DIHEdral    C4R   O4R   C1R   N1U    13.80 0  241.4   !  6.5  ! c3'-endo  S
DIHEdral    C4R   O4R   C1R   N9G    13.80 0  241.4   !  6.5  ! c3'-endo  S
DIHEdral    C4R   O4R   C1R   N9A    13.80 0  241.4   !  6.5  ! c3'-endo  S
DIHEdral    C4R   O4R   C1R   N9P    13.80 0  241.4   !  6.5  ! c3'-endo  S

!RNA c3'-endo sugar base joint torsions  (combined RNA/DNA statistics used)
DIHEdral    O4R   C1R   N1T   C2T    13.38 0  195.7   !  6.6  ! c3'-endo  S
DIHEdral    O4R   C1R   N1C   C2C    13.38 0  195.7   !  6.6  ! c3'-endo  S
DIHEdral    O4R   C1R   N1U   C2U    13.38 0  195.7   !  6.6  ! c3'-endo  S
DIHEdral    O4R   C1R   N9A   C4A     2.97 0  193.3   ! 14.0  ! c3'-endo  S
DIHEdral    O4R   C1R   N9P   C4P     2.97 0  193.3   ! 14.0  ! c3'-endo  S
DIHEdral    O4R   C1R   N9G   C4G     2.97 0  193.3   ! 14.0  ! c3'-endo  S

!DNA statistics (c2'-endo)
DIHEdral    C5D   C4D   C3D   O3R    36.44 0  145.2   ! delta ! c2'-endo  S  (4.0)
DIHEdral    O4D   C1D   C2D   C3D    24.28 0   32.8   !  4.9  ! c2'-endo  S
DIHEdral    O4D   C4D   C3D   O3R    31.53 0  265.8   !  4.3  ! c2'-endo  S
DIHEdral    C1D   C2D   C3D   C4D    44.99 0  326.9   !  3.6  ! c2'-endo  S
DIHEdral    C2D   C3D   C4D   O4D    28.79 0   22.6   !  4.5  ! c2'-endo  S
DIHEdral    C3D   C4D   O4D   C1D    15.67 0  357.7   !  6.1  ! c2'-endo  S
DIHEdral    C4D   O4D   C1D   C2D    14.69 0  340.7   !  6.3  ! c2'-endo  S
DIHEdral    C5D   C4D   C3D   C2D    34.68 0  262.0   !  4.1  ! c2'-endo  S
DIHEdral    C4D   O4D   C1D   N1T    12.99 0  217.7   !  6.7  ! c2'-endo  S
DIHEdral    C4D   O4D   C1D   N1C    12.99 0  217.7   !  6.7  ! c2'-endo  S
DIHEdral    C4D   O4D   C1D   N1U    12.99 0  217.7   !  6.7  ! c2'-endo  S
```

1 Force field parameter and topology files

```
DIHEdral    C4D   O4D   C1D   N9G    12.99 0    217.7    !  6.7   ! c2'-endo S
DIHEdral    C4D   O4D   C1D   N9A    12.99 0    217.7    !  6.7   ! c2'-endo S
DIHEdral    C4D   O4D   C1D   N9P    12.99 0    217.7    !  6.7   ! c2'-endo S

!DNA c2'-endo sugar base joint torsions  (combined RNA/DNA statistics used)
DIHEdral    O4D   C1D   N1T   C2T     1.72 0    229.8    ! 18.4   ! c2'-endo S
DIHEdral    O4D   C1D   N1C   C2C     1.72 0    229.8    ! 18.4   ! c2'-endo S
DIHEdral    O4D   C1D   N1U   C2U     1.72 0    229.8    ! 18.4   ! c2'-endo S
DIHEdral    O4D   C1D   N9A   C4A     1.00 0    237.0    ! 24.3   ! c2'-endo S
DIHEdral    O4D   C1D   N9P   C4P     1.00 0    237.0    ! 24.3   ! c2'-endo S
DIHEdral    O4D   C1D   N9G   C4G     1.00 0    237.0    ! 24.3   ! c2'-endo S

!In the case of c3'-endo conformation, the following DIHEdrals are provided
!to overwrite the c2'-endo DIHEdrals

!RNA statistics (c2'-endo)
!DIHEdral    C5R   C4R   C3R   O3R    24.28 0    147.3    ! delta ! c2'-endo S (4.9)
!DIHEdral    O4R   C1R   C2R   C3R    50.43 0     35.2    !  3.4  ! c2'-endo S
!DIHEdral    O4R   C4R   C3R   O3R    20.75 0    268.1    !  5.3  ! c2'-endo S
!DIHEdral    C1R   C2R   C3R   C4R    74.36 0    324.6    !  2.8  ! c2'-endo S
!DIHEdral    C2R   C3R   C4R   O4R    31.53 0     24.2    !  4.3  ! c2'-endo S
!DIHEdral    C3R   C4R   O4R   C1R    17.94 0    357.7    !  5.7  ! c2'-endo S
!DIHEdral    C4R   O4R   C1R   C2R    21.56 0    339.2    !  5.2  ! c2'-endo S
!DIHEdral    C5R   C4R   C3R   C2R    34.68 0    263.4    !  4.1  ! c2'-endo S
!DIHEdral    O3R   C3R   C2R   O2R    33.05 0    319.7    !  4.2  ! c2'-endo S
!DIHEdral    C4R   O4R   C1R   N1T    19.27 0    216.6    !  5.5  ! c2'-endo S
!DIHEdral    C4R   O4R   C1R   N1C    19.27 0    216.6    !  5.5  ! c2'-endo S
!DIHEdral    C4R   O4R   C1R   N1U    19.27 0    216.6    !  5.5  ! c2'-endo S
!DIHEdral    C4R   O4R   C1R   N9G    19.27 0    216.6    !  5.5  ! c2'-endo S
!DIHEdral    C4R   O4R   C1R   N9A    19.27 0    216.6    !  5.5  ! c2'-endo S

!RNA c2'-endo sugar base joint torsions  (combined RNA/DNA statistics used)
!DIHEdral    O4R   C1R   N1T   C2T     1.72 0    229.8    ! 18.4  ! c2'-endo S
!DIHEdral    O4R   C1R   N1C   C2C     1.72 0    229.8    ! 18.4  ! c2'-endo S
!DIHEdral    O4R   C1R   N1U   C2U     1.72 0    229.8    ! 18.4  ! c2'-endo S
!DIHEdral    O4R   C1R   N9A   C4A     1.00 0    237.0    ! 24.3  ! c2'-endo S
!DIHEdral    O4R   C1R   N9G   C4G     1.00 0    237.0    ! 24.3  ! c2'-endo S

!DNA statistics, c3'-endo (insuficient data, RNA values used)
!DIHEdral    C5D   C4D   C3D   O3R    30.12 0     81.1    ! delta ! c3'-endo S (4.4)
!DIHEdral    O4D   C4D   C3D   O3R    33.10 0    201.8    !  4.2  ! c3'-endo S
!DIHEdral    O4D   C1D   C2D   C3D    24.28 0    335.4    !  4.9  ! c3'-endo S
!DIHEdral    C1D   C2D   C3D   C4D    74.36 0     35.9    !  2.8  ! c3'-endo S
!DIHEdral    C2D   C3D   C4D   O4D    60.67 0    324.7    !  3.1  ! c3'-endo S
!DIHEdral    C3D   C4D   O4D   C1D    22.42 0     20.5    !  5.1  ! c3'-endo S
!DIHEdral    C4D   O4D   C1D   C2D    15.67 0      2.8    !  6.1  ! c3'-endo S
!DIHEdral    C5D   C4D   C3D   C2D    60.67 0    204.0    !  3.1  ! c3'-endo S
!DIHEdral    C4D   O4D   C1D   N1T    13.80 0    241.4    !  6.5  ! c3'-endo S
!DIHEdral    C4D   O4D   C1D   N1C    13.80 0    241.4    !  6.5  ! c3'-endo S
!DIHEdral    C4D   O4D   C1D   N1U    13.80 0    241.4    !  6.5  ! c3'-endo S
!DIHEdral    C4D   O4D   C1D   N9G    13.80 0    241.4    !  6.5  ! c3'-endo S
!DIHEdral    C4D   O4D   C1D   N9A    13.80 0    241.4    !  6.5  ! c3'-endo S

!DNA c3'-endo sugar base joint torsions  (combined RNA/DNA statistics used)
!DIHEdral    O4D   C1D   N1T   C2T    13.38 0    195.7    !  6.6  ! c3'-endo S
```

Script code

```
!DIHEdral  O4D  C1D  N1C  C2C  13.38  0  195.7  !  6.6  !  c3'-endo  S
!DIHEdral  O4D  C1D  N1U  C2U  13.38  0  195.7  !  6.6  !  c3'-endo  S
!DIHEdral  O4D  C1D  N9A  C4A   2.97  0  193.3  ! 14.0  !  c3'-endo  S
!DIHEdral  O4D  C1D  N9G  C4G   2.97  0  193.3  ! 14.0  !  c3'-endo  S
!-------------------------------------------------------------------
!-------------------------------------------------------------------
}




!Impropers taken from param11.dna , 3*kq
IMProper  C5R  X    X    C2R  94.5  0  35.260
IMProper  C5R  X    X    C1R  94.5  0  35.260
IMProper  OH   X    X    C3R  94.5  0  35.260
IMProper  OH   X    X    C4R  94.5  0  35.260
IMProper  OH   X    X    C1R  94.5  0  35.260
IMProper  O3R  X    X    C3R  94.5  0  35.260
IMProper  O5R  X    X    C1R  94.5  0  35.260
IMProper  O2R  X    X    C2R  94.5  0  35.260
IMProper  C4R  O5R  C1R  N1T  94.5  0  35.260
IMProper  C4R  O5R  C1R  N1C  94.5  0  35.260
IMProper  C4R  O5R  C1R  N9G  94.5  0  35.260
IMProper  C4R  O5R  C1R  N9A  94.5  0  35.260
IMProper  C4R  O5R  C1R  N9P  94.5  0  35.260
IMProper  C5R  O4R  C3R  C4R  94.5  0  35.260
IMProper  N1T  C2R  O4R  C1R  94.5  0  35.260
IMProper  N1C  C2R  O4R  C1R  94.5  0  35.260
IMProper  N9A  C2R  O4R  C1R  94.5  0  35.260
IMProper  N9P  C2R  O4R  C1R  94.5  0  35.260
IMProper  N9G  C2R  O4R  C1R  94.5  0  35.260
IMProper  C4R  O5R  C1R  N1U  94.5  0  35.260
IMProper  N1U  C2R  O4R  C1R  94.5  0  35.260

IMProper  C5D  X    X    C2D  94.5  0  35.260  !DNA
IMProper  C5D  X    X    C1D  94.5  0  35.260
IMProper  OH   X    X    C3D  94.5  0  35.260
IMProper  OH   X    X    C4D  94.5  0  35.260
IMProper  OH   X    X    C1D  94.5  0  35.260
IMProper  O3R  X    X    C3D  94.5  0  35.260
IMProper  O5R  X    X    C1D  94.5  0  35.260
IMProper  C4D  O5R  C1D  N1T  94.5  0  35.260
IMProper  C4D  O5R  C1D  N1C  94.5  0  35.260
IMProper  C4D  O5R  C1D  N9G  94.5  0  35.260
IMProper  C4D  O5R  C1D  N9A  94.5  0  35.260
IMProper  C4D  O5R  C1D  N9P  94.5  0  35.260
IMProper  C5D  O4D  C3D  C4D  94.5  0  35.260
IMProper  N1T  C2D  O4D  C1D  94.5  0  35.260
IMProper  N1C  C2D  O4D  C1D  94.5  0  35.260
IMProper  N9A  C2D  O4D  C1D  94.5  0  35.260
IMProper  N9P  C2D  O4D  C1D  94.5  0  35.260
```

1 Force field parameter and topology files

```
IMProper    N9G   C2D   O4D   C1D   94.5    0   35.260
IMProper    C4D   O5R   C1D   N1U   94.5    0   35.260
IMProper    N1U   C2D   O4D   C1D   94.5    0   35.260

!the following impropers were taken from param11x.dna
!the higher kq was used to enforce the ring planarity
!cytosine
IMProper    C4C   X     X     ON    2400.0  0    0.000
IMProper    C4C   X     X     N1C   250.0   0    0.000
IMProper    C6C   X     X     NC    250.0   0    0.000
IMProper    C4C   X     X     N2    2400.0  0    0.000
IMProper    C2C   X     X     ON    2400.0  0    0.000
!infer
IMProper    C1R   C2C   C6C   N1C   2400.0  0    0.000
IMProper    C1D   C2C   C6C   N1C   2400.0  0    0.000
IMProper    N4C   NC    C5C   C4C   2400.0  0    0.000
IMProper    C2C   NC    C4C   C5C   250.0   0    0.000
IMProper    C5C   C6C   N1C   C2C   250.0   0    0.000
IMProper    H2    C4C   H2    N4C   250.0   0    0.000
IMProper    C5C   C4C   N4C   H2    2000.0  0    0.000

!uracil
IMProper    C4U   X     X     ON    2400.0  0    0.000
IMProper    C4U   X     X     N1U   250.0   0    0.000
IMProper    C6U   X     X     N3U   250.0   0    0.000
IMProper    C4U   X     X     N2    2400.0  0    0.000
IMProper    C2U   X     X     ON    2400.0  0    0.000
IMProper    C1R   C2U   C6U   N1U   2400.0  0    0.000
IMProper    C1D   C2U   C6U   N1U   2400.0  0    0.000
IMProper    ON    N3U   C5U   C4U   250.0   0    0.000
IMProper    C2U   N3U   C4U   C5U   250.0   0    0.000
IMProper    C5U   C6U   N1U   C2U   250.0   0    0.000
IMProper    H2    C4U   H2    ON    250.0   0    0.000
IMProper    HN    C2U   C4U   N3U   250.0   0    0.000
IMProper    H     C3D   NX29  C1D   500.0   0   -73.5   !added for pyr, jt

!thymidine
IMProper    C4T   X     X     ON    2400.0  0    0.000
IMProper    C4T   X     X     N1T   250.0   0    0.000
IMProper    C6T   X     X     N3T   250.0   0    0.000
IMProper    C4T   X     X     N2    2400.0  0    0.000
IMProper    C2T   X     X     ON    2400.0  0    0.000
IMProper    C1R   C2T   C6T   N1T   2400.0  0    0.000
IMProper    C1D   C2T   C6T   N1T   2400.0  0    0.000
IMProper    ON    N3T   C5T   C4T   250.0   0    0.000
IMProper    C2T   N3T   C4T   C5T   250.0   0    0.000
IMProper    C5T   C6T   N1T   C2T   250.0   0    0.000
IMProper    H2    C4T   H2    ON    250.0   0    0.000
IMProper    CC3E  C4T   C6T   C5T   2400.0  0    0.000

!infer
IMProper    HN    C2T   C4T   N3T   250.0   0    0.000

! The ring-spanning impropers have been left out.
!adenine
IMProper    N2A   N3A   NC    C2A   250.0   0    0.000
```

Script code

```
IMProper    H2    C2A   H2    N2A  250.0   0   0.000
IMProper    C4A   C5A   N7A   C8A  250.0   0   0.000
IMProper    C5A   C4A   N9A   C8A  250.0   0   0.000
IMProper    C4A   X     X     NC   250.0   0   0.000
IMProper    C2A   X     X     N9A  250.0   0   0.000
IMProper    C2A   X     X     C5A  250.0   0   0.000
IMProper    C6A   C5A   C4A   N3A  250.0   0   0.000
IMProper    C5A   X     X     N9A  250.0   0   0.000
IMProper    C6A   X     X     N6A  2400.0  0   0.000
IMProper    H2    X     X     N6A  250.0   0   0.000

!infer
IMProper    C1R   C4A   C8A   N9A  2400.0  0   0.000
IMProper    C1D   C4A   C8A   N9A  2400.0  0   0.000
IMProper    N9A   C4A   C5A   N7A  250.0   0   0.000
IMProper    N7A   C8A   N9A   C4A  250.0   0   0.000
IMProper    N3A   C2A   NC    C6A  250.0   0   0.000
IMProper    C5A   C6A   N6A   H2   2000.0  0   0.000

! The ring-spanning impropers have been left out.
!purine
IMProper    N2P   N3P   NC    C2P  250.0   0   0.000
IMProper    C4P   C5P   N7P   C8P  250.0   0   0.000
IMProper    C5P   C4P   N9P   C8P  250.0   0   0.000
IMProper    C4P   X     X     NC   250.0   0   0.000
IMProper    C2P   X     X     N9P  250.0   0   0.000
IMProper    C2P   X     X     C5P  250.0   0   0.000
IMProper    C6P   C5P   C4P   N3P  250.0   0   0.000
IMProper    C5P   X     X     N9P  250.0   0   0.000

!infer
IMProper    C1R   C4P   C8P   N9P  2400.0  0   0.000
IMProper    C1D   C4P   C8P   N9P  2400.0  0   0.000
IMProper    N9P   C4P   C5P   N7P  250.0   0   0.000
IMProper    N7P   C8P   N9P   C4P  250.0   0   0.000
IMProper    N3P   C2P   NC    C6P  250.0   0   0.000

! The ring-spanning impropers have been left out.
!guanine
IMProper    C4G   C5G   N7G   C8G  250.0   0   0.000
IMProper    C5G   C4G   N9G   C8G  250.0   0   0.000
IMProper    C4G   X     X     NNA  250.0   0   0.000
IMProper    C2G   X     X     N9G  250.0   0   0.000
IMProper    C2G   X     X     C5G  250.0   0   0.000
IMProper    C6G   C5G   C4G   N3G  250.0   0   0.000
IMProper    C5G   X     X     N9G  250.0   0   0.000
IMProper    C6G   X     X     O6G  2400.0  0   0.000
IMProper    C2G   X     X     N2G  2400.0  0   0.000

!infer
IMProper    C1R   C4G   C8G   N9G  2400.0  0   0.000
IMProper    C1D   C4G   C8G   N9G  2400.0  0   0.000
IMProper    N9G   C4G   C5G   N7G  250.0   0   0.000
IMProper    N7G   C8G   N9G   C4G  250.0   0   0.000
IMProper    N3G   C2G   NNA   C6G  250.0   0   0.000
IMProper    H2    H2    C2G   N2G  250.0   0   0.000
```

1 Force field parameter and topology files

```
IMPRoper    HN   C2G  C6G   NNA  2000.0  0   0.000
IMPRoper    N3G  C2G  N2G   H2   2000.0  0   0.000

!                Lennard-Jones parameters
!                                 ------1-4------
!                  epsilon   sigma     epsilon   sigma
!                 (Kcal/mol)  (A)     (Kcal/mol)  (A)
! Taken from Rossky Karplus and Rahman BIOPOLY (1979)
! 0.05 ADDED TO RADII TO IMPRoperOVE ON NUCL. ACID STACKING/LN
!
!                   eps      sigma       eps(1:4)  sigma(1:4)

!****************** mod by anda - HNF ******************

NONBonded PX1   0.5849  3.3854   0.5849  3.3854  ! mod by anda, std value
NONBonded OX2   0.2304  2.7290   0.2304  2.7290  ! mod by anda, std value
NONBonded OX3   0.2304  2.7290   0.2304  2.7290  ! mod by anda, std value
NONBonded OX4   0.2304  2.7290   0.2304  2.7290  ! mod by anda, std value
NONBonded CX5   0.0900  3.2970   0.0900  3.2970  ! mod by anda, std value
NONBonded HX6   0.0045  2.6160   0.0045  2.6160  ! mod by anda, std value
NONBonded HX7   0.0045  2.6160   0.0045  2.6160  ! mod by anda, std value
NONBonded CX8   0.0900  3.2970   0.0900  3.2970  ! mod by anda, std value
NONBonded HX9   0.0045  2.6160   0.0045  2.6160  ! mod by anda, std value
NONBonded OX10  0.2304  2.7290   0.2304  2.7290  ! mod by anda, std value
NONBonded CX11  0.0900  3.2970   0.0900  3.2970  ! mod by anda, std value
NONBonded CX12  0.0900  3.2970   0.0900  3.2970  ! mod by anda, std value
NONBonded HX13  0.0045  2.6160   0.0045  2.6160  ! mod by anda, std value
NONBonded CX14  0.0900  3.2970   0.0900  3.2970  ! mod by anda, std value
NONBonded HX15  0.0045  2.6160   0.0045  2.6160  ! mod by anda, std value
NONBonded HX16  0.0045  2.6160   0.0045  2.6160  ! mod by anda, std value
NONBonded OX17  0.2304  2.7290   0.2304  2.7290  ! mod by anda, std value
NONBonded OX18  0.2304  2.7290   0.2304  2.7290  ! mod by anda, std value
NONBonded CX19  0.0900  3.2970   0.0900  3.2970  ! mod by anda, std value
NONBonded CX20  0.0900  3.2970   0.0900  3.2970  ! mod by anda, std value
NONBonded CX21  0.0900  3.2970   0.0900  3.2970  ! mod by anda, std value
NONBonded CX22  0.0900  3.2970   0.0900  3.2970  ! mod by anda, std value
NONBonded CX23  0.0900  3.2970   0.0900  3.2970  ! mod by anda, std value
NONBonded CX24  0.0900  3.2970   0.0900  3.2970  ! mod by anda, std value
NONBonded CX25  0.0900  3.2970   0.0900  3.2970  ! mod by anda, std value
NONBonded CX26  0.0900  3.2970   0.0900  3.2970  ! mod by anda, std value
NONBonded CX27  0.0900  3.2970   0.0900  3.2970  ! mod by anda, std value
NONBonded CX28  0.0900  3.2970   0.0900  3.2970  ! mod by anda, std value
NONBonded CX29  0.0900  3.2970   0.0900  3.2970  ! mod by anda, std value
NONBonded CX30  0.0900  3.2970   0.0900  3.2970  ! mod by anda, std value
NONBonded CX31  0.0900  3.2970   0.0900  3.2970  ! mod by anda, std value
NONBonded NX32  0.1600  2.8591   0.1600  2.8591  ! mod by anda, std value
NONBonded OX33  0.2304  2.7290   0.2304  2.7290  ! mod by anda, std value
NONBonded OX34  0.2304  2.7290   0.2304  2.7290  ! mod by anda, std value
NONBonded HX35  0.0045  2.6160   0.0045  2.6160  ! mod by anda, std value
NONBonded HX36  0.0045  2.6160   0.0045  2.6160  ! mod by anda, std value
NONBonded HX37  0.0045  2.6160   0.0045  2.6160  ! mod by anda, std value
NONBonded HX38  0.0045  2.6160   0.0045  2.6160  ! mod by anda, std value
NONBonded HX39  0.0045  2.6160   0.0045  2.6160  ! mod by anda, std value
NONBonded HX40  0.0045  2.6160   0.0045  2.6160  ! mod by anda, std value
NONBonded HX41  0.0045  2.6160   0.0045  2.6160  ! mod by anda, std value
NONBonded HX42  0.0045  2.6160   0.0045  2.6160  ! mod by anda, std value
```

Script code

```
           NONBonded  HX43  0.0045   2.6160        0.0045   2.6160  ! mod by anda, std value

!******************* end of mod by anda - HNF *************

           NONBonded  C5R    0.0900   3.2970        0.0900   3.2970
           NONBonded  C1R    0.0900   3.2970        0.0900   3.2970
           NONBonded  C2R    0.0900   3.2970        0.0900   3.2970
           NONBonded  C3R    0.0900   3.2970        0.0900   3.2970
           NONBonded  C4R    0.0900   3.2970        0.0900   3.2970

           NONBonded  C5D    0.0900   3.2970        0.0900   3.2970   !DNA
           NONBonded  C1D    0.0900   3.2970        0.0900   3.2970
           NONBonded  C2D    0.0900   3.2970        0.0900   3.2970
           NONBonded  C3D    0.0900   3.2970        0.0900   3.2970
           NONBonded  C4D    0.0900   3.2970        0.0900   3.2970

           NONBonded  HN     0.0045   2.6160        0.0045   2.6160
           NONBonded  H2     0.0045   1.6040        0.0045   1.6040
           NONBonded  H      0.0045   2.6160        0.0045   2.6160
           !
           ! give it the same as th Hn from RKR
           NONBonded  HO     0.0045   1.6040        0.0045   1.6040

           !
           ! THIS STILL IS AN EXTENDED ATOM
           NONBonded  O3R    0.2304   2.7290        0.2304   2.7290
           NONBonded  O4R    0.2304   2.7290        0.2304   2.7290
           NONBonded  O4D    0.2304   2.7290        0.2304   2.7290
           NONBonded  O5R    0.2304   2.7290        0.2304   2.7290
           NONBonded  O1P    0.2304   2.7290        0.2304   2.7290
           NONBonded  O2P    0.2304   2.7290        0.2304   2.7290
           NONBonded  P      0.5849   3.3854        0.5849   3.3854

           !bases
           NONBonded  C2     0.0900   3.2970        0.0900   3.2970
           NONBonded  C3     0.0900   3.2970        0.0900   3.2970
           NONBonded  CB     0.0900   3.2970        0.0900   3.2970
           NONBonded  CE     0.0900   3.2970        0.0900   3.2970
           NONBonded  CH     0.0900   3.2970        0.0900   3.2970

           NONBonded  N2     0.1600   2.8591        0.1600   2.8591
           NONBonded  N3U    0.1600   2.8591        0.1600   2.8591
           NONBonded  N3T    0.1600   2.8591        0.1600   2.8591
           NONBonded  NNA    0.1600   2.8591        0.1600   2.8591
           NONBonded  NB     0.1600   2.8591        0.1600   2.8591
           NONBonded  NC     0.1600   2.8591        0.1600   2.8591

           NONBonded  NH2E   0.1600   3.0291        0.1600   3.0291
           NONBonded  NS     0.1600   2.8591        0.1600   2.8591
           NONBonded  N1T    0.1600   2.8591        0.1600   2.8591
           NONBonded  N1C    0.1600   2.8591        0.1600   2.8591
           NONBonded  N9A    0.1600   2.8591        0.1600   2.8591
           NONBonded  N9P    0.1600   2.8591        0.1600   2.8591
           NONBonded  N9G    0.1600   2.8591        0.1600   2.8591
           NONBonded  N1U    0.1600   2.8591        0.1600   2.8591
           NONBonded  ON     0.2304   2.7290        0.2304   2.7290
```

1 Force field parameter and topology files

```
NONBonded  O2R   0.2304  2.7290  0.2304  2.7290
NONBonded  OH    0.2304  2.5508  0.2304  2.5508
NONBonded  SD    0.3515  2.6727  0.3515  2.6727  ! G U E S S
NONBonded  O2    0.2304  2.7290  0.2304  2.7290

! NEW
NONBonded  C6C   0.0900  3.2970  0.0900  3.2970
NONBonded  C5C   0.0900  3.2970  0.0900  3.2970
NONBonded  C4C   0.0900  3.2970  0.0900  3.2970
NONBonded  C2C   0.0900  3.2970  0.0900  3.2970
NONBonded  C6U   0.0900  3.2970  0.0900  3.2970
NONBonded  C5U   0.0900  3.2970  0.0900  3.2970
NONBonded  C4U   0.0900  3.2970  0.0900  3.2970
NONBonded  C2U   0.0900  3.2970  0.0900  3.2970
NONBonded  C8A   0.0900  3.2970  0.0900  3.2970
NONBonded  C6A   0.0900  3.2970  0.0900  3.2970
NONBonded  C5A   0.0900  3.2970  0.0900  3.2970
NONBonded  C4A   0.0900  3.2970  0.0900  3.2970
NONBonded  C2A   0.0900  3.2970  0.0900  3.2970
NONBonded  C8P   0.0900  3.2970  0.0900  3.2970
NONBonded  C6P   0.0900  3.2970  0.0900  3.2970
NONBonded  C5P   0.0900  3.2970  0.0900  3.2970
NONBonded  C4P   0.0900  3.2970  0.0900  3.2970
NONBonded  C2P   0.0900  3.2970  0.0900  3.2970
NONBonded  C8G   0.0900  3.2970  0.0900  3.2970
NONBonded  C6G   0.0900  3.2970  0.0900  3.2970
NONBonded  C5G   0.0900  3.2970  0.0900  3.2970
NONBonded  C4G   0.0900  3.2970  0.0900  3.2970
NONBonded  C2G   0.0900  3.2970  0.0900  3.2970
NONBonded  C6T   0.0900  3.2970  0.0900  3.2970
NONBonded  C5T   0.0900  3.2970  0.0900  3.2970
NONBonded  C4T   0.0900  3.2970  0.0900  3.2970
NONBonded  C2T   0.0900  3.2970  0.0900  3.2970
NONBonded  N4C   0.1600  2.8591  0.1600  2.8591
NONBonded  O4U   0.2304  2.7290  0.2304  2.7290
NONBonded  N7G   0.1600  2.8591  0.1600  2.8591
NONBonded  N3G   0.1600  2.8591  0.1600  2.8591
NONBonded  N2G   0.1600  2.8591  0.1600  2.8591
NONBonded  N3A   0.1600  2.8591  0.1600  2.8591
NONBonded  N7A   0.1600  2.8591  0.1600  2.8591
NONBonded  N6A   0.1600  2.8591  0.1600  2.8591
NONBonded  O6G   0.2304  2.7290  0.2304  2.7290
NONBonded  CC3E  0.0900  3.2970  0.0900  3.2970
NONBonded  N2A   0.1600  2.8591  0.1600  2.8591
NONBonded  N2P   0.1600  2.8591  0.1600  2.8591
NONBonded  N3P   0.1600  2.8591  0.1600  2.8591
NONBonded  N7P   0.1600  2.8591  0.1600  2.8591

! special solute-solute hydrogen bonding potential parameters
!AEXP 4
!REXP 6

!HAEX 4

!AAEX 2
! "all" possible combinations of HB-pairs in nucleic acids:
```

Script code

```
!  WELL DEPTHS DEEPENED BY 0.5 KCAL TO IMPROVE BASEPAIR ENERGIES /LN
!  AND DISTANCES INCREASED BY 0.05
!                        Emin         Rmin
!                      (Kcal/mol)      (A)
!hbond   N*    O*       -14.0         2.95
!hbond   N*    N*       -14.5         3.05
!hbond   O*    O*       -15.75        2.80
!hbond   O*    N*       -15.50        2.90

! the following NBFIXes are for DNA-DNA hydrogen bonding

! terms
!                                              --------1-4------
!                          A            B          A          B
!                   [Kcal/(mol A^12)] [Kcal/(mol A^6)]
!
nbfix   HO   ON         0.05         0.1         0.05        0.1
nbfix   HO   O3R        0.05         0.1         0.05        0.1
nbfix   HO   O5R        0.05         0.1         0.05        0.1
nbfix   HO   OH         0.05         0.1         0.05        0.1
nbfix   HO   O2R        0.05         0.1         0.05        0.1
nbfix   HO   NC         0.05         0.1         0.05        0.1


nbfix   H    ON         0.05         0.1         0.05        0.1
nbfix   H    O2         0.05         0.1         0.05        0.1
nbfix   H    O5R        0.05         0.1         0.05        0.1
nbfix   H    O4R        0.05         0.1         0.05        0.1
nbfix   H    O4D        0.05         0.1         0.05        0.1
nbfix   H    O3R        0.05         0.1         0.05        0.1
nbfix   H    O2R        0.05         0.1         0.05        0.1
nbfix   H    OH         0.05         0.1         0.05        0.1
nbfix   H    N7A        0.05         0.1         0.05        0.1
nbfix   H    N7P        0.05         0.1         0.05        0.1
nbfix   H    N7G        0.05         0.1         0.05        0.1
nbfix   H    N3A        0.05         0.1         0.05        0.1
nbfix   H    N3P        0.05         0.1         0.05        0.1
nbfix   H    N3G        0.05         0.1         0.05        0.1


nbfix   HN   ON         0.05         0.1         0.05        0.1
nbfix   HN   O2R        0.05         0.1         0.05        0.1
nbfix   HN   OH         0.05         0.1         0.05        0.1
nbfix   HN   NC         0.05         0.1         0.05        0.1

nbfix   H2   ON         0.05         0.1         0.05        0.1
nbfix   H2   O2R        0.05         0.1         0.05        0.1
nbfix   H2   OH         0.05         0.1         0.05        0.1

nbfix   H2   NC         0.05         0.1         0.05        0.1

!mod by anda————————————————————————————————

!2-AP
BOND      C2P    N2G          1110.000    1.341  !0.010  B
ANGLe     NC     C2P    N2G    449.866  116.20   !0.90   B
```

1 Force field parameter and topology files

```
ANGLe      N3P    C2P    N2G         743.656   119.900  !0.70 B
ANGLe      C2P    N2G    H2          105.000   120.000  !from param11.dna, 3*keq
DIHEdral   X      C2P    N2G    X      18.0    2    180.000
IMProper   H2     H2     C2P    N2G   250.0    0      0.000
IMProper   N3P    C2P    N2G    H2   2000.0    0      0.000
IMProper   C2P    X      X      N2G  2400.0    0      0.000

!end of mod by
    anda——————————————————————————————————————————————————————————

set echo=on message=on end


!RNA TOPOLOGY FILE 'FRAMEWORK' FROM TOPALLHDG.DNA AND ATOM NAMES
    !   FROM DNA-RNA.PARAM
    !INCLUDES ALL NONEXCHANGEABLE HYDROGENS AND TERMS FOR BOND, ANGLE, AND
    !IMPROPERS. NONEXCHANGEABLE HYDROGEN CHARGES WERE ASSIGNED 0.035.
    !CARBON CHARGES WERE REDUCED 0.035 FOR EACH ATTACHED HYDROGEN.
    !CREATED 2/24/96-- JASON P. RIFE AND PETER B. MOORE
    !   DNA-RNA-ALLATOM.TOP

    set echo=false end


    ! checkversion 1.0


    AUTOGENERATE  ANGLES=TRUE  END
    {====================================}


!******************* mod by anda - HNF *******************

    MASS PX1     30.97400  ! assuming P -> 30.97400 + 1.008 * 0 (Hs)
    MASS OX2     15.99900  ! assuming O -> 15.99900 + 1.008 * 0 (Hs)
    MASS OX3     15.99900  ! assuming O -> 15.99900 + 1.008 * 0 (Hs)
    MASS OX4     15.99900  ! assuming O -> 15.99900 + 1.008 * 0 (Hs)
    MASS CX5     12.01100  ! assuming C -> 12.01100 + 1.008 * 0 (Hs)
    MASS HX6      1.00800  ! assuming H -> 1.00800 + 1.008 * 0 (Hs)
    MASS HX7      1.00800  ! assuming H -> 1.00800 + 1.008 * 0 (Hs)
    MASS CX8     12.01100  ! assuming C -> 12.01100 + 1.008 * 0 (Hs)
    MASS HX9      1.00800  ! assuming H -> 1.00800 + 1.008 * 0 (Hs)
    MASS OX10    15.99900  ! assuming O -> 15.99900 + 1.008 * 0 (Hs)
    MASS CX11    12.01100  ! assuming C -> 12.01100 + 1.008 * 0 (Hs)
    MASS CX12    12.01100  ! assuming C -> 12.01100 + 1.008 * 0 (Hs)
    MASS HX13     1.00800  ! assuming H -> 1.00800 + 1.008 * 0 (Hs)
    MASS CX14    12.01100  ! assuming C -> 12.01100 + 1.008 * 0 (Hs)
    MASS HX15     1.00800  ! assuming H -> 1.00800 + 1.008 * 0 (Hs)
    MASS HX16     1.00800  ! assuming H -> 1.00800 + 1.008 * 0 (Hs)
    MASS OX17    15.99900  ! assuming O -> 15.99900 + 1.008 * 0 (Hs)
    MASS OX18    15.99900  ! assuming O -> 15.99900 + 1.008 * 0 (Hs)
    MASS CX19    12.01100  ! assuming C -> 12.01100 + 1.008 * 0 (Hs)
    MASS CX20    12.01100  ! assuming C -> 12.01100 + 1.008 * 0 (Hs)
    MASS CX21    12.01100  ! assuming C -> 12.01100 + 1.008 * 0 (Hs)
    MASS CX22    12.01100  ! assuming C -> 12.01100 + 1.008 * 0 (Hs)
    MASS CX23    12.01100  ! assuming C -> 12.01100 + 1.008 * 0 (Hs)
```

Script code

```
MASS CX24    12.01100 ! assuming C -> 12.01100 + 1.008 * 0 (Hs)
MASS CX25    12.01100 ! assuming C -> 12.01100 + 1.008 * 0 (Hs)
MASS CX26    12.01100 ! assuming C -> 12.01100 + 1.008 * 0 (Hs)
MASS CX27    12.01100 ! assuming C -> 12.01100 + 1.008 * 0 (Hs)
MASS CX28    12.01100 ! assuming C -> 12.01100 + 1.008 * 0 (Hs)
MASS CX29    12.01100 ! assuming C -> 12.01100 + 1.008 * 0 (Hs)
MASS CX30    12.01100 ! assuming C -> 12.01100 + 1.008 * 0 (Hs)
MASS CX31    12.01100 ! assuming C -> 12.01100 + 1.008 * 0 (Hs)
MASS NX32    14.00700 ! assuming N -> 14.00700 + 1.008 * 0 (Hs)
MASS OX33    15.99900 ! assuming O -> 15.99900 + 1.008 * 0 (Hs)
MASS OX34    15.99900 ! assuming O -> 15.99900 + 1.008 * 0 (Hs)
MASS HX35     1.00800 ! assuming H ->  1.00800 + 1.008 * 0 (Hs)
MASS HX36     1.00800 ! assuming H ->  1.00800 + 1.008 * 0 (Hs)
MASS HX37     1.00800 ! assuming H ->  1.00800 + 1.008 * 0 (Hs)
MASS HX38     1.00800 ! assuming H ->  1.00800 + 1.008 * 0 (Hs)
MASS HX39     1.00800 ! assuming H ->  1.00800 + 1.008 * 0 (Hs)
MASS HX40     1.00800 ! assuming H ->  1.00800 + 1.008 * 0 (Hs)
MASS HX41     1.00800 ! assuming H ->  1.00800 + 1.008 * 0 (Hs)
MASS HX42     1.00800 ! assuming H ->  1.00800 + 1.008 * 0 (Hs)
MASS HX43     1.00800 ! assuming H ->  1.00800 + 1.008 * 0 (Hs)

!******************* end of mod by anda - HNF *************

{* DNA/RNA default masses *}

MASS  P      30.97400! phosphorus
MASS  O1P    15.99940! O in phosphate
MASS  O2P    15.99940! O in phosphate
MASS  O5R    15.99940! ester -P-O-C-
MASS  C5R    12.011!   corresp. to CH2E
MASS  C4R    12.011!   corresp. to CH1E
MASS  C3R    12.011!   corresp. to CH1E
MASS  C2R    12.011!   corresp. to CH1E
MASS  C1R    12.011!   corresp. to CH1E
MASS  O4R    15.99940! ester -P-O-C-
MASS  O3R    15.99940! ester -P-O-C-
MASS  O2R    15.99940! ester -P-O-C-
MASS  OH     15.99940! corresp. to OH1

!DEOXY SUGAR
MASS  C5D    14.02700! corresp. to CH2E
MASS  C4D    13.01900! corresp. to CH1E
MASS  C3D    13.01900! corresp. to CH1E
MASS  C2D    13.01900! corresp. to CH1E
MASS  C1D    13.01900! corresp. to CH1E
MASS  O4D    15.99940! ester -P-O-C-
MASS  O5D    15.99940!
MASS  O3D    15.99940!

! Insert Bases
! Generic

MASS  N2     14.00670! nitrogen in -NH2
MASS  NNA    14.00670! corresp. to NH1
MASS  ON     15.99940! corresp. to O
MASS  NC     14.00670! corresp. to NR
```

1 Force field parameter and topology files

```
MASS    NS      14.00670! nitrogen in ring >N-

! Insert 4 Bases
! GUA
MASS    N9G     14.00670! nitrogen in ring >N-
MASS    C2G     12.011! (prev CE)
MASS    N3G     14.00670! (prev NC)
MASS    C4G     12.01100! (prev CB)
MASS    C5G     12.01100! (prev CB)
MASS    C6G     12.01100! (prev CN)
MASS    N7G     14.00670! (prev NB)
MASS    C8G     12.011! (prev CE)
MASS    O6G     15.99940! (prev CE)
MASS    N2G     14.00670! nitrogen in -NH2
! ADE
MASS    N9A     14.00670! nitrogen in ring >N-
MASS    C2A     12.011! (prev CE)
MASS    N3A     14.00670! (prev NC)
MASS    C4A     12.01100! (prev CB)
MASS    C5A     12.01100! (prev CB)
MASS    C6A     12.01100! (prev CA)
MASS    N7A     14.00670! (prev NB)
MASS    C8A     12.011! (prev CE)
MASS    N6A     14.00670! nitrogen in -NH2

! PUR
MASS    N9P     14.00670! nitrogen in ring >N-
MASS    C2P     12.011! (prev CE)
MASS    N3P     14.00670! (prev NC)
MASS    C4P     12.01100! (prev CB)
MASS    C5P     12.01100! (prev CB)
MASS    C6P     12.01100! (prev CA)
MASS    N7P     14.00670! (prev NB)
MASS    C8P     12.011! (prev CE)

! CYT
MASS    N1C     14.00670! nitrogen in ring >N-
MASS    C2C     12.01100! (prev CN)
MASS    C4C     12.01100! (prev CA)
MASS    C5C     12.011! (prev CF)
MASS    C6C     12.011! (prev CF)
MASS    N4C     14.00670! nitrogen in -NH2

! THY
MASS    N1T     14.00670! nitrogen in ring >N-
MASS    N3T     14.00670! nitrogen in ring >N-
MASS    C2T     12.01100! (prev CN)
MASS    C4T     12.01100! (prev CN)
MASS    C5T     12.011! (prev CS)
MASS    C6T     12.011! (prev CF)
MASS    CC3E    12.01100! (prev CF)

! END

MASS    H       1.00800! non-exchangeable Hydrogens
MASS    HN      1.00800! corresp. to H
```

Script code

```
MASS    H2      1.00800! hydrogen in -NH2
MASS    HO      1.00800! hydroxy hydrogen

!  URI
MASS    N1U     14.00670! nitrogen in ring >N-
MASS    C2U     12.01100! (prev CN)
MASS    C4U     12.01100! (prev CA)
MASS    C5U     12.011!   (prev CF)
MASS    C6U     12.011!   (prev CF)
MASS    N3U     14.00670!
```


```
!******************* mod by anda - HNF ***********************
RESIdue HNF

{ Note: electrostatics should normally not be used in }
{ crystallographic refinement since it can produce }
{ artefacts. For this reason, all charges are set to }
{ zero by default. Edit them if necessary }

GROUp
ATOM    P       TYPE PX1    CHARGE=1.20     END
ATOM    O1P     TYPE OX2    CHARGE=-0.40    END
ATOM    O2P     TYPE OX3    CHARGE=-0.40    END
ATOM    O5'     TYPE OX4    CHARGE=-0.36    END
GROUp
ATOM    C5'     TYPE CX5    CHARGE=-0.070   END
ATOM    H5'     TYPE HX6    CHARGE=0.035    END
ATOM    H5''    TYPE HX7    CHARGE=0.035    END
GROUp
ATOM    C4'     TYPE CX8    CHARGE=0.100    END !mod. to match DFT calc
ATOM    H4'     TYPE HX9    CHARGE=0.105    END !mod. to match DFT calc
ATOM    O4'     TYPE OX10   CHARGE=-0.263   END !mod. to match DFT calc, inc. by +0.13 for neutrality
ATOM    C1'     TYPE CX11   CHARGE=0.302    END !mod. to match DFT calc
ATOM    H1''    TYPE HX43   CHARGE=0.076    END !mod. to match DFT calc
ATOM    O2      TYPE OX18   CHARge=-0.244   END !reduced charge by 0.076 (residual charge of HNF)
GROUp
ATOM    C2'     TYPE CX14   CHARGE=-0.070   END
ATOM    H2'     TYPE HX15   CHARGE=0.035    END
ATOM    H2''    TYPE HX16   CHARGE=0.035    END
GROUp
ATOM    C3'     TYPE CX12   CHARGE=-0.035   END
ATOM    H3'     TYPE HX13   CHARGE=0.035    END
GROUp
ATOM    O3'     TYPE OX17   CHARGE=-0.36    END
!HNF-base
GROUp
ATOM    C2      TYPE CX19   CHARge  0.449   END
ATOM    C3      TYPE CX20   CHARge -0.372   END
ATOM    C4      TYPE CX21   CHARge -0.089   END
ATOM    C10     TYPE CX22   CHARge -0.119   END
ATOM    C11     TYPE CX23   CHARge  0.225   END
```

1 Force field parameter and topology files

```
ATOM  C1   TYPE CX24  CHARge  -0.447  END
ATOM  C13  TYPE CX25  CHARge   0.078  END
ATOM  C12  TYPE CX26  CHARge   0.134  END
ATOM  C9   TYPE CX27  CHARge  -0.191  END
ATOM  C5   TYPE CX28  CHARge  -0.154  END
ATOM  C6   TYPE CX29  CHARge  -0.230  END
ATOM  C7   TYPE CX30  CHARge   0.078  END
ATOM  C8   TYPE CX31  CHARge  -0.317  END
ATOM  N7   TYPE NX32  CHARge   0.751  END
ATOM  O71  TYPE OX33  CHARge  -0.453  END
ATOM  O72  TYPE OX34  CHARge  -0.456  END
ATOM  H4   TYPE HX35  CHARge   0.147  END
ATOM  H3   TYPE HX36  CHARge   0.156  END
ATOM  H5   TYPE HX37  CHARge   0.141  END
ATOM  H6   TYPE HX38  CHARge   0.181  END
ATOM  H8   TYPE HX39  CHARge   0.193  END
ATOM  H1   TYPE HX40  CHARge   0.102  END
ATOM  H91  TYPE HX41  CHARge   0.105  END
ATOM  H92  TYPE HX42  CHARge   0.164  END


BOND  P     O1P       BOND  P     O2P       BOND  P     O5'       BOND  O5'   C5'
BOND  C5'   H5'       BOND  C5'   H5''      BOND  C5'   C4'       BOND  C4'   H4'
BOND  C4'   O4'       BOND  C4'   C3'       BOND  O4'   C1'       BOND  C1'   C2'
BOND  C1'   O2        BOND  C1'   H1''      BOND  C3'   H3'       BOND  C3'   C2'
BOND  C3'   O3'       BOND  C2'   H2'       BOND  C2'   H2''      BOND  O2    C2
BOND  C2    C3        BOND  C2    C1        BOND  C3    C4        BOND  C3    H3
BOND  C4    C10       BOND  C4    H4        BOND  C10   C11       BOND  C10   C13
BOND  C11   C1        BOND  C11   C9        BOND  C1    H1        BOND  C13   C12
BOND  C13   C5        BOND  C12   C9        BOND  C12   C8        BOND  C9    H91
BOND  C9    H92       BOND  C5    C6        BOND  C5    H5        BOND  C6    C7
BOND  C6    H6        BOND  C7    C8        BOND  C7    N7        BOND  C8    H8
BOND  N7    O71       BOND  N7    O72

{dihedrals taken out in accord with other bases}
{
{ Note: edit these DIHEdrals if necessary }
  DIHEdral O2P  P    O5'  C5'  ! flat ? (180 degrees = trans)   172.75
! DIHEdral P    O5'  C5'  H5'  ! flexible dihedral ???          -63.48
! DIHEdral P    O5'  C5'  H5'' ! flexible dihedral ???           55.20
  DIHEdral P    O5'  C5'  C4'  ! flat ? (180 degrees = trans)   176.08
  DIHEdral O5'  C5'  C4'  H4'  ! flat ? (180 degrees = trans)   172.35
! DIHEdral O5'  C5'  C4'  O4'  ! flexible dihedral ???          -69.24
! DIHEdral O5'  C5'  C4'  C3'  ! flexible dihedral ???           52.07
! DIHEdral H5'  C5'  C4'  H4'  ! flexible dihedral ???           51.91
  DIHEdral H5'  C5'  C4'  O4'  ! flat ? (180 degrees = trans)   170.31
! DIHEdral H5'  C5'  C4'  C3'  ! flexible dihedral ???          -68.38
! DIHEdral H5'' C5'  C4'  H4'  ! flexible dihedral ???          -66.18
! DIHEdral H5'' C5'  C4'  O4'  ! flexible dihedral ???           52.23
  DIHEdral H5'' C5'  C4'  C3'  ! flat ? (180 degrees = trans)   173.54
! DIHEdral C5'  C4'  C3'  C2'  ! flexible dihedral ???         -129.37
! DIHEdral C5'  C4'  C3'  O3'  ! flexible dihedral ???          112.32
! DIHEdral H4'  C4'  C3'  C2'  ! flexible dihedral ???          110.44
  DIHEdral H4'  C4'  C3'  O3'  ! flat ? (0 degrees = cis)        -7.87
! DIHEdral O4'  C4'  C3'  H3'  ! flexible dihedral ???          111.59
  DIHEdral O4'  C4'  C3'  C2'  ! flat ? (0 degrees = cis)        -5.92
```

Script code

```
! DIHEdral  O4'  C4'  C3'  O3'  ! flexible dihedral ???   -124.23
! DIHEdral  O2   C1'  C2'  H2'  ! flexible dihedral ???     80.69
  DIHEdral  O4'  C1'  O2   C2   ! flat ? (180 degrees = trans)  177.32
! DIHEdral  C2'  C1'  O2   C2   ! flexible dihedral ???    -67.60
! DIHEdral  H1'' C1'  O2   C2   ! flexible dihedral ???     56.28
! DIHEdral  C4'  C3'  C2'  H2'  ! flexible dihedral ???    -88.75
! DIHEdral  H3'  C3'  C2'  C1'  ! flexible dihedral ???    -90.38
! DIHEdral  O3'  C3'  C2'  H2'  ! flexible dihedral ???    -93.06}
  DIHEdral  O2   C2   C3   C4   ! flat ? (180 degrees = trans)  180.39
  DIHEdral  O2   C2   C3   H3   ! flat ? (0 degrees = cis)       -0.12
  DIHEdral  C1   C2   C3   C4   ! flat ? (0 degrees = cis)       -0.46
  DIHEdral  C1   C2   C3   H3   ! flat ? (180 degrees = trans)  179.03
  DIHEdral  O2   C2   C1   C11  ! flat ? (180 degrees = trans)  179.62
  DIHEdral  O2   C2   C1   H1   ! flat ? (0 degrees = cis)       -0.53
  DIHEdral  C3   C2   C1   C11  ! flat ? (0 degrees = cis)        0.42
  DIHEdral  C3   C2   C1   H1   ! flat ? (180 degrees = trans)  180.27
  DIHEdral  C2   C3   C4   C10  ! flat ? (0 degrees = cis)        0.26
  DIHEdral  C2   C3   C4   H4   ! flat ? (180 degrees = trans)  180.16
  DIHEdral  H3   C3   C4   C10  ! flat ? (180 degrees = trans)  180.76
  DIHEdral  H3   C3   C4   H4   ! flat ? (0 degrees = cis)        0.67
  DIHEdral  C3   C4   C10  C11  ! flat ? (0 degrees = cis)       -0.02
  DIHEdral  C3   C4   C10  C13  ! flat ? (180 degrees = trans)  179.94
  DIHEdral  H4   C4   C10  C11  ! flat ? (180 degrees = trans)  180.08
  DIHEdral  H4   C4   C10  C13  ! flat ? (0 degrees = cis)        0.03
  DIHEdral  C4   C10  C11  C1   ! flat ? (0 degrees = cis)       -0.02
  DIHEdral  C4   C10  C11  C9   ! flat ? (180 degrees = trans)  179.89
  DIHEdral  C13  C10  C11  C1   ! flat ? (180 degrees = trans)  180.02
  DIHEdral  C13  C10  C11  C9   ! flat ? (0 degrees = cis)       -0.07
  DIHEdral  C4   C10  C13  C12  ! flat ? (180 degrees = trans)  180.10
  DIHEdral  C4   C10  C13  C5   ! flat ? (0 degrees = cis)        0.02
  DIHEdral  C11  C10  C13  C12  ! flat ? (0 degrees = cis)        0.06
  DIHEdral  C11  C10  C13  C5   ! flat ? (180 degrees = trans)  179.98
  DIHEdral  C10  C11  C1   C2   ! flat ? (0 degrees = cis)       -0.18
  DIHEdral  C10  C11  C1   H1   ! flat ? (180 degrees = trans)  179.97
  DIHEdral  C9   C11  C1   C2   ! flat ? (180 degrees = trans)  179.93
  DIHEdral  C9   C11  C1   H1   ! flat ? (0 degrees = cis)        0.08
  DIHEdral  C10  C11  C9   C12  ! flat ? (0 degrees = cis)        0.05
! DIHEdral  C10  C11  C9   H91  ! flexible dihedral ???    117.87
! DIHEdral  C10  C11  C9   H92  ! flexible dihedral ???   -117.81
  DIHEdral  C1   C11  C9   C12  ! flat ? (180 degrees = trans)  179.95
! DIHEdral  C1   C11  C9   H91  ! flexible dihedral ???    -62.23
! DIHEdral  C1   C11  C9   H92  ! flexible dihedral ???     62.09
  DIHEdral  C10  C13  C12  C9   ! flat ? (0 degrees = cis)       -0.03
  DIHEdral  C10  C13  C12  C8   ! flat ? (180 degrees = trans)  179.97
  DIHEdral  C5   C13  C12  C9   ! flat ? (180 degrees = trans)  180.04
  DIHEdral  C5   C13  C12  C8   ! flat ? (0 degrees = cis)        0.04
  DIHEdral  C10  C13  C5   C6   ! flat ? (180 degrees = trans)  180.08
  DIHEdral  C10  C13  C5   H5   ! flat ? (0 degrees = cis)        0.06
  DIHEdral  C12  C13  C5   C6   ! flat ? (0 degrees = cis)       -0.01
  DIHEdral  C12  C13  C5   H5   ! flat ? (180 degrees = trans)  179.97
  DIHEdral  C13  C12  C9   C11  ! flat ? (0 degrees = cis)       -0.01
! DIHEdral  C13  C12  C9   H91  ! flexible dihedral ???   -117.79
! DIHEdral  C13  C12  C9   H92  ! flexible dihedral ???    117.79
  DIHEdral  C8   C12  C9   C11  ! flat ? (180 degrees = trans)  179.99
! DIHEdral  C8   C12  C9   H91  ! flexible dihedral ???     62.22
! DIHEdral  C8   C12  C9   H92  ! flexible dihedral ???    -62.20
```

1 Force field parameter and topology files

```
DIHEdral  C13  C12  C8   C7   ! flat ? (0 degrees = cis)       -0.06
DIHEdral  C13  C12  C8   H8   ! flat ? (180 degrees = trans)   180.04
DIHEdral  C9   C12  C8   C7   ! flat ? (180 degrees = trans)   179.94
DIHEdral  C9   C12  C8   H8   ! flat ? (0 degrees = cis)         0.04
DIHEdral  C13  C5   C6   C7   ! flat ? (0 degrees = cis)         0.01
DIHEdral  C13  C5   C6   H6   ! flat ? (180 degrees = trans)   180.05
DIHEdral  H5   C5   C6   C7   ! flat ? (180 degrees = trans)   180.03
DIHEdral  H5   C5   C6   H6   ! flat ? (0 degrees = cis)         0.07
DIHEdral  C5   C6   C7   C8   ! flat ? (0 degrees = cis)        -0.03
DIHEdral  C5   C6   C7   N7   ! flat ? (180 degrees = trans)   180.02
DIHEdral  H6   C6   C7   C8   ! flat ? (180 degrees = trans)   179.93
DIHEdral  H6   C6   C7   N7   ! flat ? (0 degrees = cis)        -0.02
DIHEdral  C6   C7   C8   C12  ! flat ? (0 degrees = cis)         0.06
DIHEdral  C6   C7   C8   H8   ! flat ? (180 degrees = trans)   179.95
DIHEdral  N7   C7   C8   C12  ! flat ? (180 degrees = trans)   180.01
DIHEdral  N7   C7   C8   H8   ! flat ? (0 degrees = cis)        -0.09
DIHEdral  C6   C7   N7   O71  ! flat ? (0 degrees = cis)         0.03
DIHEdral  C6   C7   N7   O72  ! flat ? (180 degrees = trans)   179.92
DIHEdral  C8   C7   N7   O71  ! flat ? (180 degrees = trans)   180.07
DIHEdral  C8   C7   N7   O72  ! flat ? (0 degrees = cis)        -0.04


{ Note: edit these IMPRopers if necessary }
!!! IMPRoper  P    O1P  O2P  O5'  ! chirality or flatness improper   -36.22 ! taken out in
              accordance with other bases
IMPRoper  C5'  O5'  H5'  H5'' ! chirality or flatness improper   -34.21
IMPRoper  C4'  C5'  H4'  O4'  ! chirality or flatness improper   -37.05
IMPRoper  C1'  O4'  C2'  O2   ! chirality or flatness improper   -33.04
IMPRoper  C3'  C4'  H3'  C2'  ! chirality or flatness improper    41.55
IMPRoper  C2'  C1'  C3'  H2'  ! chirality or flatness improper    28.65
IMPRoper  C2   O2   C3   C1   ! chirality or flatness improper     0.51
IMPRoper  C3   C2   C4   H3   ! chirality or flatness improper    -0.27
IMPRoper  C4   C3   C10  H4   ! chirality or flatness improper    -0.05
IMPRoper  C10  C4   C11  C13  ! chirality or flatness improper    -0.03
IMPRoper  C11  C10  C1   C9   ! chirality or flatness improper    -0.06
IMPRoper  C1   C2   C11  H1   ! chirality or flatness improper    -0.08
IMPRoper  C13  C10  C12  C5   ! chirality or flatness improper    -0.04
IMPRoper  C12  C13  C9   C8   ! chirality or flatness improper     0.00
IMPRoper  C9   C11  C12  H91  ! chirality or flatness improper   -29.65
IMPRoper  C5   C13  C6   H5   ! chirality or flatness improper    -0.01
IMPRoper  C6   C5   C7   H6   ! chirality or flatness improper     0.02
IMPRoper  C7   C6   C8   N7   ! chirality or flatness improper     0.03
IMPRoper  C8   C12  C7   H8   ! chirality or flatness improper     0.05
IMPRoper  N7   C7   O71  O72  ! chirality or flatness improper    -0.06
IMPRoper  C6   C8   C1   C3   ! chirality or flatness improper    -0.01
IMPRoper  C5   C12  C11  C4   ! chirality or flatness improper     0.02
IMPRoper  O71  N7   C7   C6   ! chirality or flatness improper     0.03
IMPRoper  O72  N7   C7   C8   ! chirality or flatness improper     0.05
IMPRoper  H6   H8   H1   H3   ! chirality or flatness improper    -0.06
IMPRoper  C8   C13  C11  C3   ! mod by anda, planarity
IMPRoper  C6   C12  C10  C2   ! mod by anda, planarity
IMPRoper  C12  C13  C10  C11  ! mod by anda, planarity
IMPRoper  C5   C9   C10  C2   ! mod by anda, planarity
IMPRoper  C8   C13  C9   C4   ! mod by anda, planarity
IMPRoper  N7   C13  C9   C4   ! mod by anda, planarity
IMPRoper  C7   C13  C9   C4   ! mod by anda, planarity
```

Script code

```
IMPRoper   C8   C13   C1    C3   ! mod by anda, planarity
IMPRoper   H6   C12   C4    H1   ! mod by anda, planarity
IMPRoper   H8   C5    C11   H3   ! mod by anda, planarity
IMPRoper   C8   C5    C11   C3   ! mod by anda, planarity

!     { Note: edit any DONOrs and ACCEptors if necessary }
!   ! DONOr H?   O1P  ! only true if -OHx (x>0)
!     ACCEptor   O1P  P
!   ! DONOr H?   O2P  ! only true if -OHx (x>0)
!     ACCEptor   O2P  P
!     ACCEptor   O5'  P
!     ACCEptor   O4'  C4'
!   ! DONOr H?   O3'  ! only true if -OHx (x>0)
!     ACCEptor   O3'  C3'
!     ACCEptor   O2   C1'
!   ! DONOr H?   O71  ! only true if -OHx (x>0)
!     ACCEptor   O71  N7
!   ! DONOr H?   O72  ! only true if -OHx (x>0)
!     ACCEptor   O72  N7

END { RESIdue HNF }

!******************** end of mod by anda - HNF ********************


       RESIdue GUA
          GROUp
          ATOM P       TYPE=P      CHARGE=1.20      END
          ATOM O1P     TYPE=O1P    CHARGE=-0.40     END
          ATOM O2P     TYPE=O2P    CHARGE=-0.40     END
          ATOM O5'     TYPE=O5R    CHARGE=-0.36     END
          GROUp
          ATOM C5'     TYPE=C5R    CHARGE=-0.070    END
          ATOM H5'     TYPE=H      CHARGE=0.035     END  !JPR
          ATOM H5''    TYPE=H      CHARGE=0.035     END  !JPR
          GROUp
          ATOM C4'     TYPE=C4R    CHARGE=0.065     END
          ATOM H4'     TYPE=H      CHARGE=0.035     END  !JPR
          ATOM O4'     TYPE=O4R    CHARGE=-0.30     END
          ATOM C1'     TYPE=C1R    CHARGE=0.165     END  !JPR
          ATOM H1'     TYPE=H      CHARGE=0.035     END  !JPR

        ! Insert Base
          GROUp
          ATOM N9      TYPE=N9G    CHARGE=-0.19     END
          ATOM C4      TYPE=C4G    CHARGE=0.19      EXCLusion=( N1 )   END
          GROUp
          ATOM N3      TYPE=N3G    CHARGE=-0.35     EXCLusion=( C6 )   END
          ATOM C2      TYPE=C2G    CHARGE=0.35      EXCLusion=( C5 )   END
          GROUp
          ATOM N2      TYPE=N2G    CHARGE=-0.42     END
          ATOM H21     TYPE=H2     CHARGE=0.21      END
          ATOM H22     TYPE=H2     CHARGE=0.21      END
          GROUp
          ATOM N1      TYPE=NNA    CHARGE=-0.26     END
          ATOM H1      TYPE=HN     CHARGE=0.26      END
```

1 Force field parameter and topology files

```
GROUp
   ATOM C6     TYPE=C6G    CHARGE=0.30      END
   ATOM O6     TYPE=O6G    CHARGE=-0.30     END
GROUp
   ATOM C5     TYPE=C5G    CHARGE=0.02      END
   ATOM N7     TYPE=N7G    CHARGE=-0.25     END
   ATOM C8     TYPE=C8G    CHARGE=0.145     END
   ATOM H8     TYPE=H      CHARGE=0.035     END

!

GROUP
   ATOM C2'    TYPE=C2R    CHARGE=0.115     END
   ATOM H2'    TYPE=H      CHARGE=0.035     END
   ATOM O2'    TYPE=O2R    CHARGE=-0.40     END
   ATOM HO2'   TYPE=HO     CHARGE=0.25      END
GROUP
   ATOM C3'    TYPE=C3R    CHARGE=-0.035    END
   ATOM H3'    TYPE=H      CHARGE=0.035     END
GROUP
   ATOM O3'    TYPE=O3R    CHARGE=-0.36     END

BOND P     O1P         BOND P     O2P          BOND P     O5'

BOND O5'   C5'         BOND C5'   C4'          BOND C4'   O4'
BOND C4'   C3'         BOND O4'   C1'          BOND C1'   N9
BOND C1'   C2'         BOND N9    C4           BOND N9    C8
BOND C4    N3          BOND C4    C5           BOND N3    C2
BOND C2    N2          BOND C2    N1           BOND N2    H21

BOND N2    H22         BOND N1    H1           BOND N1    C6
BOND C6    O6          BOND C6    C5           BOND C5    N7
BOND N7    C8          BOND C2'   C3'          BOND C3'   O3'
BOND C2'   O2'         BOND C8    H8

BOND O2'   HO2'
BOND C5'   H5'         BOND C5'   H5''         BOND C4'   H4'
BOND C3'   H3'         BOND C2'   H2'          BOND C1'   H1'
{
   DIHEdral P    O5'   C5'   C4'        DIHEdral O5'   C5'   C4'   O4'
   DIHEdral O5'  C5'   C4'   C3'
}{
   DIHEdral C3'  C4'   O4'   C1'
   DIHEdral C4'  O4'   C1'   C2'        DIHEdral O4'   C1'   C2'   C3'

   DIHEdral C1'  C2'   C3'   C4'        DIHEdral O4'   C4'   C3'   O3'
   DIHEdral C5'  C4'   C3'   C2'        DIHEdral O3'   C3'   C2'   O2'
   DIHEdral O4'  C1'   N9    C4
   DIHEdral C3'  C2'   O2'   H2'
}

!
   IMPRoper N3   C2    N2    H21        IMPRoper C1'   C4    C8    N9
   IMPRoper N9   C4    C5    N7         IMPRoper C4    C5    N7    C8
   IMPRoper C5   N7    C8    N9         IMPRoper N7    C8    N9    C4
   IMPRoper C8   N9    C4    C5         IMPRoper N2    N3    N1    C2
```

181

Script code

```
IMPRoper  H1   C2   C6   N1       IMPRoper  O6   N1   C5   C6
IMPRoper  C4   N3   C2   N1       IMPRoper  N3   C2   N1   C6
IMPRoper  C2   N1   C6   C5       IMPRoper  N1   C6   C5   C4

IMPRoper  C6   C5   C4   N3       IMPRoper  C5   C4   N3   C2
IMPRoper  H22  H21  C2   N2
IMPRoper  H8   N7   N9   C8

!IMPRoper to keep the two purine rings parallel:
IMPRoper  C8   C4   C5   N1       IMPRoper  C8   C5   C4   C2
IMPRoper  N3   C4   C5   N7       IMPRoper  C6   C5   C4   N9

!RIBOSE IMPROPERS
IMPRoper       H1'   C2'   O4'   N9   !C1'
IMPRoper       H2'   C3'   C1'   O2'  !C2'
IMPRoper       H3'   C4'   C2'   O3'  !C3'
IMPRoper       H4'   C5'   C3'   O4'  !C4'
IMPRoper       H5'   O5'   H5''  C4'  !C5'


END {GUA}

! ───────────────────────────────────────────────────────────

RESIdue ADE
GROUp
    ATOM  P      TYPE=P      CHARGE=1.20     END
    ATOM  O1P    TYPE=O1P    CHARGE=-0.40    END
    ATOM  O2P    TYPE=O2P    CHARGE=-0.40    END
    ATOM  O5'    TYPE=O5R    CHARGE=-0.36    END
GROUp
    ATOM  C5'    TYPE=C5R    CHARGE=-0.070   END
    ATOM  H5'    TYPE=H      CHARGE=0.035    END
    ATOM  H5''   TYPE=H      CHARGE=0.035    END
GROUp
    ATOM  C4'    TYPE=C4R    CHARGE=0.065    END
    ATOM  H4'    TYPE=H      CHARGE=0.035    END
    ATOM  O4'    TYPE=O4R    CHARGE=-0.30    END
    ATOM  C1'    TYPE=C1R    CHARGE=0.165    END
    ATOM  H1'    TYPE=H      CHARGE=0.035    END

! Insert Base
GROUp
    ATOM  N9     TYPE=N9A    CHARGE=-0.19    END
    ATOM  C4     TYPE=C4A    CHARGE=0.19     EXCLusion=( N1 )  END
GROUp
    ATOM  N3     TYPE=N3A    CHARGE=-0.26    EXCLusion=( C6 )  END
    ATOM  C2     TYPE=C2A    CHARGE=0.225    EXCLusion=( C5 )  END
    ATOM  H2     TYPE=H      CHARGE=0.035    END
GROUp
    ATOM  N1     TYPE=NC     CHARGE=-0.28    END
    ATOM  C6     TYPE=C6A    CHARGE=0.28     END
GROUp
    ATOM  N6     TYPE=N6A    CHARGE=-0.42    END
```

1 Force field parameter and topology files

```
  ATOM H61      TYPE=H2       CHARGE=0.21      END
  ATOM H62      TYPE=H2       CHARGE=0.21      END
GROUp
  ATOM C5       TYPE=C5A      CHARGE=0.02      END
  ATOM N7       TYPE=N7A      CHARGE=-0.25     END
  ATOM C8       TYPE=C8A      CHARGE=0.195     END
  ATOM H8       TYPE=H        CHARGE=0.035     END
! END

GROUP
  ATOM C2'      TYPE=C2R      CHARGE=0.115     END
  ATOM H2'      TYPE=H        CHARGE=0.035     END
  ATOM O2'      TYPE=O2R      CHARGE=-0.40     END
  ATOM HO2'     TYPE=HO       CHARGE=0.25      END
GROUP
  ATOM C3'      TYPE=C3R      CHARGE=-0.035    END
  ATOM H3'      TYPE=H        CHARGE=0.035     END
GROUP
  ATOM O3'      TYPE=O3R      CHARGE=-0.36     END

BOND P     O1P          BOND P     O2P         BOND P     O5'
BOND O5'   C5'          BOND C5'   C4'         BOND C4'   O4'
BOND C4'   C3'          BOND O4'   C1'         BOND C1'   N9
BOND C1'   C2'          BOND N9    C4          BOND N9    C8
BOND C4    N3           BOND C4    C5          BOND N3    C2
BOND C2    N1           BOND N1    C6          BOND C6    N6

BOND N6    H61          BOND N6    H62         BOND C6    C5

BOND C5    N7           BOND N7    C8          BOND C2'   C3'
BOND C2'   O2'          BOND C3'   O3'
BOND C8    H8           BOND C2    H2
BOND O2'   HO2'
BOND C5'   H5'          BOND C5'   H5''        BOND C4'   H4'
BOND C3'   H3'          BOND C2'   H2'         BOND C1'   H1'
{
DIHEdral P    O5'   C5'   C4'           DIHEdral O5'   C5'   C4'   O4'
DIHEdral O5'  C5'   C4'   C3'
}{
DIHEdral C3'  C4'   O4'   C1'
DIHEdral C4'  O4'   C1'   C2'           DIHEdral O4'   C1'   C2'   C3'

DIHEdral C1'  C2'   C3'   C4'           DIHEdral O4'   C4'   C3'   O3'
DIHEdral C5'  C4'   C3'   C2'           DIHEdral O2'   C2'   C3'   O3'
DIHEdral O4'  C1'   N9    C4
DIHEdral C3'  C2'   O2'   H2'
}


!
IMPRoper C5   C6   N6   H61             IMPRoper C1'   C4   C8   N9
IMPRoper N9   C4   C5   N7              IMPRoper C4    C5   N7   C8
IMPRoper C5   N7   C8   N9              IMPRoper N7    C8   N9   C4
IMPRoper C8   N9   C4   C5              IMPRoper N6    N1   C5   C6
IMPRoper H62  C6   H61  N6              IMPRoper C4    N3   C2   N1
IMPRoper N3   C2   N1   C6              IMPRoper C2    N1   C6   C5
IMPRoper N1   C6   C5   C4              IMPRoper C6    C5   C4   N3
```

Script code

```
IMProper  C5    C4    N3    C2
IMProper  H2    N1    N3    C2           IMProper  H8    N7    N9    C8
! IMProper to keep the two purine rings parallel:
IMProper  C8    C4    C5    N1           IMProper  C8    C5    C4    C2
IMProper  N3    C4    C5    N7           IMProper  C6    C5    C4    N9


!RIBOSE IMPROPERS
IMProper       H1'   C2'   O4'   N9   !C1'
IMProper       H2'   C3'   C1'   O2'  !C2'
IMProper       H3'   C4'   C2'   O3'  !C3'
IMProper       H4'   C5'   C3'   O4'  !C4'
IMProper       H5'   O5'   H5''  C4'  !C5'


END {ADE}

! ----------------------------------------------------------------

RESIdue PUR
GROUp
    ATOM  P      TYPE=P     CHARGE=1.20    END
    ATOM  O1P    TYPE=O1P   CHARGE=-0.40   END
    ATOM  O2P    TYPE=O2P   CHARGE=-0.40   END
    ATOM  O5'    TYPE=O5R   CHARGE=-0.36   END
GROUp
    ATOM  C5'    TYPE=C5R   CHARGE=-0.070  END
    ATOM  H5'    TYPE=H     CHARGE=0.035   END
    ATOM  H5''   TYPE=H     CHARGE=0.035   END
GROUp
    ATOM  C4'    TYPE=C4R   CHARGE=0.065   END
    ATOM  H4'    TYPE=H     CHARGE=0.035   END
    ATOM  O4'    TYPE=O4R   CHARGE=-0.30   END
    ATOM  C1'    TYPE=C1R   CHARGE=0.165   END
    ATOM  H1'    TYPE=H     CHARGE=0.035   END
! Insert Base
GROUp
    ATOM  N9     TYPE=N9P   CHARGE=-0.19   END
    ATOM  C4     TYPE=C4P   CHARGE=0.19    EXCLusion=( N1 )  END
GROUp
    ATOM  N3     TYPE=N3P   CHARGE=-0.26   EXCLusion=( C6 )  END
    ATOM  C2     TYPE=C2P   CHARGE=0.225   EXCLusion=( C5 )  END
    ATOM  H2     TYPE=H     CHARGE=0.035   END
GROUp
    ATOM  N1     TYPE=NC    CHARGE=-0.28   END
    ATOM  C6     TYPE=C6P   CHARGE=0.28    END
    ATOM  H6     TYPE=H     CHARge= 0.035  END
GROUp
    ATOM  C5     TYPE=C5P   CHARGE=0.02    END
    ATOM  N7     TYPE=N7P   CHARGE=-0.25   END
    ATOM  C8     TYPE=C8P   CHARGE=0.195   END
    ATOM  H8     TYPE=H     CHARGE=0.035   END
! END
```

1 Force field parameter and topology files

```
GROUP
    ATOM  C2'   TYPE=C2R   CHARGE= 0.115    END
    ATOM  H2'   TYPE=H     CHARGE= 0.035    END
    ATOM  O2'   TYPE=O2R   CHARGE=-0.40     END
    ATOM  HO2'  TYPE=HO    CHARGE= 0.25     END
GROUP
    ATOM  C3'   TYPE=C3R   CHARGE=-0.035    END
    ATOM  H3'   TYPE=H     CHARGE= 0.035    END
GROUP
    ATOM  O3'   TYPE=O3R   CHARGE=-0.36     END


BOND  P    O1P           BOND  P    O2P            BOND  P    O5'
BOND  O5'  C5'           BOND  C5'  C4'            BOND  C4'  O4'
BOND  C4'  C3'           BOND  O4'  C1'            BOND  C1'  N9
BOND  C1'  C2'           BOND  N9   C4             BOND  N9   C8
BOND  C4   N3            BOND  C4   C5             BOND  N3   C2
BOND  C2   N1            BOND  N1   C6             BOND  C6   H6

BOND  C6   C5

BOND  C5   N7            BOND  N7   C8             BOND  C2'  C3'
BOND  C2'  O2'           BOND  C3'  O3'
BOND  C8   H8            BOND  C2   H2
BOND  O2'  HO2'
BOND  C5'  H5'           BOND  C5'  H5''           BOND  C4'  H4'
BOND  C3'  H3'           BOND  C2'  H2'            BOND  C1'  H1'
{
  DIHEdral  P    O5'  C5'  C4'          DIHEdral  O5'  C5'  C4'  O4'
  DIHEdral  O5'  C5'  C4'  C3'
}{
  DIHEdral  C3'  C4'  O4'  C1'
  DIHEdral  C4'  O4'  C1'  C2'          DIHEdral  O4'  C1'  C2'  C3'

  DIHEdral  C1'  C2'  C3'  C4'          DIHEdral  O4'  C4'  C3'  O3'
  DIHEdral  C5'  C4'  C3'  C2'          DIHEdral  O2'  C2'  C3'  O3'
  DIHEdral  O4'  C1'  N9   C4
  DIHEdral  C3'  C2'  O2'  H2'
}

!
  IMPRoper  H6  N1  C5  C6              IMPRoper  C1'  C4  C8  N9
  IMPRoper  N9  C4  C5  N7              IMPRoper  C4   C5  N7  C8
  IMPRoper  C5  N7  C8  N9              IMPRoper  N7   C8  N9  C4
  IMPRoper  C8  N9  C4  C5              IMPRoper  N6   N1  C5  C6
  IMPRoper  C4  N3  C2  N1
  IMPRoper  N3  C2  N1  C6              IMPRoper  C2   N1  C6  C5
  IMPRoper  N1  C6  C5  C4              IMPRoper  C6   C5  C4  N3
  IMPRoper  C5  C4  N3  C2
  IMPRoper  H2  N1  N3  C2              IMPRoper  H8   N7  N9  C8
! IMPRoper to keep the two purine rings parallel:
  IMPRoper  C8  C4  C5  N1              IMPRoper  C8   C5  C4  C2
  IMPRoper  N3  C4  C5  N7              IMPRoper  C6   C5  C4  N9


!RIBOSE IMPROPERS
  IMPRoper       C2'  C3'  C1'  O2'
```

Script code

```
IMPRoper    H1'   C2'   O4'   N9    !C1'
IMPRoper    H2'   C3'   C1'   O2'   !C2'
IMPRoper    H3'   C4'   C2'   O3'   !C3'
IMPRoper    H4'   C5'   C3'   O4'   !C4'
IMPRoper    H5'   O5'   H5''  C4'   !C5'


END {PUR}

! ────────────────────────────────────────────────────────────

RESIdue ABA
  GROUp
    ATOM  P      TYPE=P      CHARGE=1.20      END
    ATOM  O1P    TYPE=O1P    CHARGE=-0.40     END
    ATOM  O2P    TYPE=O2P    CHARGE=-0.40     END
    ATOM  O5'    TYPE=O5R    CHARGE=-0.36     END
  GROUp
    ATOM  C5'    TYPE=C5R    CHARGE=-0.070    END
    ATOM  H5'    TYPE=H      CHARGE=0.035     END
    ATOM  H5''   TYPE=H      CHARGE=0.035     END
  GROUp
    ATOM  C4'    TYPE=C4R    CHARGE=0.065     END
    ATOM  H4'    TYPE=H      CHARGE=0.035     END
    ATOM  O4'    TYPE=O4R    CHARGE=-0.30     END
    ATOM  C1'    TYPE=C1R    CHARGE=0.165     END
    ATOM  H1'    TYPE=H      CHARGE=0.018     END
    ATOM  H1''   TYPE=H      CHARGE=0.017     END

  GROUP
    ATOM  C2'    TYPE=C2R    CHARGE=0.115     END
    ATOM  H2'    TYPE=H      CHARGE=0.035     END
    ATOM  O2'    TYPE=O2R    CHARGE=-0.40     END
    ATOM  HO2'   TYPE=HO     CHARGE=0.25      END
  GROUP
    ATOM  C3'    TYPE=C3R    CHARGE=-0.035    END
    ATOM  H3'    TYPE=H      CHARGE=0.035     END
  GROUP
    ATOM  O3'    TYPE=O3R    CHARGE=-0.36     END

    BOND  P    O1P           BOND  P    O2P          BOND  P    O5'
    BOND  O5'  C5'           BOND  C5'  C4'          BOND  C4'  O4'
    BOND  C4'  C3'           BOND  O4'  C1'
    BOND  C1'  C2'           BOND  C2'  C3'
    BOND  C3'  O3'           BOND  C2'  O2'
    BOND  O2'  HO2'          BOND  C1'  H1''
    BOND  C5'  H5'           BOND  C5'  H5''         BOND  C4'  H4'
    BOND  C3'  H3'           BOND  C2'  H2'          BOND  C1'  H1'

    {
    DIHEdral  P    O5'  C5'  C4'              DIHEdral  O5'  C5'  C4'  O4'
    DIHEdral  O5'  C5'  C4'  C3'
    }{
    DIHEdral  C3'  C4'  O4'  C1'
    DIHEdral  C4'  O4'  C1'  C2'              DIHEdral  O4'  C1'  C2'  C3'
```

1 Force field parameter and topology files

```
    DIHedral C1'   C2'   C3'   C4'            DIHedral O4'   C4'   C3'   O3'
    DIHedral C5'   C4'   C3'   C2'            DIHedral O2'   C2'   C3'   O3'
    DIHedral O4'   C1'   H1''  C2
    DIHedral C3'   C2'   O2'   H2'

    ! New dihedrals
    DIHedral C5'   C4'   C3'   O3'            DIHedral C4'   O4'   C1'   H1''
    }

        !RIBOSE IMPROPERS
    !IMPRoper         H1'   C2'   O4'   H1''   !C1'  !mod by anda
    IMPRoper          H2'   C3'   C1'   O2'   !C2'
    IMPRoper          H3'   C4'   C2'   O3'   !C3'
    IMPRoper          H4'   C5'   C3'   O4'   !C4'
    IMPRoper          H5'   O5'   H5''  C4'   !C5'


    END {ABA}


!  ─────────────────────────────────────────────────────────

    RESIdue CYT
        GROUp
        ATOM P      TYPE=P      CHARGE=1.20    END
        ATOM O1P    TYPE=O1P    CHARGE=-0.40   END
        ATOM O2P    TYPE=O2P    CHARGE=-0.40   END
        ATOM O5'    TYPE=O5R    CHARGE=-0.36   END
        GROUp
        ATOM C5'    TYPE=C5R    CHARGE=-0.070  END
        ATOM H5'    TYPE=H      CHARGE=0.035   END
        ATOM H5''   TYPE=H      CHARGE=0.035   END
        GROUp
        ATOM C4'    TYPE=C4R    CHARGE=0.065   END
        ATOM H4'    TYPE=H      CHARGE=0.035   END
        ATOM O4'    TYPE=O4R    CHARGE=-0.30   END
        ATOM C1'    TYPE=C1R    CHARGE=0.165   END
        ATOM H1'    TYPE=H      CHARGE=0.035   END

    ! Insert Base

        GROUp
        ATOM N1     TYPE=N1C    CHARGE=-0.19   EXCLUSION=( C4 )   END
        ATOM C6     TYPE=C6C    CHARGE=0.155   EXCLUSION=( N3 )   END
        ATOM H6     TYPE=H      CHARGE=0.035   END
        GROUp
        ATOM C2     TYPE=C2C    CHARGE=0.30    EXCLUSION=( C5 )   END
        ATOM O2     TYPE=ON     CHARGE=-0.30   END
        GROUp
        ATOM N3     TYPE=NC     CHARGE=-0.28   END
        ATOM C4     TYPE=C4C    CHARGE=0.28    END
        GROUp
        ATOM N4     TYPE=N4C    CHARGE=-0.42   END
        ATOM H41    TYPE=H2     CHARGE=0.21    END
```

Script code

```
    ATOM  H42    TYPE=H2     CHARGE=0.21     END
GROUp
    ATOM  C5     TYPE=C5C    CHARGE=-0.035   END  !CHRG
    ATOM  H5     TYPE=H      CHARGE=0.035    END
GROUp

! END


GROUP
    ATOM  C2'    TYPE=C2R    CHARGE=0.115    END
    ATOM  H2'    TYPE=H      CHARGE=0.035    END
    ATOM  O2'    TYPE=O2R    CHARGE=-0.40    END
    ATOM  HO2'   TYPE=HO     CHARGE=0.25     END
GROUP
    ATOM  C3'    TYPE=C3R    CHARGE=-0.035   END
    ATOM  H3'    TYPE=H      CHARGE=0.035    END
GROUP
    ATOM  O3'    TYPE=O3R    CHARGE=-0.36    END

    BOND  P     O1P              BOND  P     O2P             BOND  P     O5'
    BOND  O5'   C5'              BOND  C5'   C4'             BOND  C4'   O4'
    BOND  C4'   C3'              BOND  O4'   C1'             BOND  C1'   N1
    BOND  C1'   C2'              BOND  N1    C2              BOND  N1    C6
                                 BOND  C2    N3              BOND  N3    C4
    BOND  C4    N4               BOND  N4    H41             BOND  N4    H42
    BOND  C2    O2
    BOND  C4    C5               BOND  C5    C6              BOND  C2'   C3'
    BOND  C3'   O3'              BOND  C2'   O2'
    BOND  C6    H6               BOND  C5    H5
    BOND  O2'   HO2'
    BOND  C5'   H5'              BOND  C5'   H5''            BOND  C4'   H4'
    BOND  C3'   H3'              BOND  C2'   H2'             BOND  C1'   H1'

    {
    DIHEdral  P    O5'   C5'   C4'             DIHEdral  O5'   C5'   C4'   O4'
    DIHEdral  O5'  C5'   C4'   C3'
    }{
    DIHEdral  C3'  C4'   O4'   C1'
    DIHEdral  C4'  O4'   C1'   C2'             DIHEdral  O4'   C1'   C2'   C3'

    DIHEdral  C1'  C2'   C3'   C4'             DIHEdral  O4'   C4'   C3'   O3'
    DIHEdral  C5'  C4'   C3'   C2'             DIHEdral  O2'   C2'   C3'   O3'
    DIHEdral  O4'  C1'   N1    C2
    DIHEdral  C3'  C2'   O2'   H2'

    ! New dihedrals
    DIHEdral  C5'  C4'   C3'   O3'             DIHEdral  C4'   O4'   C1'   N1
    }

    IMPRoper  C5   C4    N4    H41             IMPRoper  C1'   C2    C6    N1
    IMPRoper  O2   N1    N3    C2              IMPRoper  N4    N3    C5    C4
    IMPRoper  N1   C2    N3    C4              IMPRoper  C2    N3    C4    C5
    IMPRoper  N3   C4    C5    C6              IMPRoper  C4    C5    C6    N1
    IMPRoper  C5   C6    N1    C2              IMPRoper  C6    N1    C2    N3
    IMPRoper  H42  C4    H41   N4
```

1 Force field parameter and topology files

```
     IMPRoper  H5   C4   C6   C5              IMPRoper   H6    N1   C5    C6

    !RIBOSE IMPROPERS
     IMPRoper        H1'   C2'   O4'   N1   !C1'
     IMPRoper        H2'   C3'   C1'   O2'  !C2'
     IMPRoper        H3'   C4'   C2'   O3'  !C3'
     IMPRoper        H4'   C5'   C3'   O4'  !C4'
     IMPRoper        H5'   O5'   H5''  C4'  !C5'



    END {CYT}

   ! ─────────────────────────────────────────────────────

    RESIdue THY
      GROUp
        ATOM P      TYPE=P      CHARGE=1.20     END
        ATOM O1P    TYPE=O1P    CHARGE=-0.40    END
        ATOM O2P    TYPE=O2P    CHARGE=-0.40    END
        ATOM O5'    TYPE=O5R    CHARGE=-0.36    END
      GROUp
        ATOM C5'    TYPE=C5R    CHARGE=-0.070   END
        ATOM H5'    TYPE=H      CHARGE=0.035    END
        ATOM H5''   TYPE=H      CHARGE=0.035    END
      GROUp
        ATOM C4'    TYPE=C4R    CHARGE=0.065    END
        ATOM H4'    TYPE=H      CHARGE=0.035    END
        ATOM O4'    TYPE=O4R    CHARGE=-0.30    END
        ATOM C1'    TYPE=C1R    CHARGE=0.20     END
        ATOM H1'    TYPE=H      CHARGE=0.165    END

    ! Insert Base
      GROUp
        ATOM N1     TYPE=N1T    CHARGE=-0.19    EXCLUSION=( C4 )   END
        ATOM C6     TYPE=C6T    CHARGE=0.155    EXCLUSION=( N3 )   END
        ATOM H6     TYPE=H      CHARGE=0.035    END
      GROUp
        ATOM C2     TYPE=C2T    CHARGE=0.35     EXCLUSION=( C5 )   END
        ATOM O2     TYPE=ON     CHARGE=-0.35    END
      GROUp
        ATOM N3     TYPE=N3T    CHARGE=-0.26    END
        ATOM H3     TYPE=HN     CHARGE=0.26     END
      GROUp
        ATOM C4     TYPE=C4T    CHARGE=0.30     END
        ATOM O4     TYPE=ON     CHARGE=-0.30    END
      GROUp
        ATOM C5     TYPE=C5T    CHARGE=-0.035   END
        ATOM C7     TYPE=CC3E   CHARGE=-0.070   END ! name per IUPAC-IUB recomm.
        ATOM H71    TYPE=H      CHARGE=0.035    END ! name per IUPAC-IUB recomm.
        ATOM H72    TYPE=H      CHARGE=0.035    END ! name per IUPAC-IUB recomm.
        ATOM H73    TYPE=H      CHARGE=0.035    END ! name per IUPAC-IUB recomm.

      GROUp

    ! END
```

Script code

```
GROUP
  ATOM  C2'   TYPE=C2R   CHARGE=0.115    END
  ATOM  H2'   TYPE=H     CHARGE=0.035    END
  ATOM  O2'   TYPE=O2R   CHARGE=-0.40    END
  ATOM  HO2'  TYPE=HO    CHARGE=0.25     END
GROUP
  ATOM  C3'   TYPE=C3R   CHARGE=-0.035   END
  ATOM  H3'   TYPE=H     CHARGE=0.035    END

GROUP
  ATOM  O3'   TYPE=O3R   CHARGE=-0.36    END


  BOND  P    O1P           BOND  P    O2P           BOND  P    O5'
  BOND  O5'  C5'           BOND  C5'  C4'           BOND  C4'  O4'
  BOND  C4'  C3'           BOND  O4'  C1'           BOND  C1'  N1
  BOND  C1'  C2'           BOND  N1   C2            BOND  N1   C6
  BOND  C2   O2            BOND  C2   N3            BOND  N3   H3
  BOND  N3   C4            BOND  C4   O4            BOND  C4   C5
  BOND  C5   C7            BOND  C5   C6            BOND  C2'  C3'
  BOND  C3'  O3'           BOND  C2'  O2'
  BOND  O2'  HO2'
  BOND  C5'  H5'           BOND  C5'  H5''
  BOND  C3'  H3'                 BOND  C2'  H2'          BOND  C1'  H1'
  BOND  C4'  H4'           BOND  C7   H71          BOND  C7   H72
  BOND  C7   H73           BOND  C6   H6
{
  DIHEdral  P    O5'  C5'  C4'            DIHEdral  O5'  C5'  C4'  O4'
  DIHEdral  O5'  C5'  C4'  C3'
}{
  DIHEdral  C3'  C4'  O4'  C1'
  DIHEdral  C4'  O4'  C1'  C2'            DIHEdral  O4'  C1'  C2'  C3'

  DIHEdral  C1'  C2'  C3'  C4'            DIHEdral  O4'  C4'  C3'  O3'
  DIHEdral  C5'  C4'  C3'  C2'            DIHEdral  O2'  C2'  C3'  O3'
  DIHEdral  O4'  C1'  N1   C2
  DIHEdral  C3'  C2'  O2'  H2'


  ! New dihedrals
  DIHEdral  C5'  C4'  C3'  O3'            DIHEdral  C4'  O4'  C1'  N1
}
  IMPRoper  O4   N3   C5   C4             IMPRoper  C1'  C2   C6   N1
  IMPRoper  O2   N1   N3   C2             IMPRoper  C4   C5   C6   N1
  IMPRoper  N1   C2   N3   C4             IMPRoper  C2   N3   C4   C5
                                          IMPRoper  N3   C4   C5   C6
  IMPRoper  C5   C6   N1   C2             IMPRoper  C6   N1   C2   N3
  IMPRoper  H3   C2   C4   N3
  IMPRoper  C7   C4   C6   C5             IMPRoper  H6   N1   C5   C6

  !RIBOSE IMPROPERS
  IMPRoper       H1'  C2'  O4'  N1  !C1'
  IMPRoper       H2'  C3'  C1'  O2' !C2'
```

1 Force field parameter and topology files

```
IMPRoper      H3'   C4'   C2'   O3'   !C3'
IMPRoper      H4'   C5'   C3'   O4'   !C4'
IMPRoper      H5'   O5'   H5''  C4'   !C5'


END {THY}
```

```
RESIdue URI
GROUp
    ATOM P       TYPE=P      CHARGE=1.20      END
    ATOM O1P     TYPE=O1P    CHARGE=-0.40     END
    ATOM O2P     TYPE=O2P    CHARGE=-0.40     END
    ATOM O5'     TYPE=O5R    CHARGE=-0.36     END
GROUp
    ATOM C5'     TYPE=C5R    CHARGE=-0.070    END
    ATOM H5'     TYPE=H      CHARGE=0.035     END
    ATOM H5''    TYPE=H      CHARGE=0.035     END
GROUp
    ATOM C4'     TYPE=C4R    CHARGE=0.065     END
    ATOM H4'     TYPE=H      CHARGE=0.035     END
    ATOM O4'     TYPE=O4R    CHARGE=-0.30     END
    ATOM C1'     TYPE=C1R    CHARGE=0.165     END
    ATOM H1'     TYPE=H      CHARGE=0.035     END


GROUp
    ATOM N1      TYPE=N1U    CHARGE=-0.19     EXCLUSION=( C4 )   END
    ATOM C6      TYPE=C6U    CHARGE=0.155     EXCLUSION=( N3 )   END
    ATOM H6      TYPE=H      CHARGE=0.035     END
GROUp
    ATOM C2      TYPE=C2U    CHARGE=0.30      EXCLUSION=( C5 )   END
    ATOM O2      TYPE=ON     CHARGE=-0.30     END
GROUp
    ATOM N3      TYPE=N3U    CHARGE=-0.28     END
    ATOM H3      TYPE=HN     CHARGE=0.26      END
GROUp
    ATOM C4      TYPE=C4U    CHARGE=0.28      END
    ATOM O4      TYPE=ON     CHARGE=-0.30     END
GROUp
    ATOM C5      TYPE=C5U    CHARGE=-0.035    END !JPR
    ATOM H5      TYPE=H      CHARGE=0.035     END !JPR

GROUP
    ATOM C2'     TYPE=C2R    CHARGE=0.115     END
    ATOM H2'     TYPE=H      CHARGE=0.035     END !
    ATOM O2'     TYPE=O2R    CHARGE=-0.40     END
    ATOM HO2'    TYPE=HO     CHARGE=0.25      END
GROUP
    ATOM C3'     TYPE=C3R    CHARGE=-0.035    END
    ATOM H3'     TYPE=H      CHARGE=0.035     END
GROUP
    ATOM O3'     TYPE=O3R    CHARGE=-0.36     END
```

Script code

```
BOND  P    O1P              BOND  P    O2P              BOND  P    O5'
BOND  O5'  C5'              BOND  C5'  C4'              BOND  C4'  O4'
BOND  C4'  C3'              BOND  O4'  C1'              BOND  C1'  N1
BOND  C1'  C2'              BOND  N1   C2               BOND  N1   C6
BOND  C2   O2               BOND  C2   N3               BOND  N3   H3
BOND  N3   C4               BOND  C4   O4               BOND  C4   C5

BOND  C5   C6               BOND  C2'  C3'              BOND  C3'  O3'
BOND  C2'  O2'
BOND  C5   H5               BOND  C6   H6

BOND  O2'  HO2'
BOND  C5'  H5'        BOND  C5'  H5''          BOND  C4'  H4'
BOND  C3'  H3'        BOND  C2'  H2'           BOND  C1'  H1'
{
  DIHEdral  P    O5'  C5'  C4'              DIHEdral  O5'  C5'  C4'  O4'
  DIHEdral  O5'  C5'  C4'  C3'
}{
  DIHEdral  C3'  C4'  O4'  C1'
  DIHEdral  C4'  O4'  C1'  C2'              DIHEdral  O4'  C1'  C2'  C3'

  DIHEdral  C1'  C2'  C3'  C4'              DIHEdral  O4'  C4'  C3'  O3'
  DIHEdral  C5'  C4'  C3'  C2'              DIHEdral  O2'  C2'  C3'  O3'
  DIHEdral  O4'  C1'  N1   C2
  DIHEdral  C3'  C2'  O2'  H2'

  DIHEdral  P    O3'  C3'  C2'              DIHEdral  P    O3'  C3'  C4'
  ! New dihedrals
  DIHEdral  C5'  C4'  C3'  O3'              DIHEdral  C4'  O4'  C1'  N1
}
                                            IMPRoper  C1'  C2   C6   N1
IMPRoper  O2   N1   N3   C2                 IMPRoper  H3   C2   C4   N3
IMPRoper  O4   N3   C5   C4                 IMPRoper  N1   C2   N3   C4
IMPRoper  C2   N3   C4   C5                 IMPRoper  N3   C4   C5   C6
IMPRoper  C4   C5   C6   N1                 IMPRoper  C5   C6   N1   C2
IMPRoper  C6   N1   C2   N3
IMPRoper  H5   C4   C6   C5                 IMPRoper  H6   N1   C5   C6

!GENERAL RIBOSE IMPROPERS
IMPRoper       H1'  C2'  O4'  N1   !C1'
IMPRoper       H2'  C3'  C1'  O2'  !C2'
IMPRoper       H3'  C4'  C2'  O3'  !C3'
IMPRoper       H4'  C5'  C3'  O4'  !C4'
IMPRoper       H5'  O5'  H5'' C4'  !C5'




END {URI}
```

```
PREsidue DEOX   ! Patch to make DEOXYribose of the ribose
    DELETE ATOM O2'   END
```

1 Force field parameter and topology files

```
   DELETE ATOM HO2'     END
GROUP
   MODIFY   ATOM C2'    TYPE=C2D     CHARGE=-0.07      END
   MODIFY   ATOM C5'    TYPE=C5D     CHARGE=-0.07      END
   MODIFY   ATOM C4'    TYPE=C4D     CHARGE=0.065      END
   MODIFY   ATOM O4'    TYPE=O4D     CHARGE=-0.30      END
   MODIFY   ATOM C1'    TYPE=C1D     CHARGE=0.165      END
   MODIFY   ATOM C3'    TYPE=C3D     CHARGE=-0.035     END
   ADD      ATOM H2''   TYPE=H       CHARGE=0.035      END

   ADD BOND      C2'    H2''
   ADD ANGLE     C1'    C2'    H2'
   ADD ANGLE     C3'    C2'    H2''
   ADD ANGLE     H2'    C2'    H2''
   ADD IMPRoper  H2'    C3'    H2''   C1'! C2' chirality term
END {DEOX}

!------------------------------------------------------------


PRESidue 3TER            ! 3-terminus (without phosphate)
                         ! should be used as 'LAST 3TER HEAD - * END'
   GROUp                 ! i.e. to be patched to the last RNA residue
   MODIFY ATOM -C3'     TYPE=C3R     CHARGE=0.15    END
   MODIFY ATOM -O3'     TYPE=OH      CHARGE=-0.40   END
   ADD ATOM -H3T  TYPE=HO     CHARGE=0.25           END
   !
   ADD BOND -O3'  -H3T
   ADD ANGLe -C3'  -O3'   -H3T
   ! ADD DIHEdral -C4'  -C3'  -O3'  -H3T
END {3TER}



!------------------------------------------------------------

PRESidue 5TER            ! 5-terminus (without phosphate)
   !                     ! should be used as 'FIRST 5TER TAIL + * END'
   GROUp                 ! i.e. to be patched to the first RNA residue
   ADD ATOM +H5T  TYPE=HO     CHARGE=0.25     END
   MODIFY ATOM +O5'   TYPE=OH   CHARGE=-0.40  END
   MODIFY ATOM +C5'   TYPE=C5R  CHARGE=0.15   END
   DELETE ATOM +P    END
   DELETE ATOM +O1P  END
   DELETE ATOM +O2P  END
   !
   ADD BOND +H5T   +O5'
   ADD ANGLe +H5T   +O5'   +C5'
   ! ADD DIHEdral +H5T  +O5'  +C5'  +C4'
END {5TER}

!------------------------------------------------------------

PRESidue NUC           ! patch for nucleic acid backbone
                       ! should be used as 'LINK NUC HEAD - * TAIL + * END'
                       ! i.e. it links the previous RNA residue (-) with
                       ! the current one (+)
```

Script code

```
GROUp

  MODIFY ATOM -O3' END    !
  MODIFY ATOM +P   END    !
  MODIFY ATOM +O1P END    ! this should correctly define the electrostatic

  MODIFY ATOM +O2P END    ! group boundary

  MODIFY ATOM +O5' END    !
  ADD BOND  -O3' +P
  ADD ANGLE -C3' -O3' +P
  ADD ANGLE -O3' +P  +O1P
  ADD ANGLE -O3' +P  +O2P
  ADD ANGLE -O3' +P  +O5'
 !ADD DIHEdral -O3' +P  +O5'  +C5'

 ! ADD DIHEdral -C4' -C3' -O3'  +P
 ! ADD DIHEdral -C3' -O3' +P   +O5'

END {NUC}

!----------------------------------------------------------------

! mod by anda----------------------------------------------------
PRESidue 2AP

GROUP
  DELETE ATOM H2 END
  ADD ATOM N2   TYPE=N2G   CHARGE=-0.91   END
  ADD ATOM H21  TYPE=H2    CHARGE=0.39    END
  ADD ATOM H22  TYPE=H2    CHARGE=0.38    END

  MODIFY ATOM N1 CHARGE=-0.72 END
  MODIFY ATOM C2 CHARGE= 0.98 END
  MODIFY ATOM N3 CHARGE=-0.80 END
  MODIFY ATOM C4 CHARGE= 0.60 END
  MODIFY ATOM C5 CHARGE= 0.08 END
  MODIFY ATOM C6 CHARGE= 0.20 END
  MODIFY ATOM N7 CHARGE=-0.45 END !changed from -0.61 to -0.45 for neutrality
  MODIFY ATOM C8 CHARGE= 0.27 END
  MODIFY ATOM N9 CHARGE=-0.25 END
!   MODIFY ATOM N2 CHARGE=-0.91 END
!   MODIFY ATOM H21 CHARGE= 0.39 END
!   MODIFY ATOM H22 CHARGE= 0.38 END
  MODIFY ATOM H6 CHARGE= 0.12 END
  MODIFY ATOM H8 CHARGE= 0.11 END

  ADD BOND C2   N2
  ADD BOND N2   H21
  ADD BOND N2   H22

  ADD ANGLE N1 C2 N2
  ADD ANGLE C2 N2 H21
  ADD ANGLE C2 N2 H22
  ADD ANGLE H21 N2 H22
  ADD ANGLE N3 C2 N2
```

1 Force field parameter and topology files

```
ADD IMPRoper N3    C2    N2    H21
ADD IMPRoper N2    N3    N1    C2
ADD IMPRoper H22   H21   C2    N2
END {2AP}
! mod by anda――――――――――――――――――――――――

set echo=true end
```

Script code

2 Xplor-NIH calculation input files

Since all calculations were carried out under the same conditions, only the input files for the 13mer2AP calculation is shown. The input scripts are based on the example files of the XPLOR-NIH package (refine_full.py and sa.inp) but were substantially modified in the course of this work. The first script was used to calculate the starting structure fir the second script.

```
    language
remarks file   nmr/sa.inp
remarks  Simulated annealing protocol for NMR structure determination.
remarks  The starting structure for this protocol can be any structure with
remarks  a reasonable geometry, such as randomly assigned torsion angles or
remarks  extended strands.
remarks  Author: Michael Nilges

{====>}
evaluate ($init_t = 3000)           {*Initial simulated annealing temperature.*}
{====>}
evaluate ($high_steps= 48000 )          {*Total number of steps at high temp.*}
{====>}
evaluate ($cool_steps = 6000 )      {*Total number of steps during cooling.*}

parameter                                       {*Read the parameter file.*}
{====>}
   @13mer_2AP_c_dna.param
end

{====>}
structure @13mer_2AP_c_dna.psf end              {*Read the structure file.*}
{====>}
coordinates @13mer_2AP_c_dna.pdb           {*Read the coordinates.*}


noe
{====>}
   nres=3000        {*Estimate greater than the actual number of NOEs.*}
   class all
{====>}
   @13mer2AP_xplor_all_3.tbl                {*Read NOE distance ranges.*}
   @hbond_13mer_2AP.tbl
end

{====>}
restraints dihedral
   nass = 1000
   @dihedral_13mer_2AP_BDNA.tbl                {*Read dihedral angle
   restraints.*}
end
@plane_13mer_2AP.inp

{* Reduce the scaling factor on the force applied to disulfide           *}
```

2 Xplor-NIH calculation input files

```
{* bonds and angles from 1000.0 to 100.0 in order to reduce computation instability. *}

parameter
      bonds ( name SG ) ( name SG ) 100. TOKEN
      angle ( name CB ) ( name SG ) ( name SG ) 50. TOKEN
end


flags exclude * include bonds angle impr vdw elec noe cdih plan end

                           {*Friction coefficient for MD heatbath, in 1/ps.  *}
vector do (fbeta=10) (all)
                           {*Uniform heavy masses to speed molecular dynamics.*}
vector do (mass=100) (all)

noe                        {*Parameters for NOE effective energy term.*}
  ceiling=1000
  averaging   * cent
  potential   * soft
  scale       * 50.
  sqoffset    * 0.0
  sqconstant  * 1.0
  sqexponent  * 2
  soexponent  * 1
  asymptote   * 0.1             {*Initial value—modified later.*}
  rswitch     * 0.5
end

parameter                  {*Parameters for the repulsive energy term.*}
    nbonds
      repel=1.                  {*Initial value for repel—modified later.*}
      rexp=2 irexp=2 rcon=1.
      nbxmod=3
      wmin=0.01
      cutnb=4.5 ctonnb=2.99 ctofnb=3.
      tolerance=0.5
    end
end

restraints dihedral
      scale=5.
end

{====>}
evaluate ($end_count=100)        {*Loop through a family of 100 structures.*}

coor copy end

evaluate ($count = 0)
evaluate ($count2 = 0)
while ($count < $end_count ) loop main

   evaluate ($count=$count+1)
   evaluate ($count2=$count2+1)
```

Script code

```
coor swap end
coor copy end

{* ========================================= Initial minimization.*}
restraints dihedral   scale=5.    end
noe asymptote * 0.1  end
parameter  nbonds repel=1.    end end
constraints interaction
        (all) (all) weights * 1  vdw 0.002 end end
minimize powell nstep=50 drop=10.  nprint=25 end


{* ========================================= High-temperature dynamics.*}
constraints interaction (all) (all)
           weights * 1  angl 0.4  impr 0.1 vdw 0.002 end end

evaluate ($nstep1=int($high_steps * 2. / 3. ) )
evaluate ($nstep2=int($high_steps * 1. / 3. ) )

dynamics  verlet
    nstep=$nstep1   timestep=0.003   iasvel=maxwell   firstt=$init_t
    tcoupling=true  tbath=$init_t  nprint=50  iprfrq=0
end


{* =========== Tilt the asymptote and increase weights on geometry.*}
noe asymptote * 1.0  end

constraints interaction
         (all) (all) weights * 1  vdw 0.002   end end

{* Bring scaling factor for S-S bonds back *}

parameter
   bonds ( name SG ) ( name SG ) 1000. TOKEN
   angle ( name CB ) ( name SG ) ( name SG ) 500. TOKEN
end

dynamics  verlet
    nstep=$nstep2   timestep=0.001   iasvel=current   tcoupling=true
    tbath=$init_t  nprint=50  iprfrq=0
end

{* ========================================= Cool the system.*}

restraints dihedral   scale=200.   end

evaluate ($final_t =  25)       { K }
evaluate ($tempstep = 25)       { K }

evaluate ($ncycle = ($init_t-$final_t)/$tempstep)
evaluate ($nstep  = int($cool_steps/$ncycle))

evaluate ($ini_rad = 0.9)         evaluate ($fin_rad = 0.75)
evaluate ($ini_con= 0.003)        evaluate ($fin_con= 4.0)
```

2 Xplor-NIH calculation input files

```
evaluate ($bath   = $init_t)
evaluate ($k_vdw = $ini_con)
evaluate ($k_vdwfact = ($fin_con/$ini_con)^(1/$ncycle))
evaluate ($radius=      $ini_rad)
evaluate ($radfact = ($fin_rad/$ini_rad)^(1/$ncycle))

evaluate ($i_cool = 0)
while ($i_cool < $ncycle) loop cool
    evaluate ($i_cool=$i_cool+1)

    evaluate ($bath   = $bath  - $tempstep)
    evaluate ($k_vdw=min($fin_con,$k_vdw*$k_vdwfact))
    evaluate ($radius=max($fin_rad,$radius*$radfact))

    parameter  nbonds repel=$radius   end end
    constraints interaction (all) (all)
                weights * 1. vdw $k_vdw end end

    dynamics  verlet
       nstep=$nstep time=0.001 iasvel=current firstt=$bath
       tcoup=true tbath=$bath nprint=$nstep iprfrq=0
    end

{====>}                                        {*Abort condition.*}
    evaluate ($critical=$temp/$bath)
    if ($critical > 10. ) then
       display ****&&&& rerun job with smaller timestep (i.e., 0.003)
           stop
    end if

end loop cool
{* ================================================ Final minimization.*}

constraints interaction (all) (all) weights * 1. vdw 1. end end

parameter               {*Parameters for the repulsive energy term.*}
   nbonds
    repel=0.            {*Initial value for repel—modified later.*}
    SWITCH
    VSWITCH
    RDIE
    cutnb=11.5
    nbxmod=5
    wmin=0.01
    ctofnb=10.5
    ctonnb=9.5
    tolerance=0.5
   end
end
flags exclude * include bonds angle impr vdw elec noe cdih plan end

minimize powell nstep=3000 drop=10.0 nprint=25 end

{* ================================================ Write out the final structure(s).*}
print threshold=0.5 noe
```

Script code

```
evaluate ($rms_noe=$result)
evaluate ($violations_noe=$violations)
print threshold=0.0 noe
evaluate ($rms_noe22=$result)
print threshold=5. cdih
evaluate ($rms_cdih=$result)
evaluate ($violations_cdih=$violations)
print thres=0.05 bonds
evaluate ($rms_bonds=$result)
print thres=5. angles
evaluate ($rms_angles=$result)
print thres=5. impropers
evaluate ($rms_impropers=$result)
remarks ===============================================================
remarks              overall,bonds,angles,improper,vdw,noe,cdih,elec
remarks energies: $ener, $bond, $angl, $impr, $vdw, $noe, $cdih, $elec
remarks ===============================================================
remarks              bonds,angles,impropers,noe,cdih
remarks rms-d: $rms_bonds,$rms_angles,$rms_impropers,$rms_noe,$rms_cdih
remarks ===============================================================
remarks              noe,  cdih
remarks violations.: $violations_noe, $violations_cdih
remarks ===============================================================
remarks enviol: $ener $$violations_noe $violations_cdih
remarks ===============================================================


{===>}                       {*Name(s) of the family of final structures.*}
evaluate ($filename='z13mer_2AP_c_dna_test'+encode($count)+'.pdb')

write coordinates output =$filename end

evaluate ($filename2='z13mer_2AP_c_dna_test'+encode($count2)+'.noe')
set display=$filename2 end
@@picktbl_13mer2AP_all
close $filename2 end
set display=OUTPUT end

end loop main

stop




seed                    = 10
numberOfStructures      = 100
startStructure          = 1



# User-specific which has to be adjusted for each new sample
andaSampleName    = '13mer_2AP'                              # specify sample name
andaNumResAll     = 26                                       # Total number of
     Residues
andaDiheIdeal     = 'B'                                      # specify which
     dihedral constraint set should be used
```

2 Xplor-NIH calculation input files

```
andaDiheBound   = 30.0                                          # specify upper and
    lower bound for dihedral constraints
andaDiheBound2  = 30.0                                          # specify loose upper
    and lower bound for dihedral constraints
andaSeqStrand1  = 'GUA CYT THY GUA CYT ADE ADE ADE CYT GUA THY CYT GUA'  # Sequence of Strand I
andaSeqStrand2  = 'CYT GUA ADE CYT GUA THY THY GUA THY ADE GUA CYT'      # Sequence of Strand II
outFilename     = andaSampleName+'_STRUCTURE.pdb'               # pdb output filename
andaNOEexp      = 'NOE_13mer2AP_xplor_ohneH5_clean.tbl'         # file for reading in
    experimental NOE constraints
andaDipoInp     = 'RDC_13mer2AP_safe.inp.xplor'                 # file for reading in
    experimental RDC constraints
andaDipoInpMe   = 'RDC_13mer2AP_safe_Me.inp.xplor'              # file for reading
    in experimental Me-RDC constraints

# User-specific data, which can be adjusted for external script,
# but which can also be generated from the information given above
andaNumRes      = andaNumResAll/2
andaInitCoord   = 'start_'+andaSampleName+'.pdb'                # file created with initial
    extended structure coordinates
andaInitPSF     = 'start_'+andaSampleName+'.psf'                # file created with initial
    extended structure
andaPlan        = 'plane_13mer_2AP.inp'                         # file for reading in planar
    constraints
andaOrie        = 'dna_positional_anda.setup'                   # file for reading in ORIE
    constraints
andaHbond       = 'hbond_13mer_2AP.tbl'                         # file for reading in Hbond
    constraints
andaDihe        = None                                          # file for reading in ideal
    dihedral constraints
andaRAMA        = 'nucleic'                                     # file for reading in ideal
    RAMA constraints


# Store each residue in list andaRes!
# (Index+1) corresponds to residue number
andaRes = range(andaNumResAll)
andai   = 0
andaii  = 3
for andabla in range(andaNumRes):
    andaRes[andabla]=andaSeqStrand1[andai:andaii]
    andaRes[andabla+andaNumRes]=andaSeqStrand2[andai:andaii]
    andai  = andai+4
    andaii = andaii+4


if andaDihe==None:
    # Generate dihedral input table with ideal values for A- or B-DNA
    # input values for dihedrals
    alphaA , alphaB    = -50.0,-46.0
    betaA , betaB      = 172.0, -147.0
    gammaA , gammaB    = 41.0 ,36.0
    deltaA , deltaB    = 79.0 ,157.0
    epsA , epsB        = -146.0,155.0
    zetaA , zetaB      = -78.0,-96.0
```

Script code

```python
fileHandle = open (andaSampleName+'_dihe.tbl','w')
#
# B-DNA conformation
#
if andaDiheIdeal=='B':
    # Alpha angle
    fileHandle.write ("! Ideal dihedral values for B-DNA\n! Values taken from Roberts: NMR of Macromolecules\n\n\n!Alpha dihedral\n\n")
    for andai in range(1,andaNumRes):
        fileHandle.write ('assign ( resid %s    and name O3'  )\n        ( resid %s   and name P    )\n        ( resid %s   and name O5'  )\n        ( resid %s   and name C5'  )    1.0 %s %s 2\n\n' %
                          (andai,andai+1,andai+1,andai+1,alphaB,andaDiheBound))
    for andai in range(14,andaNumRes+13):
        fileHandle.write ('assign ( resid %s    and name O3'  )\n        ( resid %s   and name P    )\n        ( resid %s   and name O5'  )\n        ( resid %s   and name C5'  )    1.0 %s %s 2\n\n' %
                          (andai,andai+1,andai+1,andai+1,alphaB,andaDiheBound))
    # Beta angle
    fileHandle.write ('\n!Beta dihedral\n\n')
    for andai in range(2,andaNumRes+1):
        fileHandle.write ('assign ( resid %s    and name P    )\n        ( resid %s   and name O5'  )\n        ( resid %s   and name C5'  )\n        ( resid %s   and name C4'  )    1.0 %s %s 2\n\n' %
                          (andai,andai,andai,andai,betaB,andaDiheBound))
    for andai in range(15,andaNumRes+14):
        fileHandle.write ('assign ( resid %s    and name P    )\n        ( resid %s   and name O5'  )\n        ( resid %s   and name C5'  )\n        ( resid %s   and name C4'  )    1.0 %s %s 2\n\n' %
                          (andai,andai,andai,andai,betaB,andaDiheBound))
    # Gamma angle
    fileHandle.write ('\n!Gamma dihedral\n\n')
    for andai in range(1,andaNumRes+1):
        fileHandle.write ('assign ( resid %s    and name O5'  )\n        ( resid %s   and name C5'  )\n        ( resid %s   and name C4'  )\n        ( resid %s   and name C3'  )    1.0 %s %s 2\n\n' %
                          (andai,andai,andai,andai,gammaB,andaDiheBound))
    for andai in range(14,andaNumRes+14):
        fileHandle.write ('assign ( resid %s    and name O5'  )\n        ( resid %s   and name C5'  )\n        ( resid %s   and name C4'  )\n        ( resid %s   and name C3'  )    1.0 %s %s 2\n\n' %
                          (andai,andai,andai,andai,gammaB,andaDiheBound))
    # Delta angle
    fileHandle.write ('\n!Delta dihedral\n\n')
    for andai in range(1,andaNumRes+1):
        fileHandle.write ('assign ( resid %s    and name C5'  )\n        ( resid %s   and name C4'  )\n        ( resid %s   and name C3'  )\n        ( resid %s   and name O3'  )    1.0 %s %s 2\n\n' %
                          (andai,andai,andai,andai,deltaB,andaDiheBound))
    for andai in range(14,andaNumRes+14):
        fileHandle.write ('assign ( resid %s    and name C5'  )\n        ( resid %s   and name C4'  )\n        ( resid %s   and name C3'  )\n        ( resid %s   and name O3'  )    1.0 %s %s 2\n\n' %
                          (andai,andai,andai,andai,deltaB,andaDiheBound))
    # Epsilon angle, loose bound
    fileHandle.write ('\n!Epsilon dihedral\n\n')
```

2 Xplor-NIH calculation input files

```
        for andai in range(2,andaNumRes):
            fileHandle.write ("assign ( resid %s    and name C4'   )\n       ( resid %s    and
                name C3'   )\n       ( resid %s    and name O3'   )\n       ( resid %s    and
                name P     )    1.0 %s %s 2\n\n" %
                (andai,andai,andai,andai+1,epsB,andaDiheBound2))
        for andai in range(15,andaNumRes+13):
            fileHandle.write ("assign ( resid %s    and name C4'   )\n       ( resid %s    and
                name C3'   )\n       ( resid %s    and name O3'   )\n       ( resid %s    and
                name P     )    1.0 %s %s 2\n\n" %
                (andai,andai,andai,andai+1,epsB,andaDiheBound2))
    # Zeta angle, loose bound
        fileHandle.write ("\n!Zeta dihedral\n\n")
        for andai in range(2,andaNumRes):
            fileHandle.write ("assign ( resid %s    and name C3'   )\n       ( resid %s    and
                name O3'   )\n       ( resid %s    and name P     )\n       ( resid %s    and
                name O5'   )    1.0 %s %s 2\n\n" %
                (andai,andai,andai+1,andai+1,zetaB,andaDiheBound2))
        for andai in range(15,andaNumRes+13):
            fileHandle.write ("assign ( resid %s    and name C3'   )\n       ( resid %s    and
                name O3'   )\n       ( resid %s    and name P     )\n       ( resid %s    and
                name O5'   )    1.0 %s %s 2\n\n" %
                (andai,andai,andai+1,andai+1,zetaB,andaDiheBound2))
    #
    # A-BDNA conformation
    #
    elif andaDiheIdeal=="A":
    # Alpha angle
        fileHandle.write ("! Ideal dihedral values for A-DNA\nValues taken from Roberts: NMR of
            Macromolecules\n\n\n!Alpha dihedral\n\n")
        for andai in range(1,andaNumRes):
            fileHandle.write ("assign ( resid %s    and name O3'   )\n       ( resid %s    and
                name P     )\n       ( resid %s    and name O5'   )\n       ( resid %s    and
                name C5'   )    1.0 %s %s 2\n\n" %
                (andai,andai+1,andai+1,andai+1,alphaA,andaDiheBound))
        for andai in range(14,andaNumRes+13):
            fileHandle.write ("assign ( resid %s    and name O3'   )\n       ( resid %s    and
                name P     )\n       ( resid %s    and name O5'   )\n       ( resid %s    and
                name C5'   )    1.0 %s %s 2\n\n" %
                (andai,andai+1,andai+1,andai+1,alphaA,andaDiheBound))
    # Beta angle
        fileHandle.write ("\n!Beta dihedral\n\n")
        for andai in range(2,andaNumRes+1):
            fileHandle.write ("assign ( resid %s    and name P     )\n       ( resid %s    and
                name O5'   )\n       ( resid %s    and name C5'   )\n       ( resid %s    and
                name C4'   )    1.0 %s %s 2\n\n" %
                (andai,andai,andai,andai,betaA,andaDiheBound))
        for andai in range(15,andaNumRes+14):
            fileHandle.write ("assign ( resid %s    and name P     )\n       ( resid %s    and
                name O5'   )\n       ( resid %s    and name C5'   )\n       ( resid %s    and
                name C4'   )    1.0 %s %s 2\n\n" %
                (andai,andai,andai,andai,betaA,andaDiheBound))
    # Gamma angle
        fileHandle.write ("\n!Gamma dihedral\n\n")
        for andai in range(1,andaNumRes+1):
            fileHandle.write ("assign ( resid %s    and name O5'   )\n       ( resid %s    and
                name C5'   )\n       ( resid %s    and name C4'   )\n       ( resid %s    and
```

Script code

```
                        name C3' )    1.0 %s %s 2\n\n' %
                    (andai,andai,andai,andai,gammaA,andaDiheBound))
            for andai in range(14,andaNumRes+14):
                fileHandle.write ('assign ( resid %s    and name O5'  )\n    ( resid %s    and
                        name C5' )\n          ( resid %s    and name C4' )\n    ( resid %s    and
                        name C3' )    1.0 %s %s 2\n\n' %
                    (andai,andai,andai,andai,gammaA,andaDiheBound))
        # Delta angle
            fileHandle.write ('\n!Delta dihedral\n\n')
            for andai in range(1,andaNumRes+1):
                fileHandle.write ('assign ( resid %s    and name C5'  )\n    ( resid %s    and
                        name C4' )\n          ( resid %s    and name C3' )\n    ( resid %s    and
                        name O3' )    1.0 %s %s 2\n\n' %
                    (andai,andai,andai,andai,deltaA,andaDiheBound))
            for andai in range(14,andaNumRes+14):
                fileHandle.write ('assign ( resid %s    and name C5'  )\n    ( resid %s    and
                        name C4' )\n          ( resid %s    and name C3' )\n    ( resid %s    and
                        name O3' )    1.0 %s %s 2\n\n' %
                    (andai,andai,andai,andai,deltaA,andaDiheBound))
        # Epsilon angle, loose bound
            fileHandle.write ('\n!Epsilon dihedral\n\n')
            for andai in range(2,andaNumRes):
                fileHandle.write ('assign ( resid %s    and name C4'  )\n    ( resid %s    and
                        name C3' )\n          ( resid %s    and name O3' )\n    ( resid %s    and
                        name P  )    1.0 %s %s 2\n\n' %
                    (andai,andai,andai,andai+1,epsA,andaDiheBound2))
            for andai in range(15,andaNumRes+13):
                fileHandle.write ('assign ( resid %s    and name C4'  )\n    ( resid %s    and
                        name C3' )\n          ( resid %s    and name O3' )\n    ( resid %s    and
                        name P  )    1.0 %s %s 2\n\n' %
                    (andai,andai,andai,andai+1,epsA,andaDiheBound2))
        # Zeta angle, loose bound
            fileHandle.write ('\n!Zeta dihedral\n\n')
            for andai in range(2,andaNumRes):
                fileHandle.write ('assign ( resid %s    and name C3'  )\n    ( resid %s    and
                        name O3' )\n          ( resid %s    and name P  )\n    ( resid %s    and
                        name O5' )    1.0 %s %s 2\n\n' %
                    (andai,andai,andai+1,andai+1,zetaA,andaDiheBound2))
            for andai in range(15,andaNumRes+13):
                fileHandle.write ('assign ( resid %s    and name C3'  )\n    ( resid %s    and
                        name O3' )\n          ( resid %s    and name P  )\n    ( resid %s    and
                        name O5' )    1.0 %s %s 2\n\n' %
                    (andai,andai,andai+1,andai+1,zetaA,andaDiheBound2))
        fileHandle.close()
        andaDihe=andaSampleName+'_dihe.tbl'


############################################ end of anda mod
############################################################

xplor.parseArguments()                  # check for typos on the command-line

simWorld.setRandomSeed(seed)

#
# Create the PSF and initial PDB files as an extended structure
```

2 Xplor-NIH calculation input files

```
# mod by anda
#
import protocol
protocol.initParams('anda_old_nucleic')
protocol.initTopology('anda_old_nucleic')
#import psfGen
#from psfGen import pdbToPSF
#pdbToPSF(andaInitCoord,psfFilename=andaInitPSF,andaPar='anda_old_nucleic',customRename=0)
#from psfGen import seqToPSF
## generates PSF-information from Sequence
#seqToPSF(andaSeqStrand1,seqType='dna',customRename=1)
#pass
#seqToPSF(andaSeqStrand2,startResid=andaNumRes+1,seqType='dna',customRename=1)
#pass
## writes out PSF info to file
#xplor.command('write psf output=%s end' % andaInitPSF)
## generates an extended structure
#protocol.genExtendedStructure(andaInitCoord,numerator=10,verbose=1,maxFixupIters=2000)
protocol.initStruct(andaInitPSF)


#
# starting coords
#
protocol.initCoords(andaInitCoord)

# protocol.fixupCovalentGeom(verbose=1)
# list of potential terms used in refinement
from potList import PotList
potList = PotList()
crossTerms=PotList('cross terms') # can add some pot terms which are not
                                   # refined against- but included in analysis


# parameters to ramp up during the simulated annealing protocol
#
from simulationTools import MultRamp, StaticRamp, InitialParams
rampedParams=[]
highTempParams=[]

from varTensorTools import create_VarTensor, calcTensor
media={}
for medium in ['pf1']:
    media[medium] = create_VarTensor(medium)
    pass


from xplorPot import XplorPot

#planarity restraints
xplor.command('@%s' % andaPlan)
potList.append(XplorPot("plan",xplor.simulation))



#initialize the aa-aa positional database
```

Script code

```
#xplor.command('@%s' % andaOrie)
#potList.append(XplorPot('orie'))

#NOE potentials
from noePotTools import create_NOEPot
noePots = PotList("noe")
noe = create_NOEPot('noeAll',andaNOEexp)
noe.setPotType('hard')
noePots.append(noe)

# need to be satisfied by all structures
noeHB  = create_NOEPot("noeNH",andaHbond)
noeHB.setPotType('hard')
noeHB.setScale(1000)
noePots.append(noeHB)
potList.append(noePots)
rampedParams.append( StaticRamp("noePots.setScale( 50 )") )

protocol.initDihedrals(andaDihe)
potList.append(XplorPot("CDIH"))
highTempParams.append( StaticRamp("potList['CDIH'].setScale(200)") )
rampedParams.append( StaticRamp("potList['CDIH'].setScale(200)") )
#rampedParams.append( MultRamp(10,200,"potList['CDIH'].setScale(VALUE)") )

#protocol.initRamaDatabase(andaRAMA)
#potList.append(XplorPot('rama'))
#highTempParams.append( StaticRamp("potList['RAMA'].setScale(0)") )
#rampedParams.append( MultRamp(1,1.0,"potList['RAMA'].setScale(VALUE)") )

from rdcPotTools import Da_prefactor, create_RDCPot, scale_toCH

#from csaPotTools import create_CSAPot
#csaPots = PotList('csa')
#for (name,medium,force,tbl) in [('POP',  'phg3',1,"justin_csa.tbl"),
#                                ('POP2','bic2',1,"justin_csa_bcl.tbl")]:
#    term = create_CSAPot(name,oTensor=media[medium],file=tbl)
#    term.setDaScale( -term.DaScale() ) #switch sign
#    term.setScale(force)
#    csaPots.append( term )
#    pass
#
#potList.append(csaPots)
#rampedParams.append( MultRamp(0.01,0.2,"csaPots.setScale( VALUE )") )
#
#
## Rh same for jch2 and phos/pho2, but da is different
## let's add this later with a cosRatio2 potential term
#
rdcPots = PotList('rdcs')
# weight is the relative weighting of expts, as determined by expt. error
for (name,medium,weight,files) in [
    ('JCH' ,'pf1',5   ,andaDipoInp),('methyl' ,'pf1',0.5 ,andaDipoInpMe)
    ]:
    term = create_RDCPot(name,oTensor=media[medium],defThreshold=1.9)
    if type(files)==type('string'):
        files=(files,)
```

```
        pass
    for file in files:
        term.addRestraints( open(file).read() )
        pass
    term.setShowAllRestraints(1)
    term.setScale(weight)
    #term.setAveType('average')
    term.setAveType('sum')
    print name
    scale_toCH(term) #also sets useDistance
    print term.info()
    print term.gyroA()
    rdcPots.append(term)
    pass


## proton setup
#for key in ('HABS', 'HAB2'): rdcPots[key].setUseSign(0)
#
#for key in ('HABS', 'HSIG', 'HAB2', 'HSI2'):
#    rdcPots[key].setPotType('square');
#    rdcPots[key].setAveType('sum')
#    pass

potList.append(rdcPots)
rampedParams.append( MultRamp(0.01,1,"rdcPots.setScale( VALUE )") )


from rdcPotTools import Da_prefactor

print 'factor:', Da_prefactor['CH'] / Da_prefactor['NH']
for medium in media.values():
    calcTensor(medium)
    print 'medium: ', medium.instanceName(), \
          'Da: ',medium.Da(), 'Rh: ',medium.Rh()
    pass

#let's try fixing Da, Rh:
print medium
for (medium,Da,Rh) in (('pf1',-41.02,0.321),):
    medium = media[medium]
    medium.setDa(Da)
    medium.setRh(Rh)
    pass


## set up J coupling
#from jCoupPotTools import create_JCoupPot
#jCoup = create_JCoupPot('jcoup','couplings.tbl',
#                         A=15.3,B=-6.1,C=1.6,phase=0 )
#jCoup.setThreshold(0)
#jCoup.setScale(10)
#potList.append(jCoup)}

#protocol.initNBond(cutnb=4.5)
potList.append( XplorPot('VDW') )
```

Script code

```
potList.append( XplorPot('elec') )
#rampedParams.append( MultRamp(0.99,0.78,
#                              'xplor.command('param nbonds repel VALUE end end')') )
#rampedParams.append( MultRamp(.0001,4,
#                              'xplor.command('param nbonds rcon VALUE end end')') )
rampedParams.append( StaticRamp('''xplor.command('''param nbonds
                                        atom
                                        repel=0
                                        wmin=0.01
                                        nbxmod=5
                                        cutnb=58.5
                                        ctonnb=56.5
                                        ctofnb=57.5
                                        tolerance=0.5
                                        rdie
                                        vswitch
                                        switch
                              end end''')''') )
#highTempParams.append( StaticRamp('''protocol.initNBond(cutnb=100,
#                                        tolerance=45,
#                                        #repel=1.2,
#                                        #selStr='name P'
#                                        )''') )

for name in ("bond","angl","impr"):
    potList.append( XplorPot(name) )
    pass
rampedParams.append( MultRamp(0.4,1.0,"potList['ANGL'].setScale(VALUE)"))
rampedParams.append( MultRamp(0.1,1.0,"potList['IMPR'].setScale(VALUE)"))


from ivm import IVM
import varTensorTools
mini = IVM()                #initial alignment of orientation tensor axes

for medium in (('pf1'),): media[medium].setFreedom('fixDa, fixRh')
#for medium in ('bic2',):
#    media[medium].setFreedom('fixDa, fixRh, fixAxisTo bic1')
#for medium in ('phg2','phg3',):
#    media[medium].setFreedom('fixDa, fixRh, fixAxisTo phg1')
varTensorTools.topologySetup(mini,media.values())

protocol.initMinimize(mini,
                      numSteps=20)
mini.fix('not resname ANI')
mini.run()              #this initial minimization is not strictly necessary

#uncomment to allow Da, Rh to vary
for medium in (('pf1'),): media[medium].setFreedom('varyDa, varyRh')
#for medium in ('bic2',):
#    media[medium].setFreedom('varyDa, varyRh, fixAxisTo bic1')
#for medium in ('phg2','phg3',):
#    media[medium].setFreedom('varyDa, fixAxisTo phg1, fixRhTo phg1')

dyn = IVM()
protocol.initDynamics(dyn,potList=potList)
```

2 Xplor-NIH calculation input files

```
varTensorTools.topologySetup(dyn,media.values())
protocol.torsionTopology(dyn)

#
#
# Give atoms uniform weights, except for the anisotropy axis
from atomAction import SetProperty
AtomSel('not resname ANI').apply( SetProperty('mass',100.) )
varTensorTools.massSetup(media.values(),300)
AtomSel('all            ').apply( SetProperty('fric',10.) )


##
## minc used for final cartesian minimization
##
from selectTools import IVM_groupRigidSidechain
minc = IVM()
protocol.initMinimize(minc,potList=potList)
IVM_groupRigidSidechain(minc)
protocol.cartesianTopology(minc,'not resname ANI')
varTensorTools.topologySetup(minc,media.values())

init_t1 = 200000
init_t2 = 20000
init_t3 = 3000

from simulationTools import AnnealIVM
anneal1= AnnealIVM(initTemp =init_t1,
                   finalTemp=init_t2,
                   tempStep =5000,
                   ivm=dyn,
                   rampedParams = rampedParams)
anneal2= AnnealIVM(initTemp =init_t2,
                   finalTemp=init_t3,
                   tempStep =500,
                   ivm=dyn,
                   rampedParams = rampedParams)
anneal3= AnnealIVM(initTemp =init_t3,
                   finalTemp=25,
                   tempStep =25,
                   ivm=dyn,
                   rampedParams = rampedParams)


# initialize parameters for initial minimization.
InitialParams( rampedParams )
# high-temp dynamics setup - only need to specify parameters which
#    differfrom initial values in rampedParams
InitialParams( highTempParams )

# initial minimization
protocol.initMinimize(dyn,
                      potList=[potList['CDIH'],potList['IMPR']],
                      numSteps=50)
dyn.run()
```

Script code

```
# initial minimization
protocol.initMinimize(dyn,
                     potList=potList,
                     numSteps=1000)
minc.run()

#from simulationTools import testGradient
#testGradient(potList,eachTerm=1)

def calcOneStructure(loopInfo):

## mod by anda: first annealing loop, to overcome high energy barriers
#    # initialize parameters for high temp dynamics.
#    InitialParams( rampedParams )
#    # high-temp dynamics setup - only need to specify parameters which
#    #   differfrom initial values in rampedParams
#    InitialParams( highTempParams )
#
#    protocol.initDynamics(dyn,
#                          initVelocities=1,
#                          bathTemp=init_t1,
#                          potList=potList,
#                          finalTime=10)
#    dyn.setETolerance( init_t1/100 )  #used to det. stepsize. default: t/1000
#    dyn.run()
#
#    # initialize parameters for cooling loop
#    InitialParams( rampedParams )
#
#    # perform simulated annealing
#    #
#    protocol.initDynamics(dyn,
#                          finalTime=0.1, #time to integrate at a given temp.
#                          numSteps=0,    # take as many steps as necessary
#                          #eTol_minimum=0.001 # cutoff for auto-TS det.
#                          )
#    anneal1.run()

# mod by anda: second annealing loop, actual annealing
    # initialize parameters for high temp dynamics.
    InitialParams( rampedParams )
    # high-temp dynamics setup - only need to specify parameters which
    #   differfrom initial values in rampedParams
    InitialParams( highTempParams )

    protocol.initDynamics(dyn,
                          initVelocities=1,
                          bathTemp=init_t2,
                          potList=potList,
                          finalTime=50)
    dyn.setETolerance( init_t2/100 )  #used to det. stepsize. default: t/1000
    dyn.run()

    # initialize parameters for cooling loop
    InitialParams( rampedParams )
```

```
# perform simulated annealing
#
protocol.initDynamics(dyn,
                      finalTime=0.5,   #time to integrate at a given temp.
                      numSteps=0,      # take as many steps as necessary
                      #eTol_minimum=0.001 # cutoff for auto-TS det.
                      )
anneal2.run()
anneal3.run()

#
# torsion angle minimization
#
protocol.initMinimize(dyn,numSteps=5000)
dyn.run()

##
##all atom minimization
##
protocol.initMinimize(minc,potList=potList,numSteps=3000)
minc.run()

#
# perform analysis and write structure
loopInfo.writeStructure(potList,crossTerms)
pass


from simulationTools import StructureLoop
StructureLoop(numStructures=numberOfStructures,
              startStructure=startStructure,
              structLoopAction=calcOneStructure,
              pdbTemplate=outFilename,
              genViolationStats=1,
              averageFilename='average_min.pdb',
              averageFitSel='not resname ANI and not (name H71 or name H72 or name H73)',
              averageRefineSteps=15,
              averageTopFraction=0.1,
              averagePotList=potList).run()
```

Script code

3 Xplor-NIH calculation restraints files and structures

The coordinates of the 10 minimum-energy, violation-free structures can be found at the Protein Databank (PDB) with accession codes 2kh0 (13merHNF, face-down), 2kh1 (13merHNF, face-up), 2kz1 (13merRef) and 2kz2 (13mer2AP). The input files to the structure calculations have been deposited along with the structures.

4 Lua scripts written for data export from CARA

The following scripts were written for use in CARA only. Their description is given in the header.

```lua
-- script to filter through peaklist with all graded peaks and compare it
-- to normal peaklist. All peaks that are not integrated are written to
-- a new peaklist for inspection.

-- written by Andre Dallmann April-23-2007
```


--   Script --

---------------- PREPARATIONS ----------------

```lua
-- initialize tables for peak information
t={}
t.label = {}
t.id = {}
t.assx = {}
t.assy = {}
t.posx = {}
t.posy = {}
t.ampl = {}
t.grade = {}
t.vol = {}
t.atomtype = {}
t.assxlabel = {}
t.assylabel = {}


-- get Project
local ProjectNames = {}
i = 0
```

4 Lua scripts written for data export from CARA

```
for a,b in pairs(cara:getProjects()) do
        i = i + 1
        ProjectNames[ i ] = b:getName()
end
t.ProjectName = dlg.getSymbol("Select Project","", unpack( ProjectNames ))
t.project = cara:getProject( t.ProjectName )

-- get Peaklist with all graded peaks
local PeaklistNames = {}
i = 0
for a,b in pairs(t.project:getPeakLists()) do
        i = i + 1
        PeaklistNames[ i ] = b:getName()
end
t.PeaklistName = dlg.getSymbol("Select Graded Peaklist","", unpack( PeaklistNames ))
for a,b in pairs(t.project:getPeakLists()) do
        i = i + 1
        if (b:getName()==t.PeaklistName) then
                t.peaklist = t.project:getPeakList( b:getId(a) )
        end
end

-- get Peaklist as Reference
local PeaklistNames = {}
i = 0
for a,b in pairs(t.project:getPeakLists()) do
        i = i + 1
        PeaklistNames[ i ] = b:getName()
end
t.PeaklistName = dlg.getSymbol("Select Reference Peaklist","", unpack( PeaklistNames ))
for a,b in pairs(t.project:getPeakLists()) do
        i = i + 1
        if (b:getName()==t.PeaklistName) then
                t.peaklistref = t.project:getPeakList( b:getId(a) )
        end
end

-- Get Output Filename
t.Filename = dlg.getText("Enter the output filename", "", t.ProjectName)

-- open outfile
outfile = io.output( t.Filename.."_diff.peaks" )

-- write header
outfile:write ("#Number of dimensions "..t.peaklist:getDimCount().."\n")
print ("#Number of dimensions "..t.peaklist:getDimCount())
for i = 1, t.peaklist:getDimCount() do
        outfile:write ("#INAME "..i.." "..t.peaklist:getAtomType(i).."\n")
        print ("#INAME "..i.." "..t.peaklist:getAtomType(i))
end

count = 0
```
―――――――――――――――――――― Main Body ――――――――――――――――――――

Script code

```
-- read out peak information
for i,j in pairs (t.peaklistref:getPeaks()) do --cycle through all peaks of refpeaklist
        boolean = 0
        for x,y in pairs (t.peaklist:getPeaks()) do --cycle through all peaks of graded
          peaklist
                if (j:getAssig() == y:getAssig()) then
                        boolean = 1
                end
        end
        if (boolean == 0) then
                t.peak = t.peaklistref:getPeak(i)
                count = count + 1
                t.id[i] = string.format ("%9.0f", t.peak:getId())
                t.ass = {t.peak:getAssig()}
                t.assx[i] = string.format ("%9.0f", t.ass[1])
                t.assy[i] = string.format ("%9.0f", t.ass[2])
                t.label[i] = string.format ("%25.25s", t.peak:getLabel())
                t.pos = {t.peak:getPos()}
                t.posx[i] = string.format ("%9.3f", t.pos[1])
                t.posy[i] = string.format ("%9.3f", t.pos[2])
                t.ampl[i] = string.format ("%7.0f", t.peak:getAmp())
                t.vol[i] = string.format ("%15.3f", t.peak:getVol())
                outfile:write (count..t.posx[i]..t.posy[i].." 0 U
                        "..t.vol[i]..t.ampl[i].."    0"..t.assx[i]..t.assy[i].." 0\n#
                        "..t.label[i].."\n")
                print (count..t.posx[i]..t.posy[i].." 0 U         "..t.vol[i]..t.ampl[i].." -
                        0"..t.assx[i]..t.assy[i].." 0\n# "..t.label[i])
        end
end -- of first for loop

------------------------ End of Main Body ------------------------

i = 0
t = nil

-- close outfiles
outfile:close()

------------------------ End of Script ------------------------

print ( "\ncheck_intpeaks_byanda is done." )
print ( "Have a nice day!" )


-- script to export cara peaklist to Sparky

-- written by Andre Dallmann April-23-2007
```

4 Lua scripts written for data export from CARA

---   Script   ---

--- PREPARATIONS ---

```lua
-- initialize tables for peak information
t={}
t.label = {}
t.id = {}
t.assx = {}
t.assy = {}
t.posx = {}
t.posy = {}
t.ampl = {}
t.grade = {}
t.vol = {}
t.atomtype = {}
t.assxlabel = {}
t.assylabel = {}

-- get Project
local ProjectNames = {}
i = 0
for a,b in pairs(cara:getProjects()) do
        i = i + 1
        ProjectNames[ i ] = b:getName()
end
t.ProjectName = dlg.getSymbol('Select Project','', unpack( ProjectNames ))
t.project = cara:getProject( t.ProjectName )

-- get Peaklist with all graded peaks
local PeaklistNames = {}
i = 0
for a,b in pairs(t.project:getPeakLists()) do
        i = i + 1
        PeaklistNames[ i ] = b:getName()
end
t.PeaklistName = dlg.getSymbol('Select Peaklist','', unpack( PeaklistNames ))
for a,b in pairs(t.project:getPeakLists()) do
        i = i + 1
        if (b:getName()==t.PeaklistName) then
                t.peaklist = t.project:getPeakList( b:getId(a) )
        end
end

-- Get Output Filename
t.Filename = dlg.getText('Enter the output filename', '', t.ProjectName)

-- open outfile
outfile = io.output( t.Filename..'_sparky.peaks' )

-- write header
```

Script code

```
outfile:write (string.format ('%25.25s','Assignment')..string.format
    ('%9.9s','w1')..string.format ('%9.9s','w1')..'\n\n')
print (string.format ('%25.25s','Assignment')..string.format ('%9.9s','w1')..string.format
    ('%9.9s','w1')..'\n')

count = 0
```

----------------------------------- Main Body -----------------------------------

```
-- read out peak information
for i,peak in pairs(t.peaklist:getPeaks()) do  --cycle through all peaks
        t.peak = t.peaklist:getPeak(i)
        t.label[i] = string.format ('%25.25s', t.peak:getLabel())
        t.ass = {t.peak:getAssig()}
        t.assx[i] = string.format ('%9.0f', t.ass[1])
        t.assy[i] = string.format ('%9.0f', t.ass[2])
        t.pos = {t.peak:getPos()}
        t.posx[i] = string.format ('%9.3f', t.pos[1])
        t.posy[i] = string.format ('%9.3f', t.pos[2])
        t.ampl[i] = string.format ('%7.0f', t.peak:getAmp())
        t.vol[i] = string.format ('%15.3f', t.peak:getVol())
        assx = t.project:getSpin(t.assx[i]):getLabel()
        assy = t.project:getSpin(t.assy[i]):getLabel()
        resxid = t.project:getSpin(t.assx[i]):getSystem():getId()
        resyid = t.project:getSpin(t.assy[i]):getSystem():getId()
        resx =
            t.project:getSpin(t.assx[i]):getSystem():getResidue():getType():getShort()
        resy =
            t.project:getSpin(t.assy[i]):getSystem():getResidue():getType():getShort()
        if (assx==assy) then
            outfile:write(resx..resxid..assx..' - '..assy..t.posx[i]..t.posy[i]..'\n')
            print (resx..resxid..assx..' - '..assy..t.posx[i]..t.posy[i])
        else
            outfile:write(resx..resxid..assx..' - '..resy..resyid..assy..t.posx[i]..t.posy[i]..'\n'
            print
                (resx..resxid..assx..' - '..resy..resyid..assy..t.posx[i]..t.posy[i])
        end
    end -- of second for loop
```

-------------------------------- End of Main Body --------------------------------

```
i = 0
t = nil

-- close outfiles
outfile:close()
```

---------------------------------- End of Script ----------------------------------

```
print ( "\ntestscript_byanda is done." )
print ( "Have a nice day!" )
```

4 Lua scripts written for data export from CARA

```
-- script to filter through peaklist which where exported from homoscope to
-- monoscope
-- produces two files, containing d2o and h2o peaks respectively

-- written by Andre Dallmann April-11-2007
```

---   Script

──────── PREPARATIONS ────────

```
-- initialize tables for peak information
t={}
t.label = {}
t.id = {}
t.assx = {}
t.assy = {}
t.posx = {}
t.posy = {}
t.ampl = {}
t.grade = {}
t.vol = {}
t.atomtype = {}
t.assxlabel = {}
t.assylabel = {}

-- Get Project
local ProjectNames = {}
i = 0
for a,b in pairs(cara:getProjects()) do
        i = i + 1
        ProjectNames[ i ] = b:getName()
end
t.ProjectName = dlg.getSymbol('Select Project','', unpack( ProjectNames ))
t.project = cara:getProject( t.ProjectName )

-- Get Peaklist
local PeaklistNames = {}
i = 0
for a,b in pairs(t.project:getPeakLists()) do
        i = i + 1
        PeaklistNames[ i ] = b:getName()
end
t.PeaklistName = dlg.getSymbol('Select Peaklist','', unpack( PeaklistNames ))
for a,b in pairs(t.project:getPeakLists()) do
        i = i + 1
        if (b:getName()==t.PeaklistName) then
                t.peaklist = t.project:getPeakList( b:getId(a) )
```

Script code

```
                end
        end

        -- Get Output Filename
        t.Filename = dlg.getText('Enter the output filename', '', t.ProjectName)

        -- open outfile
        outfile_h2o = io.output( t.Filename..'_h2o.peaks' )
        outfile_d2o = io.output( t.Filename..'_d2o.peaks' )


        ----------------------------------------------------------------
        ------------------------- Main Body -----------------------------
        ----------------------------------------------------------------

        -- read out peak information
        for i,peak in pairs (t.peaklist:getPeaks()) do --cycle through all peaks
                t.peak = t.peaklist:getPeak(i)
                t.id[i] = string.format ('%9.0f', t.peak:getId())
                t.ass = {t.peak:getAssig()}
                t.assx[i] = string.format ('%9.0f', t.ass[1])
                t.assy[i] = string.format ('%9.0f', t.ass[2])
                t.label[i] = string.format ('%25.25s', t.peak:getLabel())
                t.pos = {t.peak:getPos()}
                t.posx[i] = string.format ('%9.3f', t.pos[1])
                t.posy[i] = string.format ('%9.3f', t.pos[2])
                t.ampl[i] = string.format ('%7.0f', t.peak:getAmp())
                t.vol[i] = string.format ('%15.3f', t.peak:getVol())
        end -- of first for loop

        -- write header
        outfile_h2o:write ('#Number of dimensions '..t.peaklist:getDimCount()..'\n')
        outfile_d2o:write ('#Number of dimensions '..t.peaklist:getDimCount()..'\n')
        --print ('#Number of dimensions '..t.peaklist:getDimCount())
        for i = 1, t.peaklist:getDimCount() do
                outfile_h2o:write ('#INAME '..i..' '..t.peaklist:getAtomType(i)..'\n')
                outfile_d2o:write ('#INAME '..i..' '..t.peaklist:getAtomType(i)..'\n')
                --print ('#INAME '..i..' '..t.peaklist:getAtomType(i))
        end

        -- initialize counter variables
        c_h2o = 0
        c_d2o = 0

        -- write peaklists
        for x,assx in pairs (t.assx) do
                t.assxlabel[x] = t.project:getSpin(t.assx[x]):getLabel() -- get Peaklabel
                t.assylabel[x] = t.project:getSpin(t.assy[x]):getLabel() -- get Peaklabel
                if ((((t.assxlabel[x]=='H41') or (t.assxlabel[x]=='H42') or (t.assxlabel[x]=='H1') or
                        (t.assxlabel[x]=='H3')) or ((t.assylabel[x]=='H41') or (t.assylabel[x]=='H42') or
                        (t.assylabel[x]=='H1') or (t.assylabel[x]=='H3'))) and (t.assx[x]~=t.assy[x]))
                        then
                                c_h2o = c_h2o + 1
                                outfile_h2o:write (c_h2o..t.posx[x]..t.posy[x]..' 0 U
                                        '..t.vol[x]..t.ampl[x]..'   -    0'..t.assx[x]..t.assy[x]..' 0\n#
                                        '..t.label[x]..'\n')
                                --print (c_h2o..t.posx[x]..t.posy[x]..' 0 U       '..t.vol[x]..t.ampl[x]..'   -
```

4 Lua scripts written for data export from CARA

```
                        0"..t.assx[x]..t.assy[x].." 0\n# "..t.label[x].."    h2o")
        else
                        if (((t.assxlabel[x]=="H2") and  ((t.assylabel[x]=="H2'") or
                            (t.assylabel[x]=="H2''") or (t.assylabel[x]=="H3'") or
                            (t.assylabel[x]=="H4'") or (t.assylabel[x]=="H5'") or
                            (t.assylabel[x]=="H5''") or (t.assylabel[x]=="H8'"))) or
                            ((t.assylabel[x]=="H2") and  ((t.assxlabel[x]=="H2'") or
                            (t.assxlabel[x]=="H2''") or (t.assxlabel[x]=="H3'") or
                            (t.assxlabel[x]=="H4'") or (t.assxlabel[x]=="H5'") or
                            (t.assxlabel[x]=="H5''") or (t.assxlabel[x]=="H8'")))) then
                        --- leaves H6 connections because of 2AP
                                i = i + 1 --- dummy statement
                        else
                                if (t.assx[x]~=t.assy[x]) then
                                        c_d2o = c_d2o + 1
                                        outfile_d2o:write (c_d2o..t.posx[x]..t.posy[x].." 0 U
                                             "..t.vol[x]..t.ampl[x].."  -    0"..t.assx[x]..t.assy[x].."
                                             0\n# "..t.label[x].."\n")
                                        ---print (c_d2o..t.posx[x]..t.posy[x].." 0 U
                                             "..t.vol[x]..t.ampl[x].."  -    0"..t.assx[x]..t.assy[x].."
                                             0\n# "..t.label[x].."    d2o")
                                end
                        end
                end
        end
end
---------------------------------- End of Main Body ----------------------------------

--- close outfiles
outfile_h2o:close()
outfile_d2o:close()

t = nil

---------------------------------- End of Script ----------------------------------

print ( "\nfilterpeaks_byanda is done." )
print ( "Have a nice day!" )

---------------------------------- End filterpeaksbyanda



--- script to filter through peaklist which where exported from homoscope to
--- monoscope
--- produces several peaklists, containing d2o and h2o peaks respectively
--- adjustments need to be made for unnatural nucleobases, peakwidths, and home spectrum for
       h2o and d2o!!!

--- written by Andre Dallmann May-11-2007
```

Script code

--- Script ---

--- PREPARATIONS ---

```
-- initialize tables for peak information
t={}
t.label = {}
t.id = {}
t.assx = {}
t.assy = {}
t.posx = {}
t.posy = {}
t.ampl = {}
t.grade = {}
t.vol = {}
t.atomtype = {}
t.assxlabel = {}
t.assylabel = {}

-- Get Project
local ProjectNames = {}
i = 0
for a,b in pairs(cara:getProjects()) do
        i = i + 1
        ProjectNames[ i ] = b:getName()
end
t.ProjectName = dlg.getSymbol('Select Project','', unpack( ProjectNames ))
t.project = cara:getProject( t.ProjectName )

-- Get Peaklist
local PeaklistNames = {}
i = 0
for a,b in pairs(t.project:getPeakLists()) do
        i = i + 1
        PeaklistNames[ i ] = b:getName()
end
t.PeaklistName = dlg.getSymbol('Select Peaklist H2O','', unpack( PeaklistNames ))
for a,b in pairs(t.project:getPeakLists()) do
        i = i + 1
        if (b:getName()==t.PeaklistName) then
                t.peaklist_h2o = t.project:getPeakList( b:getId(a) )
        end
end
i=0
t.PeaklistName = dlg.getSymbol('Select Peaklist D2O','', unpack( PeaklistNames ))
for a,b in pairs(t.project:getPeakLists()) do
        i = i + 1
        if (b:getName()==t.PeaklistName) then
                t.peaklist_d2o = t.project:getPeakList( b:getId(a) )
        end
```

4 Lua scripts written for data export from CARA

end

---------------------------- Main Body ----------------------------

---------------------------- h2o peaks ----------------------------

```
-- initialize variables
local c = 0
local d = 0
local e = 0
local f = 0
local waminox = 0.027 -- peak width for amino
local wrestx = 0.027 -- peak width for rest in x dim
local wally = 0.027 -- peak width for y dim
local h2o_spectrum = 6 -- home spectrum needs to be adjusted !!!


for i,p in pairs (t.peaklist_h2o:getPeaks()) do --cycle through all peaks
        t.posx_ref,t.posy_ref = p:getPos() -- get Positions of Reference peak
        t.assx,t.assy = p:getAssig() -- get Assignment of Reference peak
        t.x = t.project:getSpin(t.assx):getLabel() -- get Peaklabel
        t.y = t.project:getSpin(t.assy):getLabel() -- get Peaklabel
        t.HNFx = t.project:getSpin(t.assx):getSystem():getResidue():getId() -- get
            Residuenumber
        t.HNFy = t.project:getSpin(t.assy):getSystem():getResidue():getId() -- get
            Residuenumber
            if (((t.x=='H41') and (t.y=='H42')) or ((t.x=='H42') and (t.y=='H41'))) and not
                ((t.HNFx==2) or (t.HNFx==12) or (t.HNFy==2) or (t.HNFy==12) or (t.HNFx==14) or
                (t.HNFx==26) or (t.HNFy==14) or (t.HNFy==26)) then
            -- selects all H41/H42 pairs which are not subjet to base pair fraying
                e = e + 1
                    if (e==1) then -- for first match establish peaklist
                        peaklist3 = spec.createPeakList('1H','1H')
                        peaklist3:setName(t.project:getName().."_h41h42_h2o")
                        peaklist3:setHome(t.project:getSpectrum(h2o_spectrum))
                        t.project:addPeakList(peaklist3)
                        peaklist3:getModel(0):setWidth(1,waminox)
                        peaklist3:getModel(0):setWidth(2,wally)
                        peaklist3:setAttr('WidthX',waminox)
                        peaklist3:setAttr('WidthY',wally)
                    end
                peak = peaklist3:createPeak(p:getPos()) -- create Peaks in Peaklist
                peak:setAssig(p:getAssig())
                peak:setLabel(p:getLabel())
                    for xx,yy in pairs (t.peaklist_h2o:getPeaks()) do --cycle through all peaks
                        t.posx,t.posy = yy:getPos() -- get Positions of peak
                        t.assxx,t.assyy = yy:getAssig() -- get Assignment of peak
                            if (((((t.posx_ref-t.posx)^2)+((t.posy_ref-t.posy)^2))^(1/2)) >
                                (2*wally)) then -- if next peak is more than double peakwidth
                                away, isolated peak
                                    if ((peak:getAttr('grade')~='b') and
                                        (peak:getAttr('grade')~='c')) then
```

Script code

```
                                    peak:setAttr('grade','a')
                            end
                    elseif (wally <
                            ((((t.posx_ref-t.posx)^2)+((t.posy_ref-t.posy)^2))^(1/2))) and
                            (((((t.posx_ref-t.posx)^2)+((t.posy_ref-t.posy)^2))^(1/2)) <
                            (2*wally)) then
                            if (peak:getAttr('grade')-='c') then
                                    peak:setAttr('grade','b')
                            end
                    elseif (((((t.posx_ref-t.posx)^2)+((t.posy_ref-t.posy)^2))^(1/2)) <
                            wally) and (i-=xx) then -- very close proximity to next peak,
                            great error
                                    peak:setAttr('grade','c')
                    end
            end
-- this procedure is repeated for all peaks of interest
-- different peaklists are required for peaks with different peakwidths
-- for h2o peaks only one half of the diagonal is used
elseif ((t.x=='H41')) and ((t.y=='H1') or (t.y=='H3') or (t.y=='H5') or (t.HNFy==7))
    and (t.assx-=t.assy) then
    -- selects all relevant H41-peaks
        c = c + 1
        if (c==1) then
                peaklist = spec.createPeakList('1H','1H')
                peaklist:setName(t.project:getName().._H41_h2o')
                peaklist:setHome(t.project:getSpectrum(h2o_spectrum))
                t.project:addPeakList(peaklist)
                peaklist:getModel(0):setWidth(1,waminox)
                peaklist:getModel(0):setWidth(2,wally)
                peaklist:setAttr('WidthX',waminox)
                peaklist:setAttr('WidthY',wally)
        end
        peak = peaklist:createPeak(p:getPos())
        peak:setAssig(p:getAssig())
        peak:setLabel(p:getLabel())
        for xx,yy in pairs (t.peaklist_h2o:getPeaks()) do --cycle through all peaks
                t.posx,t.posy = yy:getPos()
                t.assxx,t.assyy = yy:getAssig()
                if (((((t.posx_ref-t.posx)^2)+((t.posy_ref-t.posy)^2))^(1/2)) >
                    (2*wally)) then -- if peak next peak is more than double
                    peakwidth away, isolated peak
                            if ((peak:getAttr('grade')-='b') and
                                (peak:getAttr('grade')-='c')) then
                                    peak:setAttr('grade','a')
                            end
                    elseif (wally <
                            ((((t.posx_ref-t.posx)^2)+((t.posy_ref-t.posy)^2))^(1/2))) and
                            (((((t.posx_ref-t.posx)^2)+((t.posy_ref-t.posy)^2))^(1/2)) <
                            (2*wally)) then
                            if (peak:getAttr('grade')-='c') then
                                    peak:setAttr('grade','b')
                            end
                    elseif (((((t.posx_ref-t.posx)^2)+((t.posy_ref-t.posy)^2))^(1/2)) <
                            wally) and (i-=xx) then -- very close proximity to next peak,
                            great error
                                    peak:setAttr('grade','c')
```

4 Lua scripts written for data export from CARA

```lua
                end
         end
elseif ((t.x=="H42")) and ((t.y=="H1") or (t.y=="H3") or (t.y=="H5") or (t.HNFy==7))
    and (t.assx-=t.assy) then
    -- selects all relevant H42 peaks
        f = f + 1
        if (f==1) then
                peaklist4 = spec:createPeakList("1H","1H")
                peaklist4:setName(t.project:getName().."_H42_h2o")
                peaklist4:setHome(t.project:getSpectrum(h2o_spectrum))
                t.project:addPeakList(peaklist4)
                peaklist4:getModel(0):setWidth(1,waminox)
                peaklist4:getModel(0):setWidth(2,wally)
                peaklist4:setAttr("WidthX",waminox)
                peaklist4:setAttr("WidthY",wally)
        end
        peak = peaklist4:createPeak(p:getPos())
        peak:setAssig(p:getAssig())
        peak:setLabel(p:getLabel())
        for xx,yy in pairs (t.peaklist_h2o:getPeaks()) do --cycle through all peaks
                t.posx,t.posy = yy:getPos()
                t.assxx,t.assyy = yy:getAssig()
                if (((((t.posx_ref-t.posx)^2)+((t.posy_ref-t.posy)^2))^(1/2)) >
                    (2*wally)) then -- if peak next peak is more than double
                    peakwidth away, isolated peak
                        if ((peak:getAttr("grade")~="b") and
                            (peak:getAttr("grade")~="c")) then
                                peak:setAttr("grade","a")
                        end
                elseif (wally <
                    (((((t.posx_ref-t.posx)^2)+((t.posy_ref-t.posy)^2))^(1/2))) and
                    (((((t.posx_ref-t.posx)^2)+((t.posy_ref-t.posy)^2))^(1/2)) <
                    (2*wally)) then
                        if (peak:getAttr("grade")~="c") then
                                peak:setAttr("grade","b")
                        end
                elseif (((((t.posx_ref-t.posx)^2)+((t.posy_ref-t.posy)^2))^(1/2)) <
                    wally) and (i~=xx) then -- very close proximity to next peak,
                    great error
                        peak:setAttr("grade","c")
                end
        end
elseif ((t.x=="H1" or t.x=="H3" or t.x=="H2") and (t.y=="H1" or t.y=="H3" or
    t.y=="H5" or t.HNFy==7)) and (t.assx-=t.assy) then
    -- selects all other exchangable proton peaks plus HNF-crosspeaks
        d = d + 1
        if (d==1) then
                peaklist2 = spec:createPeakList("1H","1H")
                peaklist2:setName(t.project:getName().."_rest_h2o")
                peaklist2:setHome(t.project:getSpectrum(h2o_spectrum))
                t.project:addPeakList(peaklist2)
                peaklist2:getModel(0):setWidth(1,wrestx)
                peaklist2:getModel(0):setWidth(2,wally)
                peaklist2:setAttr("WidthX",wrestx)
                peaklist2:setAttr("WidthY",wally)
        end
```

Script code

```
                peak = peaklist2:createPeak(p:getPos())
                peak:setAssig(p:getAssig())
                peak:setLabel(p:getLabel())
                for xx,yy in pairs (t.peaklist_h2o:getPeaks()) do --cycle through all peaks
                        t.posx,t.posy = yy:getPos()
                        t.assxx,t.assyy = yy:getAssig()
                        if (((((t.posx_ref-t.posx)^2)+((t.posy_ref-t.posy)^2))^(1/2)) >
                            (2*wally)) then -- if peak next peak is more than double
                            peakwidth away, isolated peak
                                if ((peak:getAttr('grade')~='b') and
                                    (peak:getAttr('grade')~='c')) then
                                        peak:setAttr('grade','a')
                                end
                        elseif (wally <
                            (((((t.posx_ref-t.posx)^2)+((t.posy_ref-t.posy)^2))^(1/2))) and
                            (((((t.posx_ref-t.posx)^2)+((t.posy_ref-t.posy)^2))^(1/2)) <
                            (2*wally)) then
                                if (peak:getAttr('grade')~='c') then
                                        peak:setAttr('grade','b')
                                end
                        elseif (((((t.posx_ref-t.posx)^2)+((t.posy_ref-t.posy)^2))^(1/2)) <
                            wally) and (i~=xx) then -- very close proximity to next peak,
                            great error
                                        peak:setAttr('grade','c')
                        end
                end
        end
end -- of first for loop




------------------------------- d2o peaks -------------------------------

function find2x (index) -- function to format the atomlabels
        local Boolean = false
        local Booleanx = false
        local Booleany = false
        for x in string.gfind(t.x,"H[2345]'[']*") do
                Booleanx=true
        end
        for y in string.gfind(t.y,"H[25]$") do
                Booleany=true
        end
        if (Booleanx==true) and (Booleany==true) then
                Boolean=true
        else
                Boolean=false
        end
        return Boolean
end

function find2y (index) -- function to filter atoms
        local Boolean = false
        local Booleanx = false
        local Booleany = false
```

4 Lua scripts written for data export from CARA

```
            for x in string.gfind(t.x,"H[25]$") do
                    Booleanx=true
            end
            for y in string.gfind(t.y,"H[2345]'['']*") do
                    Booleany=true
            end
            if (Booleanx==true) and (Booleany==true) then
                    Boolean=true
            else
                    Boolean=false
            end
            return Boolean
end


--- initialize variables
local c = 0
local d = 0
local e = 0
local f = 0
local wh2 = 0.040 --- peak width for H2' or H2''
local wrest = 0.025 --- peak width for rest
local d2o_spectrum = 1

for i,p in pairs (t.peaklist_d2o:getPeaks()) do ---cycle through all peaks
        t.posx_ref,t.posy_ref = p:getPos()
        t.assx,t.assy = p:getAssig()
        t.x = t.project:getSpin(t.assx):getLabel() --- get Peaklabel
        t.y = t.project:getSpin(t.assy):getLabel() --- get Peaklabel
        if (t.x=="H41") or (t.x=="H42") or (t.x=="H1") or (t.x=="H3") or (t.y=="H41") or
            (t.y=="H42") or (t.y=="H1") or (t.y=="H3") then
            t.peaklist_d2o:removePeak(p) --- remove all h2o-peaks

--- selection process for different peaklists as in h2o, repeated several times

        elseif ((t.x=="H2'") or (t.x=="H2''")) and (t.assx~=t.assy) and (find2x(i)==false) and
            (find2y(i)==false) then
                --- selects all peaks including H2' to H2'' spins on x-axis (differentiation
                    necessary because of different peakwidth for y and x axis)
                    d = d +1
                    if (d==1) then
                            peaklist2 = spec.createPeakList("1H","1H")
                            peaklist2:setName(t.project:getName().."_h2inx_d2o")
                            peaklist2:setHome(t.project:getSpectrum(d2o_spectrum))
                            t.project:addPeakList(peaklist2)
                            peaklist2:getModel(0):setWidth(1,wh2)
                            peaklist2:getModel(0):setWidth(2,wrest)
                            peaklist:setAttr("WidthX",wh2)
                            peaklist:setAttr("WidthY",wrest)
                    end
                    peak = peaklist2:createPeak(p:getPos())
                    peak:setAssig(p:getAssig())
                    peak:setLabel(p:getLabel())
                    for xx,yy in pairs (t.peaklist_d2o:getPeaks()) do ---cycle through all peaks
                            t.posx,t.posy = yy:getPos()
                            t.assxx,t.assyy = yy:getAssig()
```

Script code

```
                if (((((t.posx_ref-t.posx)^2)+((t.posy_ref-t.posy)^2))^(1/2)) >
                    (2*wh2)) then -- if peak next peak is more than double peakwidth
                    away, isolated peak
                        if ((peak:getAttr('grade')~='b') and
                            (peak:getAttr('grade')~='c')) then
                                    peak:setAttr('grade','a')
                        end
                elseif (wh2 <
                    ((((t.posx_ref-t.posx)^2)+((t.posy_ref-t.posy)^2))^(1/2))) and
                    (((((t.posx_ref-t.posx)^2)+((t.posy_ref-t.posy)^2))^(1/2)) <
                    (2*wh2)) then
                        if (peak:getAttr('grade')~='c') then
                                    peak:setAttr('grade','b')
                        end
                elseif (((((t.posx_ref-t.posx)^2)+((t.posy_ref-t.posy)^2))^(1/2)) <
                    wh2) and (i~=xx) then -- very close proximity to next peak, great
                    error
                            peak:setAttr('grade','c')
                end
        end
elseif ((t.y=='H2'') or (t.y=='H2''')) and (t.assx~=t.assy) and (find2x(i)==false) and
    (find2y(i)==false) then
    -- selects all peaks including H2' to H2'' spins on y-axis (differentiation
    necessary because of different peakwidth for y and x axis)
        e = e +1
        if (e==1) then
                    peaklist3 = spec.createPeakList('1H','1H')
                    peaklist3:setName(t.project:getName()..'_h2iny_d2o')
                    peaklist3:setHome(t.project:getSpectrum(d2o_spectrum))
                    t.project:addPeakList(peaklist3)
                    peaklist3:getModel(0):setWidth(1,wrest)
                    peaklist3:getModel(0):setWidth(2,wh2)
                    peaklist:setAttr('WidthX',wrest)
                    peaklist:setAttr('WidthY',wh2)
        end
        peak = peaklist3:createPeak(p:getPos())
        peak:setAssig(p:getAssig())
        peak:setLabel(p:getLabel())
        for xx,yy in pairs (t.peaklist_d2o:getPeaks()) do --cycle through all peaks
                    t.posx,t.posy = yy:getPos()
                    t.assxx,t.assyy = yy:getAssig()
                    if (((((t.posx_ref-t.posx)^2)+((t.posy_ref-t.posy)^2))^(1/2)) >
                        (2*wh2)) then -- if peak next peak is more than double peakwidth
                        away, isolated peak
                            if ((peak:getAttr('grade')~='b') and
                                (peak:getAttr('grade')~='c')) then
                                        peak:setAttr('grade','a')
                            end
                    elseif (wh2 <
                        ((((t.posx_ref-t.posx)^2)+((t.posy_ref-t.posy)^2))^(1/2))) and
                        (((((t.posx_ref-t.posx)^2)+((t.posy_ref-t.posy)^2))^(1/2)) <
                        (2*wh2)) then
                            if (peak:getAttr('grade')~='c') then
                                        peak:setAttr('grade','b')
                            end
                    elseif (((((t.posx_ref-t.posx)^2)+((t.posy_ref-t.posy)^2))^(1/2)) <
```

4 Lua scripts written for data export from CARA

```
                        wh2) and (i~=xx) then  -- very close proximity to next peak, great
                        error
                                peak:setAttr('grade','c')
                        end
                end
        elseif (t.assx~=t.assy) and (find2x(i)==false) and (find2y(i)==false) then
                -- select all other peaks which are not sugar to H2'* or H5'* peaks
                f = f +1
                if (f==1) then
                        peaklist4 = spec.createPeakList('1H','1H')
                        peaklist4:setName(t.project:getName().."_rest_d2o")
                        peaklist4:setHome(t.project:getSpectrum(d2o_spectrum))
                        t.project:addPeakList(peaklist4)
                        peaklist4:getModel(0):setWidth(1,wrest)
                        peaklist4:getModel(0):setWidth(2,wrest)
                        peaklist:setAttr('WidthX',wrest)
                        peaklist:setAttr('WidthY',wrest)
                end
                peak = peaklist4:createPeak(p:getPos())
                peak:setAssig(p:getAssig())
                peak:setLabel(p:getLabel())
                for xx,yy in pairs (t.peaklist_d2o:getPeaks()) do --cycle through all peaks
                        t.posx,t.posy = yy:getPos()
                        t.assxx,t.assyy = yy:getAssig()
                        if (((((t.posx_ref-t.posx)^2)+((t.posy_ref-t.posy)^2))^(1/2)) >
                                (2*wrest)) then
                                        if ((peak:getAttr('grade')~='b') and
                                            (peak:getAttr('grade')~='c')) then
                                                peak:setAttr('grade','a')
                                        end
                                elseif (wrest <
                                        ((((t.posx_ref-t.posx)^2)+((t.posy_ref-t.posy)^2))^(1/2))) and
                                        (((((t.posx_ref-t.posx)^2)+((t.posy_ref-t.posy)^2))^(1/2)) <
                                        (2*wrest)) then
                                                if (peak:getAttr('grade')~='c') then
                                                        peak:setAttr('grade','b')
                                                end
                                        elseif (((((t.posx_ref-t.posx)^2)+((t.posy_ref-t.posy)^2))^(1/2)) <
                                                wrest) and (i~=xx) then
                                                        peak:setAttr('grade','c')
                                                -- print (p:getLabel().."  "..yy:getLabel().."
                                                        "..(((((t.posx_ref-t.posx)^2)+((t.posy_ref-t.posy)^2))^(1/2)))
                                end
                        end
                end
end  -- of first for loop

--------------------------- End of Main Body ---------------------------

-- t.Project:removePeakList(peaklist_d2o)
-- t.Project:removePeakList(peaklist_h2o)
t = nil


--------------------------- End of Script ---------------------------
```

Script code

```
print ( '\nfilterpeaks_byanda is done.' )
print ( 'Have a nice day!' )
```

---------------------------------- End filterpeaksbyanda

--- script to generate ppm-file for input to GIFA

---   Script   ---

---------------------------------- PREPARATIONS ----------------------------------

```
-- initialize tables for peak information
t={}
t.label = {}
t.id = {}
t.assx = {}
t.assy = {}
t.posx = {}
t.posy = {}
t.ampl = {}
t.grade = {}
t.vol = {}
t.atomtype = {}
t.assxlabel = {}
t.assylabel = {}

-- get Project
local ProjectNames = {}
local i = 0
for a,b in pairs(cara:getProjects()) do
        i = i + 1
        ProjectNames[ i ] = b:getName()
end
t.ProjectName = dlg.getSymbol('Select Project','', unpack( ProjectNames ))
t.project = cara:getProject( t.ProjectName )

-- get Input-Filename
--SparkyList = dlg.getText( 'Enter input filename', '', t.ProjectName'_sparky.peaks')

-- open input file
--io.open (SparkyList, r)

-- Get Output Filename
t.Filename = dlg.getText('Enter the output filename', '', t.ProjectName..'_gifa.ppm')
```

4 Lua scripts written for data export from CARA

```lua
-- open outfile
outfile = io.output( t.Filename )

-- write header

for i, spin in pairs (t.project:getSpins()) do
    t.label[i]=spin:getLabel()
    if (t.label[i]=='H5') then
        t.label[i]='HQ5'
    elseif (t.label[i]=='H2') then
        t.label[i]='HQ2'
    end
end
for i, spin in pairs (t.project:getSpins()) do
    if (t.label[i]=='H7') then
        outfile:write("PPM "..spin:getSystem():getResidue():getType():getShort().."
        "..spin:getSystem():getResidue():getId().." "..t.label[i].."1
        "..spin:getShift().." 1\n")
        outfile:write("PPM "..spin:getSystem():getResidue():getType():getShort().."
        "..spin:getSystem():getResidue():getId().." "..t.label[i].."2
        "..spin:getShift().." 1\n")
        outfile:write("PPM "..spin:getSystem():getResidue():getType():getShort().."
        "..spin:getSystem():getResidue():getId().." "..t.label[i].."3
        "..spin:getShift().." 1\n")
    else
        outfile:write("PPM "..spin:getSystem():getResidue():getType():getShort().."
        "..spin:getSystem():getResidue():getId().." "..t.label[i].."
        "..spin:getShift().." 1\n")
    end
end

outfile:close()
t={}

print ('done with gifappm')



-- script to remove peaklists, peaklists are picked by their names

-- written by Andre Dallmann

for i,p in pairs(cara:getProject():getPeakLists()) do
    temp = string.gfind(p:getName(),'13merHNF')
    for y in temp do
        cara:getProject():removePeakList(p)
        print (y)
    end
end

print ('peaklists removed!')



-- writes 1H shifts from different spins to separate columns in an external file
```

Script code

```
-- written by Andre Dallmann

function Format( Number )  -- function to format the chemical shifts
        FormattedNumber = string.format( '%7.3f', Number )
        return FormattedNumber
end
function Format2( String )  -- function to format the chemical shifts
        FormattedString = string.format( '%7.7s', String )
        return FormattedString
end

-- User editable parameters are below: ======

-- Spacer between elements of table to write out:
Spacer  = ' '
Spacer2 = '  '

-- Table of spin labels whose shifts should be written to a column

SpinsInColumns = {}
SpinsInColumns[ 1  ] = 'H1''
SpinsInColumns[ 2  ] = 'H2''
SpinsInColumns[ 3  ] = 'H2'''
SpinsInColumns[ 4  ] = 'H3''
SpinsInColumns[ 5  ] = 'H4''
SpinsInColumns[ 6  ] = 'H5''
SpinsInColumns[ 7  ] = 'H5'''
SpinsInColumns[ 8  ] = 'H1'
SpinsInColumns[ 9  ] = 'H2'
SpinsInColumns[ 10 ] = 'H3'
SpinsInColumns[ 11 ] = 'H41'
SpinsInColumns[ 12 ] = 'H42'
SpinsInColumns[ 13 ] = 'H5'
SpinsInColumns[ 14 ] = 'H6'
SpinsInColumns[ 15 ] = 'H7'
SpinsInColumns[ 16 ] = 'H8'
SpinsInColumns[ 17 ] = 'H4'
SpinsInColumns[ 18 ] = 'H1'''

-- End of user editable section =======
-- define a table of temporary script variables
t={}

-- Get Parameters from User

-- 1. Get Project: -----------------------------------------
local projectnames = {}
i=0
for ProjName,Proj in pairs( cara:getProjects() ) do
        i = i + 1
        projectnames[ i ] = ProjName
end

if i==1 then
        t.ProjectName = projectnames[ i ]
```

4 Lua scripts written for data export from CARA

```
        else
                t.ProjectName = dlg.getSymbol( 'Choose project', 'select one', unpack( projectnames ) )
        end
        if not t.ProjectName then
                error( 'No project name defined')
        else
                t.P = cara:getProject( t.ProjectName )
        end

        -- 2. Get Output filename:  --------------------------------------
        t.FileName = dlg.getText( 'Enter output filename', 'output filename','shiftstable.txt' )

        --3. Get Label of Spin to write out chemical shifts from
        -- I replaced this step with a table , see the top of the script
        --t.Label = dlg.getText( 'Enter label of the spins whose chemical shifts you want to write out
             : ', 'Enter Label of spins whose shifts will be written out (e.g. HA): ')


        -- loop through the sequence and for each residue, create a Line
        -- then for each Line look for each column entry in turn
        -- add it to the end of the growing line

        Seq = t.P:getSequence()
        j = 0
        Lines = {}
        for ResId,Res in pairs( Seq ) do
                Sys = Res:getSystem()
                if Sys then -- if residue is assigned
                        SpinsInSys = Sys:getSpins()
                        j = j + 1
                        Lines[ j ] = string.format ('%-7.7s',ResId)
                        for k = 1,table.getn( SpinsInColumns ) do
                                LabelToFind = SpinsInColumns[ k ]
                                MatchingSpin = nil -- reset to none found
                                for SpinId ,Spin in pairs( SpinsInSys ) do -- search for a match to
                                    LabelToFind
                                        if Spin:getLabel() == LabelToFind then
                                                MatchingSpin = Spin
                                        end -- if Spins label matches LabelToFind
                                end -- for all Spins in System (look for match to this Label)
                                if MatchingSpin then
                                        FormShift = Format( MatchingSpin:getShift() )
                                        Lines[ j ] = Lines[ j ]..FormShift
                                else
                                        Formzero = Format2( '-' )
                                        Lines[ j ] = Lines[ j ]..Formzero -- no shift assigned, leave
                                            empty
                                end
                        end -- for all elements k of SpinsInColumns (try to find a shift for this
                            label)
                        --Lines[ j ] = Lines[ j ]..'\n' -- next line
                end -- if System is assigned
        end -- for all residues in sequence

        --create string 'Table' with lines
```

231

Script code

```
--create the first line of table
for m = 1,table.getn( SpinsInColumns ) do
        if m == 1 then
                Labels = string.format ("%7.7s", SpinsInColumns[ 1 ])
        else
                Formlabels = string.format ("%7.7s", SpinsInColumns[ m ] )
                Labels = Labels..Formlabels
        end
end
res = string.format ("%-7.7s",'Res')
Header = res..Labels

for l=1,table.getn( Lines ) do
        if l==1 then
                Table = Header.."\n"..Lines[ 1 ]
        else
                Table = Table.."\n"..Lines[ l ]
        end
end

-- Now write out all lines to a file
file = io.open( t.FileName, 'w' )
file:write( Table )
file:flush()
file:close()


print("Wrote out "..table.getn( Lines ).." lines to file "..t.FileName )
print( "script WriteShiftsInColumns is done" )
t = nil



-- writes 1H shifts from different spins to separate columns in an external file

-- written by Andre Dallmann

function Format( Number ) -- function to format the chemical shifts
        FormattedNumber = string.format( "%11.3f", Number )
        return FormattedNumber
end
function Format2( String ) -- function to format the chemical shifts
        FormattedString = string.format( "%11.11s", String )
        return FormattedString
end

-- User editable parameters are below: ==================

-- Spacer between elements of table to write out:
Spacer  = " "
Spacer2 = "  "

-- Table of spin labels whose shifts should be written to a column

SpinsInColumns = {}
SpinsInColumns[ 1 ] = "C1'"
```

4 Lua scripts written for data export from CARA

```
SpinsInColumns[  2 ] = "C2'"
SpinsInColumns[  3 ] = "C3'"
SpinsInColumns[  4 ] = "C4'"
SpinsInColumns[  5 ] = "C5'"
SpinsInColumns[  6 ] = "C5"
SpinsInColumns[  7 ] = "C6"
SpinsInColumns[  8 ] = "C8"
SpinsInColumns[  9 ] = "C1"
SpinsInColumns[ 10 ] = "C2"
SpinsInColumns[ 11 ] = "C3"
SpinsInColumns[ 12 ] = "C4"


-- End of user editable section ===================
-- define a table of temporary script variables
t={}

-- Get Parameters from User

-- 1. Get Project: --------------------------------
local projectnames = {}
i=0
for ProjName,Proj in pairs( cara:getProjects() ) do
        i = i + 1
        projectnames[ i ] = ProjName
end

if i==1 then
        t.ProjectName = projectnames[ i ]
else
        t.ProjectName = dlg.getSymbol( "Choose project", "select one", unpack( projectnames ) )
end
if not t.ProjectName then
        error( "No project name defined")
else
        t.P = cara:getProject( t.ProjectName )
end

-- 2. Get Output filename: -------------------------
t.FileName = dlg.getText( "Enter output filename", "output filename","shiftstable_C.txt" )

--3. Get Label of Spin to write out chemical shifts from
-- I replaced this step with a table , see the top of the script
--t.Label = dlg.getText( "Enter label of the spins whose chemical shifts you want to write out
        : ", "Enter Label of spins whose shifts will be written out (e.g. HA): ")


-- loop through the sequence and for each residue, create a Line
-- then for each Line look for each column entry in turn
-- add it to the end of the growing line

Seq = t.P:getSequence()
j = 0
Lines = {}
for ResId,Res in pairs( Seq ) do
        Sys = Res:getSystem()
```

Script code

```
            if Sys then -- if residue is assigned
                SpinsInSys = Sys:getSpins()
                j = j + 1
                Lines[ j ] = string.format ("%-11.11s",ResId)
                for k = 1,table.getn( SpinsInColumns ) do
                        LabelToFind = SpinsInColumns[ k ]
                        MatchingSpin = nil -- reset to none found
                        for SpinId,Spin in pairs( SpinsInSys ) do -- search for a match to
                                LabelToFind
                                if Spin:getLabel() == LabelToFind then
                                        MatchingSpin = Spin
                                end -- if Spins label matches LabelToFind
                        end -- for all Spins in System (look for match to this Label)
                        if MatchingSpin then
                                FormShift = Format( MatchingSpin:getShift() )
                                Lines[ j ] = Lines[ j ]..FormShift
                        else
                                Formzero = Format2( '-' )
                                Lines[ j ] = Lines[ j ]..Formzero -- no shift assigned, leave
                                        empty
                        end
                end -- for all elements k of SpinsInColumns (try to find a shift for this
                        label)
                --Lines[ j ] = Lines[ j ].."\n" -- next line
        end -- if System is assigned
end -- for all residues in sequence

--create string 'Table' with lines

--create the first line of table
for m = 1,table.getn( SpinsInColumns ) do
        if m == 1 then
                Labels = string.format ("%11.11s", SpinsInColumns[ 1 ])
        else
                Formlabels = string.format ("%11.11s", SpinsInColumns[ m ] )
                Labels = Labels..Formlabels
        end
end
res = string.format ("%-11.11s",'Res')
Header = res..Labels

for l=1,table.getn( Lines ) do
        if l==1 then
                Table = Header.."\n"..Lines[ l ]
        else
                Table = Table.."\n"..Lines[ l ]
        end
end

-- Now write out all lines to a file
file = io.open( t.FileName, 'w' )
file:write( Table )
file:flush()
file:close()
```

4 Lua scripts written for data export from CARA

```
print("Wrote out "..table.getn( Lines ).." lines to file "..t.FileName )
print( "script WriteShiftsInColumns is done" )
t = nil
```

```
-- First part: Script to output all chosen and integrated peaks from one project
-- and combine them in one peaklist.

-- Second part: Choose the best integrated peak among same ones or average over
-- equivalently rated peaks

-- Third part: Convert peak volumes to distances. Tricky is here the
-- differentiation of d2o and h2o and methyl peaks (all have different
-- reference peaks)!

-- Fourth part: An XPLOR-inputfile is generated where the distance information
-- and some predefined upper and lower limits (deduced from the maximum
-- deviation of the standard peaks) are used

-- written by Andre Dallmann April-05-2007, mod. May-11-2007
```


--   FIRST PART   --

------------------------ PREPARATIONS ------------------------

```
t = {} -- table for all the variables used in the script

-- choosing one project
local ProjectNames = {}
local i = 0
for a,b in pairs(cara:getProjects()) do
        i = i + 1
        ProjectNames[ i ] = b:getName()
end
t.ProjectName=dlg.getSymbol('Select Project',"", unpack( ProjectNames ) )
t.project = cara:getProject( t.ProjectName )

-- Get Output Filename
t.Filename = dlg.getText('Enter the output filename', "", t.ProjectName)

-- open outfile
outfile = io.output( t.Filename.."_all.peaks" )

-- Write header to peaklist
local label = string.format ("%25.25s", 'Peaklabel')
local id = string.format ("%9.9s", 'PeakID')
local assx = string.format ("%9.9s", 'ID(X)')
local assy = string.format ("%9.9s", 'ID(Y)')
local posx = string.format ("%9.9s", 'PPM(X)')
local posy = string.format ("%9.9s", 'PPM(Y)')
```

Script code

```
local ampl = string.format ("%7.7s", 'Ampl')
local grade = string.format ("%9.9s", 'Grade')
local vol = string.format ("%15.15s", 'VolumeInt')
outfile:write("IDnew"..id..label..assx..assy..posx..posy..ampl..vol..grade.."\n")


-- generate tables for information
local count = 0
local i = 0
t.label = {}
t.id_old = {}
t.assx = {}
t.assy = {}
t.posx = {}
t.posy = {}
t.ampl = {}
t.grade = {}
t.vol = {}


-------------------------------------------------------------------------
------------------------------- Main Body -------------------------------
-------------------------------------------------------------------------


-- generate list of all peaks graded abc of all peaklists in specified project
for peaklistid, peaklist in pairs(t.project:getPeakLists()) do --cycle through all peaklists
        t.peaklist = t.project:getPeakList(peaklistid)
        for peakid,peak in pairs(t.peaklist:getPeaks()) do --cycle through all peaks
                t.peak = t.peaklist:getPeak(peakid)
                if ((t.peak:getAttr('grade')== 'a') or (t.peak:getAttr('grade')== 'b') or
                    (t.peak:getAttr('grade')== 'c')) then -- choose only peaks with grade abc
                        i = i + 1 -- this is the index for all the tables, corresponds to new
                                  peakid
                        t.label[i] = string.format ("%25.25s", t.peak:getLabel())
                        t.id_old[i] = string.format ("%9.0f", t.peak:getId())
                        t.ass = {t.peak:getAssig()}
                        t.assx[i] = string.format ("%9.0f", t.ass[1])
                        t.assy[i] = string.format ("%9.0f", t.ass[2])
                        t.pos = {t.peak:getPos()}
                        t.posx[i] = string.format ("%9.3f", t.pos[1])
                        t.posy[i] = string.format ("%9.3f", t.pos[2])
                        t.ampl[i] = string.format ("%7.0f", t.peak:getAmp())
                        t.grade[i] = string.format ("%7.7s", t.peak:getAttr('grade'))
                        t.vol[i] = string.format ("%15.3f", t.peak:getVol())
                        outfile:write(i.."
                        "..t.id_old[i]..t.label[i]..t.assx[i]..t.assy[i]..t.posx[i]..t.posy[i]
..t.ampl[i]..t.vol[i]..t.grade[i].."\n")
                end --of if loop
        end -- of second for loop
end -- of first for loop


-------------------------------- End of Main Body --------------------------------
---------------------------------------------------------------------------------


-- close outfile
outfile:close()
```

4 Lua scripts written for data export from CARA

---------- End of FIRST PART ----------

--- SECOND PART ---

---------- PREPARATIONS ----------

```
-- open outfile
outfile = io.output( t.Filename.."_combo.peaks" )
outfile2 = io.output( t.Filename.."_negvol.peaks" )

-- initialize variables
local x = 0
local counter = 1
local a = string.format ("%7.7s","a")
local b = string.format ("%7.7s","b")
local c = string.format ("%7.7s","c")
```

---------- Main Body ----------

---------- Preparing combination ----------

```
for i,assx in pairs (t.assx) do
    for j,assy in pairs (t.assy) do
        if ((((t.assx[i]==t.assx[j]) and (t.assy[i]==t.assy[j])) or
            ((t.assx[i]==t.assy[j]) and (t.assy[i] == t.assx[j]))) and not (j==i))
        then
            -- select all peaks that have the same assignment (including cross-diagonal
               peaks)
            counter = counter + 1
            if (t.grade[i]==t.grade[j]) then
                t.vol[i] = string.format ("%15.3f",(t.vol[i] + t.vol[j])) --
                    average volumes, rest stays
                t.label[j] = nil -- set jth peak to nil
                t.assx[j] = nil
                t.assy[j] = nil
                t.posx[j] = nil
                t.posy[j] = nil
                t.ampl[j] = nil
                t.grade[j] = nil
                t.vol[j] = nil
                t.id_old[j] = nil
```

Script code

```
                                    end
                                    if (((a==t.grade[i]) and ((t.grade[j]==b) or (t.grade[j]==c))) or
                                        ((b==t.grade[i]) and (t.grade[j]==c))) then
                                        counter = 1
                                        t.label[j] = nil  -- set jth peak to nil
                                        t.assx[j] = nil
                                        t.assy[j] = nil
                                        t.posx[j] = nil
                                        t.posy[j] = nil
                                        t.ampl[j] = nil
                                        t.grade[j] = nil
                                        t.vol[j] = nil
                                        t.id_old[j] = nil
                                    end
                                    if (((a==t.grade[j]) and ((t.grade[i]==b) or (t.grade[i]==b))) or
                                        ((b==t.grade[j]) and (t.grade[i]==c))) then
                                        counter = 1
                                        t.vol[i] = t.vol[j]  --transfer volume and grade of better
                                            integrated peak (j)
                                        t.grade[i] = t.grade[j]
                                        t.label[j] = nil  -- set jth peak to nil
                                        t.assx[j] = nil
                                        t.assy[j] = nil
                                        t.posx[j] = nil
                                        t.posy[j] = nil
                                        t.ampl[j] = nil
                                        t.grade[j] = nil
                                        t.vol[j] = nil
                                        t.id_old[j] = nil
                                    end
                            end  -- if loop
                    end  -- second for loop
                    if (counter > 1) then  -- only valid if grades are the same and averaging is needed
                            t.vol[i] = string.format ("%15.3f",(t.vol[i]/counter))
                    end
                    counter = 1
            end  -- first for loop


------------------------------- Generating new combined peaklist -------------------------------

-- initialize new tables for the combined peaklist
t.labelnew = {}
t.assxnew = {}
t.assynew = {}
t.assxlabel = {}
t.assylabel = {}
t.assxresid = {}
t.assyresid = {}
t.gradenew= {}
t.volnew = {}

for i,assx in pairs (t.assx) do  -- generate new table with combined peaks
        t.volnum=tonumber(t.vol[i])
        if (t.volnum>0) then
```

4 Lua scripts written for data export from CARA

```
                    x = x + 1
                    t.labelnew[x] = t.label[i]
                    t.assxnew[x] = t.assx[i]
                    t.assynew[x] = t.assy[i]
                    t.assxlabel[x] = string.format ('%7.7s',
                            t.project:getSpin(t.assx[i]):getLabel()) -- get Peaklabel
                    t.assylabel[x] = string.format ('%7.7s',
                            t.project:getSpin(t.assy[i]):getLabel()) -- get Peaklabel
                    t.assxresid[x] = string.format ('%5.5s',
                            t.project:getSpin(t.assx[i]):getSystem():getResidue():getId()) -- get
                            residue id
                    t.assyresid[x] = string.format ('%5.5s',
                            t.project:getSpin(t.assy[i]):getSystem():getResidue():getId()) -- get
                            residue id
                    t.gradenew[x] = t.grade[i]
                    t.volnew[x] = t.vol[i]
                    outfile:write (x..'
                            '..t.labelnew[x]..t.assxlabel[x]..t.assylabel[x]..t.gradenew[x]..t.volnew[x]..'\n')
            else
                    outfile2:write (i..'
                            '..t.label[i]..t.assx[i]..t.assy[i]..t.grade[i]..t.vol[i]..'\n')
            end
    end
end

-- loop to correct for base rectangle sum method error
-- for peaks with grade c or b divide volume by 2 or 1.5 respectively
-- this is a very rough approximation!!!
for i,vol in pairs (t.volnew) do
        if (t.gradenew[i]==c) then
                t.volnew[i]=string.format('%15.3f',vol/2)
        elseif (t.gradenew[i]==b) then
                t.volnew[i]=string.format('%15.3f',vol/1.5)
        end
end

------------------------------ End of Main Body ------------------------------

-- close outfile
outfile:close()
outfile2:close()

local i = 0

------------------------------ End of SECOND PART ------------------------------



--                            THIRD PART                                   --
```

Script code

---------------------------------- PREPARATIONS ----------------------------------

```
-- open outfile
outfile = io.output( t.Filename.."'_dist.peaks' )
outfile2 = io.output( t.Filename.."'_notused.peaks' )

-- initialize variables
local sumcyt = 0
local summet = 0
local sumcytamino = 0
local sumcyth42h5 = 0
local sumcyth41h5 = 0
local countcyt = 0
local countmet = 0
local countcytamino = 0
local countcyth42h5 = 0
local countcyth41h5 = 0
```

---------------------------------- Main Body ----------------------------------

---------------------- Setting up Reference Volumes and Distances ----------------------

```
-- sum up reference peaks
for j,assx in pairs( t.assxnew) do
    for y in string.gfind (t.labelnew[j],"H[56]/H[56] [0-9]:C[0-9]+") do -- establish
        reference for d2o peaks
            countcyt = countcyt + 1
            sumcyt = sumcyt + t.volnew[j]
    end
    for y in string.gfind (t.labelnew[j],"H[67]/H[67] [0-9]:T[0-9]+") do -- establish
        reference for methyl peaks
            countmet = countmet + 1
            summet = summet + t.volnew[j]
    end
    for y in string.gfind (t.labelnew[j],"H4[12]/H4[12] [0-9]:C[0-9]+") do -- establish
        reference for h2o exchangeable peaks
            countcytamino = countcytamino + 1
            sumcytamino = sumcytamino + t.volnew[j]
    end
    for y in string.gfind (t.labelnew[j],"H42/H5 [0-9]:C[0-9]+") do -- establish reference
        for h2o exchangeable-non-exchangeable peaks
            countcyth42h5 = countcyth42h5 + 1
            sumcyth42h5 = sumcyth42h5 + t.volnew[j]
    end
    for y in string.gfind (t.labelnew[j],"H5/H42 [0-9]:C[0-9]+") do -- establish reference
        for h2o exchangeable-non-exchangeable peaks
            countcyth42h5 = countcyth42h5 + 1
            sumcyth42h5 = sumcyth42h5 + t.volnew[j]
    end
    for y in string.gfind (t.labelnew[j],"H41/H5 [0-9]:C[0-9]+") do -- establish reference
        for h2o exchangeable-non-exchangeable with H41 peaks
            countcyth41h5 = countcyth41h5 + 1
```

4 Lua scripts written for data export from CARA

```
                    sumcyth41h5 = sumcyth41h5 + t.volnew[j]
            end
            for y in string.gfind (t.labelnew[j],"H5/H41 [0-9]:C[0-9]+") do -- establish reference
                for h2o exchangeable-non-exchangeable with H41 peaks
                    countcyth41h5 = countcyth41h5 + 1
                    sumcyth41h5 = sumcyth41h5 + t.volnew[j]
            end
    end
end

refvolcyt = string.format ("%13.3f", sumcyt / countcyt) -- average volume of CYT H5-H6
refdistcyt = 2.48 -- distance of CYT H5-H6
refvolmet = string.format ("%13.3f", summet / countmet) -- average volume of THY H6-H7
refdistmet = 3.09 -- distance of THY H6-H7
refvolcytamino = string.format ("%13.3f", sumcytamino / countcytamino) -- average volume of
    CYT H41-H42
refdistcytamino = 1.70 -- distance of CYT H41-H42
refvolcyth42h5 = string.format ("%13.3f", sumcyth42h5 / countcyth42h5) -- average volume of
    CYT H42-H5
refdistcyth42h5 = 2.40 -- distance of CYT H42-H5
refvolcyth41h5 = string.format ("%13.3f", sumcyth41h5 / countcyth41h5) -- average volume of
    CYT H41-H5
refdistcyth41h5 = 3.62 -- distance of CYT H41-H5


---------------------- Prepare standard deviations for references ----------------------

-- initialize variables
local stddevsumcyt = 0
local stddevsummet = 0
local stddevsumcytamino = 0
local stddevsumcyth42h5 = 0
local stddevsumcyth41h5 = 0
local maxdev1 = 0
local maxdev2 = 0
local maxdev3 = 0
local maxdev4 = 0
local maxdev5 = 0

for j,assx in pairs (t.assxnew) do
    for y in string.gfind (t.labelnew[j],"H[56]/H[56] [0-9]:C[0-9]+") do
        stddevsumcyt = stddevsumcyt + (t.volnew[j]-refvolcyt)^2 -- standard deviation
        dummy1 = math.abs(t.volnew[j]-refvolcyt) -- dummy for maximum deviation
        if (dummy1 > maxdev1) then
            maxdev1 = dummy1
        end
    end
    for y in string.gfind (t.labelnew[j],"H[67]/H[67] [0-9]:T[0-9]+") do -- establish
        reference for methyl peaks
        stddevsummet = stddevsummet + (t.volnew[j]-refvolmet)^2 -- standard deviation
        dummy2 = math.abs(t.volnew[j]-refvolmet) -- dummy for maximum deviation
        if (dummy2 > maxdev2) then
            maxdev2 = dummy2
        end
    end
    for y in string.gfind (t.labelnew[j],"H4[12]/H4[12] [0-9]:C[0-9]+") do -- establish
        reference for h2o exchangeable peaks
```

Script code

```
                stddevsumcytamino = stddevsumcytamino + (t.volnew[j]-refvolcytamino)^2 --
                    standard deviation
                dummy3 = math.abs(t.volnew[j]-refvolcytamino) -- dummy for maximum deviation
                if (dummy3 > maxdev3) then
                    maxdev3 = dummy3
                end
        end
        for y in string.gfind (t.labelnew[j],'H42/H5 [0-9]:C[0-9]+') do -- establish reference
            for h2o exchangeable-non-exchangeable peaks (appears twice because of selection
            reasons)
                stddevsumcyth42h5 = stddevsumcyth42h5 + (t.volnew[j]-refvolcyth42h5)^2 --
                    standard deviation
                dummy4 = math.abs(t.volnew[j]-refvolcyth42h5) -- dummy for maximum deviation
                if (dummy4 > maxdev4) then
                    maxdev4 = dummy4
                end
        end
        for y in string.gfind (t.labelnew[j],'H5/H42 [0-9]:C[0-9]+') do -- establish reference
            for h2o exchangeable-non-exchangeable peaks
                stddevsumcyth42h5 = stddevsumcyth42h5 + (t.volnew[j]-refvolcyth42h5)^2 --
                    standard deviation
                dummy4 = math.abs(t.volnew[j]-refvolcyth42h5) -- dummy for maximum deviation
                if (dummy4 > maxdev4) then
                    maxdev4 = dummy4
                end
        end
        for y in string.gfind (t.labelnew[j],'H41/H5 [0-9]:C[0-9]+') do -- establish reference
            for h2o exchangeable-non-exchangeable peaks (appears twice because of selection
            reasons)
                stddevsumcyth41h5 = stddevsumcyth41h5 + (t.volnew[j]-refvolcyth41h5)^2 --
                    standard deviation
                dummy5 = math.abs(t.volnew[j]-refvolcyth41h5) -- dummy for maximum deviation
                if (dummy5 > maxdev5) then
                    maxdev5 = dummy5
                end
        end
        for y in string.gfind (t.labelnew[j],'H5/H41 [0-9]:C[0-9]+') do -- establish reference
            for h2o exchangeable-non-exchangeable peaks
                stddevsumcyth41h5 = stddevsumcyth41h5 + (t.volnew[j]-refvolcyth41h5)^2 --
                    standard deviation
                dummy5 = math.abs(t.volnew[j]-refvolcyth41h5) -- dummy for maximum deviation
                if (dummy5 > maxdev5) then
                    maxdev5 = dummy5
                end
        end

end

----------------- Calculate standard deviations for references --------------------

stddevcyt       = string.format ('%13.3f', (stddevsumcyt / countcyt)^(1/2))
stddevmet       = string.format ('%13.3f', (stddevsummet / countmet)^(1/2))
stddevcytamino  = string.format ('%13.3f', (stddevsumcytamino / countcytamino)^(1/2))
stddevcyth42h5  = string.format ('%13.3f', (stddevsumcyth42h5 / countcyth42h5)^(1/2))
stddevcyth41h5  = string.format ('%13.3f', (stddevsumcyth41h5 / countcyth41h5)^(1/2))
```

4 Lua scripts written for data export from CARA

`---------- Prepare maximum deviations in percent for references ----------`

```
if (maxdev1==nil) then
        maxdevcyt = string.format ("%13.3f", dummy1/refvolcyt)
else
        maxdevcyt = string.format ("%13.3f", maxdev1/refvolcyt)
end
if (maxdev2==nil) then
        maxdevmet = string.format ("%13.3f", dummy2/refvolmet)
else
        maxdevmet = string.format ("%13.3f", maxdev2/refvolmet)
end
if (maxdev3==nil) then
        maxdevcytamino = string.format ("%13.3f", dummy3/refvolcytamino)
else
        maxdevcytamino = string.format ("%13.3f", maxdev3/refvolcytamino)
end
if (maxdev4==nil) then
        maxdevcyth42h5 = string.format ("%13.3f", dummy4/refvolcyth42h5)
else
        maxdevcyth42h5 = string.format ("%13.3f", maxdev4/refvolcyth42h5)
end
if (maxdev5==nil) then
        maxdevcyth41h5 = string.format ("%13.3f", dummy5/refvolcyth41h5)
else
        maxdevcyth41h5 = string.format ("%13.3f", maxdev5/refvolcyth41h5)
end
```

`---------- Prepare for distance calculation ----------`

```
function f ( String ) -- function to format the atomlabels
        FormattedString = string.format( "%7.7s", String )
        return FormattedString
end
function f2 ( String ) -- function to format the atomlabels
        FormattedString = string.format( "%9.9s", String )
        return FormattedString
end
function d2ox ( index ) -- function to find non-exchangeable protons on x axis
        local Boolean=false
        for x in string.gfind(t.assxlabel[index],"H[12345]'[']*") do
                Boolean=true
        end
        for y in string.gfind(t.assxlabel[index],"H[2568]") do
                Boolean=true
        end
        return Boolean
end
function d2oy ( index ) -- function to find non-exchangeable protons on x axis
        local Boolean=false
        for x in string.gfind(t.assylabel[index],"H[12345]'[']*") do
                Boolean=true
        end
        for y in string.gfind(t.assylabel[index],"H[2568]") do
                Boolean=true
```

Script code

```
                end
                return Boolean
end
function h2ox ( index ) -- function to find non-exchangeable protons on x axis
        local Boolean=false
        for x in string.gfind(t.assxlabel[index],"H[13]") do
                Boolean=true
        end
        return Boolean
end
function h2oy ( index ) -- function to find non-exchangeable protons on x axis
        local Boolean=false
        for x in string.gfind(t.assylabel[index],"H[13]") do
                Boolean=true
        end
        return Boolean
end
function HNFx ( index ) -- function to find non-exchangeable protons on x axis
        local Boolean=false
        local resid=t.project:getSpin(t.assxnew[index]):getSystem():getResidue():getId()
        if resid==7 or resid==20 then
                Boolean=true
        end
        return Boolean
end
function HNFy ( index ) -- function to find non-exchangeable protons on x axis
        local Boolean=false
        local resid=t.project:getSpin(t.assynew[index]):getSystem():getResidue():getId()
        if resid==7 or resid==20 then
                Boolean=true
        end
        return Boolean
end

-- initialize new tables for distance and the lower and upper limit (same)
t.distance = {}
t.limit = {}
local i = nil
local assx = nil

-------------------------- Distance and Limit calculation --------------------------

-- select atom pairs corresponding to references and calculate distances and
-- limits and write out new peaklist to file

-- limits are calculated by taking the maximum deviation of the corresponding
--reference peak times the distance

for i,assx in pairs (t.assxlabel) do
    if (((HNFx(i)==true) or (d2ox(i)==true)) and ((HNFy(i)==true) or (d2oy(i)==true))) then
                t.distance[i] = string.format ("%7.2f",
                        refdistcyt*(refvolcyt/t.volnew[i])^(1/6))
                if (t.gradenew[i]==a) then -- error bounds scaled by grading of integration
                        t.limit[i] = string.format ("%7.1f", t.distance[i]*maxdevcyt)
                else
```

4 Lua scripts written for data export from CARA

```
                    if (t.gradenew[i]==b) then
                        t.limit[i] = string.format ("%7.1f",
                            t.distance[i]*maxdevcyt*1.2)
                    else
                        t.limit[i] = string.format ("%7.1f",
                            t.distance[i]*maxdevcyt*1.4)
                    end
                end
                outfile:write (f2("d2o:   ")..f(i)..t.labelnew[i]..assx..t.assylabel[i]
..t.gradenew[i]..t.volnew[i]..t.distance[i]..t.limit[i].."\n")
            elseif ((assx==f("H7")) or (t.assylabel[i]==f("H7"))) then
                t.distance[i] = string.format ("%7.2f",
                    refdistmet*(refvolmet/t.volnew[i])^(1/6))
                if (t.gradenew[i]==a) then -- error bounds scaled by grading of integration
                    t.limit[i] = string.format ("%7.1f", t.distance[i]*maxdevmet)
                else
                    if (t.gradenew[i]==b) then
                        t.limit[i] = string.format ("%7.1f",
                            t.distance[i]*maxdevmet*1.2)
                    else
                        t.limit[i] = string.format ("%7.1f",
                            t.distance[i]*maxdevmet*1.4)
                    end
                end
                outfile:write (f2("methyl:  ")..f(i)..t.labelnew[i]..assx
..t.assylabel[i]..t.gradenew[i]..t.volnew[i]..t.distance[i]..t.limit[i].."\n")
            elseif (((HNFx(i)==true) or (h2ox(i)==true) or (d2ox(i)==true)) and ((h2oy(i)==true)
                or (t.assylabel[i]==f("H42")))) or (((h2ox(i)==true) or (assx==f("H42"))) and
                ((HNFy(i)==true) or (h2oy(i)==true) or (d2oy(i)==true))) then
                t.distance[i] = string.format ("%7.2f",
                    refdistcyth42h5*(refvolcyth42h5/t.volnew[i])^(1/6))
                if (t.gradenew[i]==a) then -- error bounds scaled by grading of integration
                    t.limit[i] = string.format ("%7.1f", t.distance[i]*maxdevcyth42h5)
                else
                    if (t.gradenew[i]==b) then
                        t.limit[i] = string.format ("%7.1f",
                            t.distance[i]*maxdevcyth42h5*1.2)
                    else
                        t.limit[i] = string.format ("%7.1f",
                            t.distance[i]*maxdevcyth42h5*1.4)
                    end
                end
                outfile:write (f2("d2o_h2o:   ")..f(i)..t.labelnew[i]..assx..
t.assylabel[i]..t.gradenew[i]..t.volnew[i]..t.distance[i]..t.limit[i].."\n")

            elseif (((HNFx(i)==true) or (h2ox(i)==true) or (d2ox(i)==true)) and
                (t.assylabel[i]==f("H41"))) or ((assx==f("H41")) and ((HNFy(i)==true) or
                (h2oy(i)==true) or (d2oy(i)==true))) then
                t.distance[i] = string.format ("%7.2f",
                    refdistcyth41h5*(refvolcyth41h5/t.volnew[i])^(1/6))
                if (t.gradenew[i]==a) then -- error bounds scaled by grading of
                    integration
                    t.limit[i] = string.format ("%7.1f",
                        t.distance[i]*maxdevcyth41h5)
                else
                    if (t.gradenew[i]==b) then
```

Script code

```
                            t.limit[i] = string.format ('%7.1f',
                                t.distance[i]*maxdevcyth41h5*1.2)
                        else
                            t.limit[i] = string.format ('%7.1f',
                                t.distance[i]*maxdevcyth41h5*1.4)
                        end
                    end
                    outfile:write (f2('H41:
                        ')..f(i)..t.labelnew[i]..assx..t.assylabel[i]..t.gradenew[i]
..t.volnew[i]..t.distance[i]..t.limit[i].."\n")
                else -- write out peaks not used to file !!!
                    t.distance[i] = string.format ('%7.2f',
                        refdistcyt*(refvolcyt/t.volnew[i])^(1/6))
                    if (t.gradenew[i]==a) then -- error bounds scaled by grading of integration
                        t.limit[i] = string.format ('%7.1f', t.distance[i]*maxdevcyt)
                    else
                        if (t.gradenew[i]==b) then
                            t.limit[i] = string.format ('%7.1f',
                                t.distance[i]*maxdevcyt*1.2)
                        else
                            t.limit[i] = string.format ('%7.1f',
                                t.distance[i]*maxdevcyt*1.4)
                        end
                    end
                    outfile2:write (f2('not used (ref on C H5-H6):
                        ')..f(i)..t.labelnew[i]..assx..t.assylabel[i]..
t.gradenew[i]..t.volnew[i]..t.distance[i]..t.limit[i].."\n")
                end
            end
        end
```

---------------------- End of Main Body ----------------------

```
-- close outfile
outfile:close()
outfile2:close()

local i = 0
```

---------------------- End of THIRD PART ----------------------

--   FOURTH PART   --

---------------------- PREPARATIONS ----------------------

```
-- open outfile
outfile = io.output( t.Filename..'_reference.peaks' )
```

4 Lua scripts written for data export from CARA

```
function f ( String )  -- function to format the atomlabels
    FormattedString = string.format( '%7.2f', String )
    return FormattedString
end
```

-- Main Body --

```
outfile:write
    ('\n----------------------------------------\n\nReference
    for non-exchangeable proton cross-peaks: CYT H5-H6\n\nreference_vol   ref_dist   standard_dev
    maximum_dev(%)\n'..refvolcyt..f(refdistcyt)..stddevcyt..maxdevcyt..'\n\n                Peaklabel      Volume        Dist      Dev\n')
for j,assx in pairs (t.assxnew) do
    for y in string.gfind (t.labelnew[j],'H[56]/H[56] [0-9]:C[0-9]+') do  -- reference for d2o peaks
        outfile:write
            (t.labelnew[j]..t.volnew[j]..t.distance[j]..f(t.distance[j]-refdistcyt)..'\n')
    end
end
outfile:write
    ('\n----------------------------------------\n\nReference
    for methyl proton cross-peaks: MET H6-H7\n\nreference_vol   ref_dist   standard_dev
    maximum_dev(%)\n'..refvolmet..f(refdistmet)..stddevmet..maxdevmet..'\n'       Peaklabel      Volume        Dist      Dev\n')
for j,assx in pairs (t.assxnew) do
    for y in string.gfind (t.labelnew[j],'H[67]/H[67] [0-9]:T[0-9]+') do  -- reference for methyl peaks
        outfile:write
            (t.labelnew[j]..t.volnew[j]..t.distance[j]..f(t.distance[j]-refdistmet)..'\n')
    end
end
outfile:write
    ('\n----------------------------------------\n\nReference
    for exchangeable proton cross-peaks: CYT H41-H42\n\nreference_vol   ref_dist   standard_dev
    maximum_dev(%)\n'..refvolcytamino..f(refdistcytamino)..stddevcytamino..maxdevcytamino..'\n'        Peaklabel      Volume        Dist      Dev\n')
for j,assx in pairs (t.assxnew) do
    for y in string.gfind (t.labelnew[j],'H4[12]/H4[12] [0-9]:C[0-9]+') do  -- reference for h2o exchangeable peaks
        outfile:write
            (t.labelnew[j]..t.volnew[j]..t.distance[j]..f(t.distance[j]-refdistcytamino)..'\n')
    end
end
outfile:write
    ('\n----------------------------------------\n\nReference
    for non-exchangeable/exchangeable proton cross-peaks: CYT H42-H5\n\nreference_vol   ref_dist   standard_dev
    maximum_dev(%)\n'..refvolcyth42h5..f(refdistcyth42h5)..stddevcyth42h5..maxdevcyth42h5..'\n\n       Peaklabel      Volume        Dist      Dev\n')
```

Script code

```
for j,assx in pairs (t.assxnew) do
    for y in string.gfind (t.labelnew[j],"H42/H5 [0-9]:C[0-9]+") do -- reference for h2o
        exchangeable-non-exchangeable peaks (appears twice because of selection reasons)
        outfile:write
            (t.labelnew[j]..t.volnew[j]..t.distance[j]..f(t.distance[j]-refdistcyth42h5).."\n")
    end
end
for j,assx in pairs (t.assxnew) do
    for y in string.gfind (t.labelnew[j],"H5/H42 [0-9]:C[0-9]+") do -- reference for h2o
        exchangeable-non-exchangeable peaks
        outfile:write
            (t.labelnew[j]..t.volnew[j]..t.distance[j]..f(t.distance[j]-refdistcyth42h5).."\n")
    end
end
outfile:write
    ("\n--------------------------------------------------------\n\nReference
    for H41 proton cross-peaks: CYT H41-H5\n\nreference_vol  ref_dist  standard_dev
    maximum_dev(%)\n"..refvolcyth41h5..f(refdistcyth41h5)..stddevcyth41h5..maxdevcyth41h5.."\n\n
            Peaklabel    Volume    Dist    Dev\n")
for j,assx in pairs (t.assxnew) do
    for y in string.gfind (t.labelnew[j],"H5/H41 [0-9]:C[0-9]+") do -- reference for
        H41-containing peaks
        outfile:write
            (t.labelnew[j]..t.volnew[j]..t.distance[j]..f(t.distance[j]-refdistcyth41h5).."\n")
    end
end
for j,assx in pairs (t.assxnew) do
    for y in string.gfind (t.labelnew[j],"H41/H5 [0-9]:C[0-9]+") do -- reference for
        H41-containing peaks
        outfile:write
            (t.labelnew[j]..t.volnew[j]..t.distance[j]..f(t.distance[j]-refdistcyth41h5).."\n")
    end
end

------------------------ End of Main Body ------------------------

-- close outfile
outfile:close()

i = 0
```

------------------------ End of FOURTH PART ------------------------
---

--   FIFTH PART   --

------------------------ PREPARATIONS ------------------------

4 Lua scripts written for data export from CARA

```
--- open outfile
outfile   = io.output( t.Filename.."_xplor.list" )
outfile2  = io.output( t.Filename.."_xplor_all.list" )
outfile3  = io.output( 'picktbl_'..t.Filename)
outfile4  = io.output( t.Filename.."_xplor.noe" )
outfile5  = io.output( t.Filename.."_xplor_all.noe" )
outfile6  = io.output( 'picktbl_all_'..t.Filename)
outfile7  = io.output( t.Filename.."_xplor_longdist.list" )
outfile8  = io.output( t.Filename.."_xplor_longdist.noe" )
outfile9  = io.output( t.Filename.."_xplor_H1H5sug.list" )
outfile10 = io.output( t.Filename.."_xplor_H1H5sug.noe" )
outfile11 = io.output( 'picktbl_longdist_'..t.Filename)
outfile12 = io.output( 'picktbl_H1H5sug_'..t.Filename)

function find (index) -- function to find atomlabels
        local Boolean = false
        local Booleanx = false
        local Booleany = false
        for x in string.gfind(t.assxlabel[index],"H[2345]'[']*") do
                Booleanx=true
        end
        for y in string.gfind(t.assylabel[index],"H[2345]'[']*") do
                Booleany=true
        end
        if (Booleanx==true) and (Booleany==true) then
                Boolean=true
        else
                Boolean=false
        end
        return Boolean
end
function find_h1 (index) -- function to format the atomlabels
        local Boolean = false
        local Booleanx = false
        local Booleany = false
        for x in string.gfind(t.assxlabel[index],"H[12345]'[']*") do
                Booleanx=true
        end
        for y in string.gfind(t.assylabel[index],"H[12345]'[']*") do
                Booleany=true
        end
        if (Booleanx==true) and (Booleany==true) then
                Boolean=true
        else
                Boolean=false
        end
        return Boolean
end
function find_h1h5sug (index) -- function to format the atomlabels
        local Boolean = false
        local Booleanx = false
        local Booleany = false
        for x in string.gfind(t.assxlabel[index],"H[15]'[']*") do
                Booleanx=true
        end
```

Script code

```
            for y in string.gfind(t.assylabel[index],"H[15]'[']*") do
                    Booleany=true
            end
            if (Booleanx==true) and (Booleany==true) then
                    Boolean=true
            else
                    Boolean=false
            end
            return Boolean
end

for i,assx in pairs (t.assxlabel) do -- iterate over all peaks
        if (t.assxlabel[i]=="H5") then
                t.assxlabel[i]="HQ5"
        elseif (t.assxlabel[i]=="H2") then
                t.assxlabel[i]="HQ2"
        elseif (t.assxlabel[i]=="H7") then
                t.assxlabel[i]="H7#"
        end
end
for i,assy in pairs (t.assylabel) do -- iterate over all peaks
        if (t.assylabel[i]=="H5") then
                t.assylabel[i]="HQ5"
        elseif (t.assylabel[i]=="H2") then
                t.assylabel[i]="HQ2"
        elseif (t.assylabel[i]=="H7") then
                t.assylabel[i]="H7#"
        end
end
for i,assx in pairs (t.assxnew) do -- iterate over all peaks
        if (t.distance[i]) and (find(i)==false) then -- filter out negative volume peaks
                if (find_h1(i)==false) and ((t.distance[i]/1) < 4.5) then
                        outfile:write ("assign (resid "..t.assxresid[i].." and
                            name"..t.assxlabel[i]..") (resid "..t.assyresid[i].." and
                            name"..t.assylabel[i]..") "..t.distance[i]..t.limit[i]..t.limit[i].."\n")
                        outfile2:write ("assign (resid "..t.assxresid[i].." and
                            name"..t.assxlabel[i]..") (resid "..t.assyresid[i].." and
                            name"..t.assylabel[i]..") "..t.distance[i]..t.limit[i]..t.limit[i].."\n")
                        outfile3:write ("pick bond (resid "..t.assxresid[i].." and
                            name"..t.assxlabel[i]..") (resid "..t.assyresid[i].." and
                            name"..t.assylabel[i]..") ".." geometry\ndisplay \$result".."\n")
                        outfile4:write (t.assxresid[i].. t.assxlabel[i]..t.assyresid[i]
..t.assylabel[i]..t.labelnew[i]..t.gradenew[i]..t.distance[i].."\n")
                        outfile5:write (t.assxresid[i].. t.assxlabel[i]..t.assyresid[i]..
t.assylabel[i]..t.labelnew[i]..t.gradenew[i]..t.distance[i].."\n")
                elseif (find_h1(i)==false) then
                        outfile2:write ("assign (resid "..t.assxresid[i].." and
                            name"..t.assxlabel[i]..") (resid "..t.assyresid[i].." and
                            name"..t.assylabel[i]..") "..t.distance[i]..t.limit[i]..t.limit[i].."
                            !added!\n")
                        outfile5:write (t.assxresid[i].. t.assxlabel[i]..t.assyresid[i]..
t.assylabel[i]..t.labelnew[i]..t.gradenew[i]..t.distance[i].."\n")
                        outfile6:write ("pick bond (resid "..t.assxresid[i].." and
                            name"..t.assxlabel[i]..") (resid "..t.assyresid[i].." and
                            name"..t.assylabel[i]..") ".." geometry\ndisplay \$result".."\n")
                        outfile7:write ("assign (resid "..t.assxresid[i].." and
```

4 Lua scripts written for data export from CARA

```
                                    name"..t.assxlabel[i].."') (resid'"..t.assyresid[i].."' and
                                    name'"..t.assylabel[i].."')'"..t.distance[i]..t.limit[i]..t.limit[i].."'
                                    !added!\n")
                        outfile8:write (t.assxresid[i].. t.assxlabel[i]..t.assyresid[i]
..t.assylabel[i]..t.labelnew[i]..t.gradenew[i]..t.distance[i].."'\n")
                        outfile11:write ("pick bond (resid'"..t.assxresid[i].."' and
                                    name"..t.assxlabel[i].."') (resid'"..t.assyresid[i].."' and
                                    name"..t.assylabel[i].."')'..' geometry\ndisplay \$result".."'\n")
                    elseif (find_h1h5sug(i)==true) then
                        outfile2:write ("assign (resid'"..t.assxresid[i].."' and
                                    name"..t.assxlabel[i].."') (resid'"..t.assyresid[i].."' and
                                    name'"..t.assylabel[i]..")'"..t.distance[i]..t.limit[i]..t.limit[i].."'
                                    !added!\n")
                        outfile5:write (t.assxresid[i].. t.assxlabel[i]..t.assyresid[i]..
t.assylabel[i]..t.labelnew[i]..t.distance[i].."'\n")
                        outfile6:write ("pick bond (resid'"..t.assxresid[i].."' and
                                    name"..t.assxlabel[i].."') (resid'"..t.assyresid[i].."' and
                                    name'"..t.assylabel[i].."')'..' geometry\ndisplay \$result".."'\n")
                        outfile9:write ("assign (resid'"..t.assxresid[i].."' and
                                    name"..t.assxlabel[i].."') (resid'"..t.assyresid[i].."' and
                                    name"..t.assylabel[i].."')'"..t.distance[i]..t.limit[i]..t.limit[i].."'
                                    !added!\n")
                        outfile10:write (t.assxresid[i].. t.assxlabel[i]..t.assyresid[i]..
t.assylabel[i]..t.labelnew[i]..t.gradenew[i]..t.distance[i].."'\n")
                        outfile12:write ("pick bond (resid'"..t.assxresid[i].."' and
                                    name"..t.assxlabel[i].."') (resid'"..t.assyresid[i].."' and
                                    name"..t.assylabel[i].."')'..' geometry\ndisplay \$result".."'\n")
                    else
                        outfile2:write ("assign (resid'"..t.assxresid[i].."' and
                                    name"..t.assxlabel[i].."') (resid'"..t.assyresid[i].."' and
                                    name"..t.assylabel[i].."')'"..t.distance[i]..t.limit[i]..t.limit[i].."'
                                    !added!\n")
                        outfile5:write (t.assxresid[i]..
                                    t.assxlabel[i]..t.assyresid[i]..t.assylabel[i]
..t.labelnew[i]..t.gradenew[i]..t.distance[i].."'\n")
                        outfile6:write ("pick bond (resid'"..t.assxresid[i].."' and
                                    name"..t.assxlabel[i].."') (resid'"..t.assyresid[i].."' and
                                    name'"..t.assylabel[i].."')'..' geometry\ndisplay \$result".."'\n")
                    end
            end
end

------------------------------ End of Main Body ------------------------------

--- close outfile
outfile:close()
outfile2:close()
outfile3:close()
outfile4:close()
outfile5:close()
outfile6:close()
outfile7:close()
outfile8:close()
outfile9:close()
outfile10:close()
```

Script code

```
outfile11:close()
outfile12:close()

i = 0
t = nil
```

———————————————————— End of FIFTH PART ————————————————————

```
print ( '\ngenerateinput_byanda is done.' )
print ( 'Have a nice day!' )
```

———————————————————— End generateinput_byanda ————————————————————

5 Utility scripts

In this section a short collection of the most frequently used utility scripts is presented. They are grouped according to their function and a small description is supplied.

Scripts for converting program inputs/outputs for further use in other programs

The next two scripts are used to convert a SPARKY resonance table to a CARA atomlist.

```
language
#!/bin/csh
#
# Skript um aus Sparky-resonance-table eine CARA-atomlist zu machen
#
awk 'NR==1 {print "#Number of dimensions 2"; print "INAME 1 H"; print
    "INAME 2 H"} {print NR-2" "$2" "$3" 0 U    0.000    0 - 0 "
```

```
#!/bin/csh
#
# Skript um aus Sparky-resonance-table eine CARA-atomlist zu machen
#
awk 'NR==1 {gsub (/h2$/,"HQ2",$2); gsub (/h5$/,"HQ5",$7); for (i=1;i<=(NF-1);i++) {printf
    "%5s",toupper($i)}} /me/ {printf "%5s%5s%5s\n","H71","H72","H73"} {if (NR!=1) {print $0"
    $NF" "$NF}}' $1".shifts" > bla

awk 'BEGIN {COUNT=0} {if (NR==1) {for (i=1;i<=NF;i++) a[i]=$i} else {for (i=2;i<=NF;i++)
    {COUNT++; printf "%s\n",COUNT" "$i" 0.005 "a[i-1]" "NR-1}}}' bla > bla2

awk '{gsub(/g/,"GUA"); gsub(/c/,"CYT"); gsub(/a/,"ADE"); gsub(/t/,"THY"); gsub(/x/,"PUR");
    gsub(/-/,"999"); print}' bla2 > $1_cara.prot
```

The next three scripts are all needed for the NOESY back-calculation with the program GIFA (the first to modify the psf-file, the second to generate the ppm-file needed as input and the last to convert the XPLOR-NIH back-calculation output into GIFA-format.

```
language
#!/bin/csh
#
# Dateiname des PSF-Files ohne Extension muss uebergeben werden.
# Es werden nur CYT,Thy,Ade,Gua selektiert , wenn andere
# Modifikationen vorhanden entsprechend mit einfuehren.
#
awk '{if (($1~/[0-9]+/)&&($3~/^[a-zA-Z]+$/)&&($3~/~GUA|ADE|CYT|THY|PUR$/)) {$6="..000000";
    $7="1.008"; printf "%14s%-5s%-5s%-5s%-16s%-7s%12s\n",$1"     ",$2,$3,$4,$5,$6,$7,$8;
    NR=1} else print}' $1.psf > $1"_spect.psf"
#
#
#BEGIN {getline VAR < "'"$1"'.ppm"; split (VAR,a)}
```

Script code

```
# if ($4~/^H/) { if ((a[4]==$4)&&(a[2]==$3)) {getline VAR < "'"$1"'.ppm"; split (VAR,a);
#else {$6=".000000"; $7='1.008'; printf "%14s%-5s%-5s%-5s%-6s%-16s%-7s%12s\n",$1"
    ",$2,$3,$4,$5,$6,$7,$8}} else print

#!/bin/csh
#
# Uebergabe des filenamens ohne Extension von pdb-file und shifts-file
# Viel Modifizieren nötig, im shifts-file klein h durch goss H ersetzen
# nur A, C, T, G selektiert!!!
# in letzter awk-zeile ist die i<=Zahl anzupassen auf die Zeilenanzahl des ppm-files
awk 'NR==1 {gsub (/h2$/,"HQ2",$2); gsub (/h5$/,"HQ5",$7); for (i=1;i<=(NF-1);i++) {printf
    "%5s",toupper($i)}} /me/ {printf "%5s%5s%5s\n","H71","H72","H73"} {if (NR!=1) {print $0"
    "$NF" "$NF}}' $2"_2.shifts" > bla
awk '{if (NR==1) {for (i=1;i<=NF;i++) a[i]=$i} else {for (i=2;i<=NF;i++) printf "%s\n","PPM"
    "$1" "NR-1" "a[i-1]" "$i" 1"}}' bla > bla2
awk '{gsub(/g[0-9]+/,"GUA"); gsub(/c[0-9]+/,"CYT"); gsub(/a[0-9]+/,"ADE");
    gsub(/t[0-9]+/,"THY"); gsub(/x[0-9]+/,"PUR"); print}' bla2 > bla3
awk '{if ($(NF-1)!~/^\-$/) print}' bla3 > $1.ppm
#awk '{getline VAR < "'"$1"'.pdb"; split (VAR,a); store=a[1]; if (a[1]=="ATOM") {store=a[1];
    if ((a[3]==$3)&&($4!~/\-/)) {printf "%s\n",$1" 'a[4]' '$2' '$3' '$4' 1"; i=1}; getline
    VAR < "'"$1"'.pdb"; split (VAR,a)}}' bla4> $1.ppm
#awk '{if ($1=="ATOM") {if ($3!~/H/) {print} else {getline VAR < "'"$1"'.ppm"; split (VAR,a);
    for (i=1;i<=211;i++) {if (($4==a[2])&&($5==a[3])&&($3==a[4])) {print; close
    ("'"$1"'.ppm"); break} else {getline VAR < "'"$1"'.ppm"; split (VAR,a)}}; close
    ("'"$1"'.ppm")}} else print}' $1.pdb > $1_gifa.pdb
awk '{getline VAR < "'"$1"'.ppm"; split (VAR,a); for (i=1;i<=211;i++) {if
    (($2==a[2])&&($1==a[3])&&($3==a[4])) {close ("'"$1"'.ppm"); for (i=1;i<=211;i++) {if
    (($5==a[2])&&($4==a[3])&&($6==a[4])) {print "INT" '$2' '$1' '$3' '$5' '$4' '$6' '$7; close
    ("'"$1"'.ppm"); break} else {getline VAR < "'"$1"'.ppm"; split (VAR,a)}} break} else
    {getline VAR < "'"$1"'.ppm"; split (VAR,a)}}; close ("'"$1"'.ppm")}' $2.spect >
    $2_gifa.spect'
rm bla*

#!/bin/csh
#
# Uebergabe des .spect files ohne Extension
#
set VAR='echo $1'
awk '{if ((NF==9)&&($3~/^(H1[^0-9a-zA-Z]|H2[^0-9a-zA-Z]*|HQ2|HQ5|H3[^0-9a-zA-Z]|H4[^0-9a-zA-Z]|
    H5[^0-9a-zA-Z]*|H6|H7[0-9]|H8)$/)&&($6~/^(H1[^0-9a-zA-Z]|H2[^0-9a-zA-Z]*|HQ2|HQ5|H3[^0-9a-zA-Z]|
    H4[^0-9a-zA-Z]|H5[^0-9a-zA-Z]*|H6|H7[0-9]|H8)$/)) {printf
    "%-4s%-4s%-3s%-5s%-4s%-3s%-5s%-10s\n","INT",$2,$1,$3,$5,$4,$6,$7}}' $VAR.spect >
    $VAR'_gifa.spect'
```

The next script is used to generate a latex table from a sparky shifts-file.

```
language
#!/bin/csh
#
# Skript um aus sparky assignment tables tabellen für Latex zu machen
# Es wird der Dateiname des shifts-files mit extension übergeben.
#
awk '{if (NR==1) {print " \\arrayrulewidth0.5pt"; print " \\doublerulesep0pt"; print "
    \\begin\{tabular\}\{c\|\|\|\*\{'(NF-1)'\}\{\>\
{\\PBS\\centering\}p\{0.9cm\}\|\}\>\{\\PBS\\centering\}p\{0.9cm\}\}"; printf "%s","&"; for
    (i=1;i<=NF;i++) {if (i==NF) {printf "%s","\\textbf\{'$i'\}"; printf
```

254

5 Utility scripts

```
'%s\n%s\n',"\\\\",'\\hline\\hline\\hline'} else printf "%s","\\textbf\{"$i"\}\&"}} else
{for (i=1;i<=NF;i++) {if (i==1) {printf "%s","\\textbf\{"$i"\}\&"} else {if (i==NF)
{printf "%s",$i; printf "%s\n","\\\\"} else printf "%s",$i"\&"}}}' $1 > $1"_tab.tex"
```

The next script is used to generate and XPLOR-NIH RDC restraints file from the input to the PALES-program.

```
#!/bin/csh
# used to convert pales input to xplor RDC input (with axes2.pdb and axis_500.psf)
awk '{if (NR==1) {print "! RDC table"}; if ((NR>3)&&(NF==10)) {print "assign ( resid 500 and
  name OO     )\n   ( resid 500  and name Z    )\n       ( resid 500  and name X
  )\n     ( resid 500  and name Y   )\n    ( resid "$1"     and name "$3"             )\n
       ( resid "$1"       and name "$6"        ) "$(NF-3)" "$(NF-2)" "$(NF-1)"\n\n"}}' $1 >
$1".xplor"
```

The next two scripts are used for generating random data subsets from RDC and NOE restraints files, respectively.

```
#!/bin/csh
# script to pick out every fourth (or fifth etc) NOE point from input
set VAR='echo $1'
awk '{if ($1~/^assign/) print}' $1 > bla
awk 'BEGIN {DUMMY=5} {if (NR==DUMMY) {DUMMY=DUMMY+3; print > "'$VAR'".unused"} else print}'
   bla > $1.third
rm bla
```

```
#!/bin/csh
# script to pick out every fourth (or fifth etc) RDC point from PALES input
set VAR='echo $1'
awk 'BEGIN {DUMMY=6;print "VARS   RESID_I RESNAME_I ATOMNAME_I RESID_J RESNAME_J ATOMNAME_J D
   DD W\nFORMAT \%5d \%6s \%6s \%5d \%6s \%6s \%9.3f \%9.3f \%.2f" > "'$VAR'".unused"} {if
   ((NR>3)&&(NR==DUMMY)) {DUMMY=DUMMY+3; print > "'$VAR'".unused"} else print}' $1 >
   $1.third
```

The next script is used for generating a single pdb-file from the average and the 10 minimum-energy structures for submission to the PDB databank.

```
#!/bin/csh
# script for generating pdb submission file with averaged,minimized structures as model 1
# and the 10 minimum-energy structures as models 2-11. TER cards, chainIDs are inserted
# and ANI residues deleted.
#
set pdbs = 'awk '{if ($1~/pdb/) {printf "%s",$1" "}}' *_##.pdb.stats''
cat average_min.pdb $pdbs > bla
awk 'BEGIN {print "MODEL      1";COUNT1=0;COUNT2=2} {if ($1=="REMARK") print; if
   (($1=="ATOM")&&($5<=13)&&($4!="ANI")) {printf
   "%-5s%5s%5s%4s%2s%4s%12s%8s%8s%6s%6s%10s\n",$1,$2,$3,$4,"A",$5,$6,$7,$8,$9,$10,"
   ";next}; if (($1=="ATOM")&&($5>=14)&&($4!="ANI")) {if (COUNT1==0) {print "TER"};printf
   "%-5s%5s%5s%4s%2s%4s%12s%8s%8s%6s%6s%10s\n",$1,$2,$3,$4,"B",$5,$6,$7,$8,$9,$10,"
   ";COUNT1=1;next}; if (($1=="END")&&(COUNT2<12)) {printf
   "%s\n%s\n%s%9s\n","TER","ENDMDL","MODEL",COUNT2;COUNT2=COUNT2+1;COUNT1=0;next}; if
```

255

Script code

```
                    ((  $1=='END')&&(COUNT2==12))  {printf "%s\n%s\n%s\n","TER","ENDMDL","END"}}' bla >
            submit.pdb
#awk 'BEGIN {print "MODEL 1";COUNT1=0;COUNT2=1} {if ($1=="REMARK") print;  if
           ((  $1=='ATOM')&&($5<=13)&&($4!='ANI')) {printf
          '%5s%-6s%-5s%-4s%-2s%-4s%-12s%-8s%-8s%-6s%-6s%-10s\n',$1,$2,$3,$4,"A",$5,$6,$7,$8,$9,$10,"
           '}}' bla > bla2
rm bla
unset pdbs
```

The next script calls the program XPLOR-VMD with the 10 minimum-energy structures of a calculation for display.

```
    language
#!/bin/csh
# script for displaying top 10 structures of xplor-nih python script in xvmd
set pdbs = 'awk '{if ($1~/pdb/) {printf "%s",$1 '}}' *'_##.pdb.stats''
vmd-xplor $pdbs
unset pdbs
```

The next script collects the CURVES helical parameter calculation out of the 3DNA output files that were generated for the 10 minimum-energy structures of a XPLOR-NIH calculation and prints them to one file.

```
    language
#!/bin/csh
#
# Einmaliges Skript zum umnummerieren von schon vorhandenen Dateien
set i=1
while ($i<= 10)
    grep -A 17 'Curves' $i/cf_7methods.par >> comp_hel_par.tab
    @ i++
end
unset i
```

The next script selects just a subset of NOEs from an NOE input file, the rest is commented out.

```
    language
#!/bin/csh
#
# uebergabe des .tbl files ohne extension
#
awk '{if ((($6~/[hH]1.\)/)&&($11~/[hH]6\)|[hH]8\)/))
||(($6~/[hH]6\)|[hH]8\)/)&&($11~/[hH]1.\)/))||((($6~/^[hH]2[^0-9a-zA-Z]\)$/)
&&(($11=="H6)")||($11=="H8)"))&&($3==88))||(($6~/
[hH]2[^0-9a-zA-Z][^0-9a-zA-Z]\)/)&&(($11=="H6)")||($11=="H8)"))&&($3!=88))))
    {if ($1=="assign") {printf "%s\n",$0} else {$1="assign"; printf "%s\n",$0}} else
        {$1="\!assign"; printf "%s\n",$0}}' $1_2.tbl > bla
awk '$1==$1 {printf "%-9s%-10s%-3s%-4s%-5s%-6s%-10s%-3s%-4s%-5s%-12s%-12s%-8s%-5s%-8s%-4s\n",
$1,$2,$3,$4,$5,$6,$7,$8,$9,$10,$11,$12,$13,$14,$15,$16}' bla > $1'_3.tbl'
rm bla*
#(($6~/[hH]6|[hH]8/)&&($11~/[hH]6|[hH]8/))||
```

5 Utility scripts

Scripts for renaming atoms in files

The script is used to rename certain atoms in an input file to XPLOR-NIH calculations.

```
language
#!/bin/csh
sort $2 > bla
sort $1.list > bla2
awk '$1!="display" {gsub("H5")","HQ5)",$0);gsub("H2")","HQ2)",$0);gsub("H7")","H7#)",$0);print
    $0}'\ndisplay \$result'}' bla > $2
awk '{gsub("H5")","HQ5)",$0);gsub("H2")","HQ2)",$0);gsub("H7")","H7#)",$0);print $0}' bla2 >
    $1.tbl
rm bla*
```

The script is used to rename certain atoms in a whole family of calculated structures.

```
language
#!/bin/csh
#
#set i=1
#while ($i<= 100)
foreach i (*.pdb)
cp $i bla$i
#   sed 's/H2'\'''\'' HNF/H2bb HNF/g' $i > 'bla'$i
#   sed 's/ H2'\'' HNF/H2'\'''\'' HNF/g' 'bla'$i > 'blabla'$i
#   sed 's/H2bb HNF/ H2'\'' HNF/g' 'blabla'$i > 'bla'$i
  awk '{if (($1=="REMARK")||(($2<208)||($2>235))) {print} else {if ($2==208)
    {gsub($2,$2+24);DUMMY1=$0;next};if ($2==209) {gsub($2,$2+24);DUMMY2=$0;next};if
    ($2==210) {gsub($2,$2+24);DUMMY3=$0;next};if ($2==211)
    {gsub($2,$2+24);DUMMY4=$0;next};if (($2>211)&&($2<235)) {gsub($2,$2-4);print;next};if
    ($2==235) {gsub($2,$2-4);print;print DUMMY1;print DUMMY2;print DUMMY3;print
    DUMMY4;next}}}' bla$i > $i
#   sed 's/H2'\'''\'' HNF/H2bb HNF/g' '13mer_HNF_'$i'.pdb' > 'bla'$i
#   sed 's/ H2'\'' HNF/H2'\'''\'' HNF/g' 'bla'$i > 'blabla'$i
#   sed 's/H2bb HNF/ H2'\'' HNF/g' 'blabla'$i > '13mer_HNF_'$i'.pdb'
#   @ i++
end
rm bla*
```

Scripts for displaying specific information content from files

This script prints out the number of intra-, interresidual and total number of NOEs, as well as the NOEs to the modification site.

```
language
#!/bin/csh
#
# Skript gibt Anzahl der NOE, der intra- und interresidualen der NOEs zur Modifikation
    (Achtung nur Position 7) aus
awk 'BEGIN {INTER=0;INTRA=0;MOD=0} {if ($1!="assign") {DUMMY=0} else {if ($3==$8)
    {INTRA=INTRA+1} else {if (($3==7)||($8==7)) {INTER=INTER+1;MOD=MOD+1} else
    {INTER=INTER+1}}}} END {TOTAL=INTER+INTRA;print " Total= "TOTAL;print "    intraresidual=
    "INTRA;print "    interresidual= "INTER;print " NOE to modification site = "MOD}' $1
#||($3==20)||($8==20)
```

Script code

This script prints out the Da and Rh values from all pdb-files in the working directory.

```
language
#!/bin/csh
# script for displaying Da values of top 10 structures of xplor-nih python script
set pdbs = `awk '{if ($1~/pdb/) {printf "%s",$1' '}}' *_##.pdb.stats'`
grep 'Da:' $pdbs
unset pdbs
```

This script prints out the sorted energies from all pdb-files in the working directory.

```
language
#!/bin/csh
grep 'summary total' *.pdb | awk '{print $4' '$5' ' $1}' | sort -n
```

This script prints out the sorted energies from just a subset of all pdb-files in the working directory.

```
language
#!/bin/csh
#
# Skript um die 10 niedrigsten Energien auszugeben.
#
set i=1
grep 'total' refine*.pdb* | awk '{print $5' '$1' '$2' '$3' '$4}' | sort -n | head -50 | awk
    '{print $5' '$1' '$2' '$3' '$4}' | sort -n | head -50 | awk '{print $3' '$4' '$5' '$1'
    '$2}' > 50minen
cat 50minen
while ($i<= 50)
  set VAR=`awk 'BEGIN {FS="."} {if (NR==''$i'') printf "%s",$1}' 50minen`
  cp $VAR.pdb "min10_'$i'.pdb"
  @ i++
end
#rm bla
```

This script generates files which contain the NOE restraints information (which was used as input for the structure calculations) with the corresponding distances in the 10 minimum-energy structures. Thus not only NOE violations but also too tight NOE restraints can be easily identified.

```
language
#!/bin/csh
#
# Skript um die 10 niedrigsten Energien auszugeben.
# Anschliessend werden die Bindungslaengen der 10 energieniedrigsten Strukturen für die
    noe-constraints
# hinter die eingegebenen noe-constraints ausgeschrieben.
# Es muss der Name des noe-constraint-files ohne die extension tbl angegeben werden.
# Es werden mehrere Files ausgegeben die die Auswertung der Strukturen erleichtern
#
set i=1
set VAR=`echo $1`
```

5 Utility scripts

```
grep 'enviol' *.pdb | awk '{print $3' '$1' '$2' '$4' '$5}' | sort -n | awk '{print $2' '$3'
    '$1' '$4' '$5}' | head -10
grep 'enviol' *.pdb | awk '{print $3' '$1' '$2' '$4' '$5}' | sort -n | awk '{print $2' '$3'
    '$1' '$4' '$5}' | head -10 > blablabla
awk '{printf "%20s%5s%5s%5s",$3'-('$6'\/'$8'-('$11,$12,"-'$(NF-1),'+'$NF; if ($1~/^\!/)
    {printf "%5s\n",'off'} else printf "%5s\n",'on'}' $VAR > bla1
cat bla1 > blabla1
while ($i<= 10)
    set FILE=`awk 'BEGIN {FS="."} {if (NR=="'$i'") print $1}' blablabla`.noe
    echo $FILE
    awk '{printf "%s",$0; getline < "'$FILE'"; printf "%8.4f\n",$0}' bla$i > bonds.log
    awk '{printf "%s",$0; REF=$3; getline < "'$FILE'"; DEV=REF-$1; printf "%8.4f\n",DEV}'
        blabla$i > noedev.log
    @ i++
    mv bonds.log bla$i
    mv noedev.log blabla$i
end
awk '{printf "%s\n\n",$0}' bla$i > bonds.log
awk '{printf "%s\n\n",$0}' blabla$i > noedev.log
awk '{AVE=($(NF-9)+$(NF-8)+$(NF-7)+$(NF-6)+$(NF-5)+$(NF-4)+$(NF-3)+$(NF-2)+$(NF-1)+$NF)/10;
    RMS=((($(NF-9)-AVE)^2+($(NF-8)-AVE)^2+($(NF-7)-AVE)^2+($(NF-6)-AVE)^2+($(NF-5)-AVE)^2+
($(NF-4)-AVE)^2+($(NF-3)-AVE)^2+($(NF-2)-AVE)^2+($(NF-1)-AVE)^2+($NF-AVE)^2)/9)^(1/2);
    RMSNOE=((($(NF-9)-$3)^2+($(NF-8)-$3)^2+($(NF-7)-$3)^2+($(NF-6)-$3)^2+($(NF-5)-$3)^2+($(NF-4)-$3)^2
+($(NF-3)-$3)^2+($(NF-2)-$3)^2+($(NF-1)-$3)^2+($NF-$3)^2)/9)^(1/2); getline < 'bla1'; printf
    '%s%-29s%-19s\n\n',$0,'    RMSD_AVE: 'RMS,"RMSD_NOE: 'RMSNOE}' bla$i > rmsd.log
rm bla*
```

Chemical shifts

In the following the chemical shifts are listed. All proton chemical shifts are referenced to the HOD signal at 4.80 ppm.

6 Chemical shifts of 13merHNF

The chemical shifts which are unique to HNF and/or the abasic site are not listed in the table and thus are given below in ppm.

HNF (Res 7): H1" 5.32, H4 6.78, C1 109.8, C3 113.7, C4 121.2

Abasic site (Res 20): H1" 4.05

Chemical shifts

Tabelle 2: ^1H chemical shifts of sugar protons of 13merHNF

Res	H1'	H2'	H2"	H3'	H4'
G1	6,00	2,67	2,79	4,86	4,28
C2	6,09	2,14	2,54	4,85	4,27
T3	5,75	2,13	2,45	4,89	4,16
G4	5,84	2,65	2,69	5,00	4,38
C5	5,47	1,97	2,31	4,82	4,16
A6	6,34	2,82	2,78	5,06	4,40
A7	5,32	2,30	2,38	4,88	4,65
A8	6,09	2,51	2,81	4,98	4,43
C9	5,50	1,95	2,34	4,82	4,15
G10	5,96	2,62	2,78	4,96	4,37
T11	6,05	2,10	2,47	4,88	4,23
C12	5,72	2,04	2,38	4,86	4,14
G13	6,18	2,64	2,40	4,71	4,20
C14	5,72	1,85	2,36	4,70	4,07
G15	5,45	2,73	2,80	5,01	4,33
A16	6,27	2,72	2,93	5,08	4,50
C17	5,58	1,95	2,32	4,83	4,16
G18	5,94	2,60	2,76	4,98	4,37
T19	6,11	2,35	2,53	4,94	4,23
T20	4,15	2,26	2,27	4,77	4,11
T21	5,79	2,00	2,45	4,91	4,33
G22	5,75	2,58	2,66	4,97	4,33
C23	5,45	1,97	2,31	4,82	4,16
A24	6,01	2,74	2,90	5,05	4,39
G25	5,82	2,48	2,66	4,97	4,36
C26	6,10	2,15	2,22	4,47	4,05

Tabelle 3: ^1H chemical shifts of base protons of 13merHNF

Res	H1	H2	H3	H41	H42	H5	H6	H7	H8
G1	-	-	-	-	-	-	-	-	7,97
C2	-	-	-	8,32	6,62	5,36	7,53	-	-
T3	-	-	13,96	-	-	-	7,34	1,64	-
G4	12,70	-	-	-	-	-	-	-	7,89
C5	-	-	-	8,21	6,40	5,43	7,37	-	-
A6	-	7,10	-	-	-	-	-	-	8,36
A7	5,97	-	6,00	-	-	7,04	7,45	-	7,71
A8	-	7,29	-	-	-	-	-	-	7,99
C9	-	-	-	7,98	6,38	5,09	7,16	-	-
G10	12,73	-	-	-	-	-	-	-	7,83
T11	-	-	13,76	-	-	-	7,29	1,41	-
C12	-	-	-	8,63	7,03	5,72	7,51	-	-
G13	-	-	-	-	-	-	-	-	7,96
C14	-	-	-	8,17	7,01	5,89	7,60	-	-
G15	12,95	-	-	-	-	-	-	-	7,96
A16	-	7,90	-	-	-	-	-	-	8,23
C17	-	-	-	8,09	6,42	5,22	7,19	-	-
G18	12,60	-	-	-	-	-	-	-	7,84
T19	-	-	13,21	-	-	-	7,29	1,50	-
T20	-	-	-	-	-	-	-	-	-
T21	-	-	12,77	-	-	-	7,29	1,51	-
G22	12,54	-	-	-	-	-	-	-	7,83
C23	-	-	-	8,26	6,35	5,38	7,35	-	-
A24	-	7,68	-	-	-	-	-	-	8,18
G25	12,93	-	-	-	-	-	-	-	7,68
C26	-	-	-	8,14	6,57	5,24	7,33	-	-

Chemical shifts

Tabelle 4: ^{13}C chemical shifts of 13merHNF

Res	C1'	C3'	C2	C5	C6	C8
G1	82.2	76.6	-	-	-	135.7
C2	83.9	73.9	-	95.5	140.0	-
T3	82.8	75.3	-	-	136.4	-
G4	81.4	76.5	-	-	-	135.2
C5	83.3	73.7	-	95.4	139.5	-
A6	82.3	76.7	151.3	-	-	139.1
A7	102.0	76.5 -	118.1	121.9	119.2	-
A8	82.7	75.2	150.5	-	-	138.3
C9	83.1	73.8	-	95.1	139.1	-
G10	81.9	76.4	-	-	-	135.2
T11	82.7	75.0	-	-	135.8	-
C12	85.1	75.2	-	96.0	140.7	-
G13	81.8	70.5	-	-	-	136.2
C14	83.7	75.0	-	96.6	140.2	-
G15	81.2	76.6	-	-	-	135.9
A16	82.1	76.6	152.6	-	-	138.1
C17	83.1	73.7	-	95.4	138.9	-
G18	82.1	76.3	-	-	-	135.3
T19	82.7	73.7	-	-	135.8	-
T20	67.2	68.8	-	-	-	-
T21	83.5	75.2	-	-	135.9	-
G22	81.2	76.4	-	-	-	135.1
C23	83.5	73.6	-	95.4	139.9	-
A24	82.1	76.6	151.4	-	-	138.5
G25	81.2	76.3	-	-	-	134.4
C26	84.0	68.6	-	95.1	139.7	-

7 Chemical shifts of 13merRefGC

Tabelle 5: ^1H chemical shifts of sugar protons of 13merRefGC

Res	H1'	H2'	H2"	H3'	H4'		
G1	6,02	2,68	2,81	4,87	4,29	3,76	3,76
C2	6,11	2,15	2,56	4,87	4,29	4,21	4,18
T3	5,77	2,17	2,49	4,91	4,18	4,17	4,13
G4	5,88	2,65	2,72	5,01	4,39	4,08	4,15
C5	5,48	1,93	2,30	4,82	4,15	4,13	4,19
A6	5,94	2,70	2,87	5,04	4,37	4,00	4,12
A7	5,46	2,57	2,69	5,00	4,36	4,15	4,17
A8	6,18	2,60	2,89	5,02	4,45	4,15	4,21
C9	5,58	2,00	2,35	4,81	4,15	4,24	4,19
G10	5,97	2,61	2,79	4,94	4,37	4,08	4,15
T11	6,06	2,10	2,48	4,89	4,24	4,10	4,15
C12	5,73	2,04	2,39	4,86	4,14	4,08	4,10
G13	6,19	2,65	2,40	4,71	4,21	4,15	4,11
C14	5,74	1,88	2,37	4,71	4,08	3,73	3,73
G15	5,47	2,74	2,82	5,02	4,33	3,99	4,11
A16	6,28	2,73	2,94	5,08	4,50	4,19	4,25
C17	5,60	2,03	2,38	4,83	4,18	4,15	4,28
G18	5,97	2,61	2,80	4,94	4,38	4,10	4,16
T19	6,04	2,19	2,55	4,89	4,27	4,24	4,16
T20	6,00	2,11	2,51	4,81	4,19	4,24	4,15
T21	5,71	2,11	2,46	4,88	4,15	4,09	4,13
G22	5,85	2,66	2,71	4,99	4,37	4,13	4,15
C23	5,49	1,98	2,32	4,83	4,14	4,12	4,19
A24	6,01	2,74	2,89	5,05	4,39	4,01	4,13
G25	5,83	2,49	2,66	4,97	4,36	4,15	4,20
C26	6,13	2,15	2,22	4,47	4,06	4,25	4,06

Chemical shifts

Tabelle 6: ^1H *chemical shifts of base protons of 13merRefGC*

Res	H1	H2	H3	H41	H42	H5	H6	H7	H8
G1	-	-	-	-	-	-	-	-	8,00
C2	-	-	-	8,33	6,57	5,39	7,56	-	-
T3	-	-	13,92	-	-	-	7,37	1,67	-
G4	12,92	-	-	-	-	-	-	-	7,91
C5	-	-	-	8,33	6,64	5,40	7,34	-	-
A6	-	7,52	-	-	-	-	-	-	8,15
A7	12,59	-	-	-	-	-	-	-	7,69
A8	-	7,76	-	-	-	-	-	-	8,09
C9	-	-	-	8,06	6,40	5,17	7,18	-	-
G10	12,70	-	-	-	-	-	-	-	7,82
T11	-	-	13,76	-	-	-	7,29	1,40	-
C12	-	-	-	8,61	7,00	5,72	7,52	-	-
G13	-	-	-	-	-	-	-	-	7,97
C14	-	-	-	-	-	5,91	7,62	-	-
G15	12,94	-	-	-	-	-	-	-	7,98
A16	-	7,89	-	-	-	-	-	-	8,23
C17	-	-	-	8,12	6,45	5,24	7,22	-	-
G18	12,74	-	-	-	-	-	-	-	7,82
T19	-	-	13,66	-	-	-	7,27	1,37	-
T20	-	-	-	8,33	6,89	5,60	7,59	-	-
T21	-	-	13,79	-	-	-	7,34	1,67	-
G22	12,69	-	-	-	-	-	-	-	7,90
C23	-	-	-	8,29	6,56	5,40	7,37	-	-
A24	-	7,64	-	-	-	-	-	-	8,19
G25	12,72	-	-	-	-	-	-	-	7,69
C26	-	-	-	8,30	6,57	5,32	7,38	-	-

8 Chemical shifts of 13merRef

Tabelle 7: ^1H chemical shifts of sugar protons of 13merRef

Res	H1'	H2'	H2"	H3'	H4'
G1	6.02	2.68	2.81	4.87	4.28
C2	6.11	2.15	2.55	4.86	4.28
T3	5.76	2.15	2.47	4.90	4.17
G4	5.86	2.63	2.70	4.99	4.38
C5	5.41	1.84	2.22	4.79	4.11
A6	5.75	2.71	2.84	5.03	4.36
A7	5.84	2.63	2.84	5.05	4.43
A8	6.05	2.54	2.83	5.00	4.44
C9	5.56	1.92	2.32	4.77	4.14
G10	5.95	2.61	2.78	4.95	4.36
T11	6.05	2.10	2.47	4.88	4.23
C12	5.72	2.03	2.38	4.85	4.14
G13	6.18	2.64	2.40	4.70	4.20
C14	5.73	1.87	2.37	4.70	4.07
G15	5.46	2.73	2.80	5.01	4.33
A16	6.27	2.71	2.92	5.08	4.49
C17	5.56	1.99	2.36	4.82	4.17
G18	6.00	2.62	2.82	4.96	4.39
T19	6.04	2.14	2.61	4.87	4.28
T20	6.13	2.17	2.62	4.91	4.19
T21	5.84	2.10	2.48	4.91	4.15
G22	5.82	2.65	2.67	5.00	4.38
C23	5.49	1.98	2.32	4.83	4.16
A24	6.01	2.74	2.89	5.04	4.38
G25	5.82	2.48	2.66	4.96	4.36
C26	6.13	2.14	2.20	4.47	4.05

Chemical shifts

Tabelle 8: ^1H chemical shifts of base protons of 13merRef

Res	H1	H2	H3	H41	H42	H5	H6	H7	H8
G1	-	-	-	-	-	-	-	-	7.99
C2	-	-	-	8.33	6.62	5.39	7.55	-	-
T3	-	-	13.95	-	-	-	7.36	1.66	-
G4	12.71	-	-	-	-	-	-	-	7.89
C5	-	-	-	8.34	6.31	5.40	7.30	-	-
A6	-	7.17	-	-	-	-	-	-	8.19
A7	-	7.13	-	-	-	-	-	-	8.11
A8	-	7.58	-	-	-	-	-	-	8.05
C9	-	-	-	7.91	6.36	5.10	7.12	-	-
G10	12.67	-	-	-	-	-	-	-	7.81
T11	-	-	13.75	-	-	-	7.27	1.39	-
C12	-	-	-	8.62	7.02	5.72	7.51	-	-
G13	-	-	-	-	-	-	-	-	7.97
C14	-	-	-	-	-	5.91	7.62	-	-
G15	12.95	-	-	-	-	-	-	-	7.97
A16	-	7.91	-	-	-	-	-	-	8.23
C17	-	-	-	8.11	6.44	5.23	7.21	-	-
G18	12.76	-	-	-	-	-	-	-	7.83
T19	-	-	13.91	-	-	-	7.25	1.37	-
T20	-	-	13.90	-	-	-	7.47	1.62	-
T21	-	-	13.70	-	-	-	7.33	1.69	-
G22	12.60	-	-	-	-	-	-	-	7.89
C23	-	-	-	8.32	6.34	5.39	7.37	-	-
A24	-	7.66	-	-	-	-	-	-	8.19
G25	12.94	-	-	-	-	-	-	-	7.69
C26	-	-	-	-	-	5.33	7.38	-	-

Tabelle 9: ^{13}C *chemical shifts of 13merRef*

Res	C1'	C3'	C2	C5	C6	C8
G1	82.2	75.3	-	-	-	135.9
C2	83.9	76.6	-	95.7	140.1	-
T3	82.8	75.3	-	-	136.6	-
G4	81.4	76.5	-	-	-	135.5
C5	83.3	76.1	-	95.5	139.6	-
A6	81.7	76.9	151.4	-	-	138.7
A7	82.2	76.1	151.1	-	-	138.2
A8	81.9	76.6	152.0	-	-	138.0
C9	83.1	74.0	-	95.1	139.1	-
G10	82.0	76.9	-	-	-	135.5
T11	82.7	75.1	-	-	135.8	-
C12	83.8	76.5	-	96.1	141.0	-
G13	82.0	70.6	-	-	-	135.5
C14	85.2	75.1	-	96.8	140.4	-
G15	81.3	76.7	-	-	-	136.4
A16	82.1	76.9	152.7	-	-	138.4
C17	83.2	74.0	-	95.4	139.2	-
G18	82.1	75.2	-	-	-	135.5
T19	82.7	75.1	-	-	135.8	-
T20	82.7	74.8	-	-	137.2	-
T21	81.5	74.7	-	-	136.7	-
G22	81.3	76.6	-	-	-	135.5
C23	83.5	75.3	-	95.5	139.9	-
A24	82.1	77.0	151.5	-	-	138.8
G25	81.2	75.0	-	-	-	134.5
C26	84.0	68.8	-	95.3	140.2	-

9 Chemical shifts of 13mer2AP

Tabelle 10: ^1H chemical shifts of sugar protons of 13mer2AP

Res	H1'	H2'	H2"	H3'	H4'
G1	6.02	2.68	2.81	4.88	4.30
C2	6.11	2.15	2.55	4.88	4.29
T3	5.77	2.15	2.48	4.91	4.17
G4	5.87	2.66	2.72	5.01	4.39
C5	5.49	1.92	2.29	4.83	4.15
A6	5.92	2.70	2.89	5.05	4.36
X7	5.55	2.57	2.70	5.01	4.38
A8	6.17	2.60	2.89	5.03	4.47
C9	5.56	1.97	2.34	4.83	4.14
G10	5.96	2.63	2.79	4.96	4.38
T11	6.06	2.11	2.49	4.89	4.24
C12	5.73	2.05	2.40	4.87	4.15
G13	6.19	2.66	2.42	4.72	4.22
C14	5.73	1.87	2.38	4.71	4.09
G15	5.47	2.74	2.82	5.03	4.34
A16	6.29	2.74	2.95	5.09	4.51
C17	5.60	2.03	2.38	4.85	4.19
G18	6.00	2.66	2.82	4.98	4.40
T19	6.07	2.14	2.64	4.89	4.27
T20	6.05	2.21	2.54	4.89	4.21
T21	5.73	2.14	2.47	4.91	4.15
G22	5.86	2.64	2.72	5.00	4.38
C23	5.48	1.97	2.32	4.84	4.15
A24	6.01	2.75	2.90	5.05	4.39
G25	5.83	2.49	2.66	4.98	4.37
C26	6.12	2.16	2.23	4.48	4.06

9 Chemical shifts of 13mer2AP

Tabelle 11: ^1H chemical shifts of base protons of 13mer2AP

Res	H1	H2	H3	H41	H42	H5	H6	H7	H8
G1	-	-	-	-	-	-	-	-	7.99
C2	-	-	-	8.33	6.62	5.38	7.55	-	-
T3	-	-	13.95	-	-	-	7.36	1.68	-
G4	12.72	-	-	-	-	-	-	-	7.92
C5	-	-	-	8.30	6.32	5.42	7.34	-	-
A6	-	7.33	-	-	-	-	-	-	8.20
X7	-	-	-	-	-	-	7.80	-	7.95
A8	-	7.72	-	-	-	-	-	-	8.09
C9	-	-	-	8.02	6.40	5.16	7.15	-	-
G10	12.72	-	-	-	-	-	-	-	7.83
T11	-	-	13.77	-	-	-	7.29	1.42	-
C12	-	-	-	8.64	7.04	5.73	7.52	-	-
G13	-	-	-	-	-	-	-	-	7.97
C14	-	-	-	-	-	5.90	7.62	-	-
G15	12.97	-	-	-	-	-	-	-	7.98
A16	-	7.92	-	-	-	-	-	-	8.24
C17	-	-	-	8.11	6.46	5.25	7.23	-	-
G18	12.72	-	-	-	-	-	-	-	7.86
T19	-	-	13.69	-	-	-	7.29	1.41	-
T20	-	-	13.38	-	-	-	7.46	1.65	-
T21	-	-	13.57	-	-	-	7.34	1.69	-
G22	12.66	-	-	-	-	-	-	-	7.89
C23	-	-	-	8.34	6.35	5.40	7.36	-	-
A24	-	7.68	-	-	-	-	-	-	8.19
G25	12.94	-	-	-	-	-	-	-	7.69
C26	-	-	-	-	-	5.26	7.35	-	-

Chemical shifts

Tabelle 12: ^{13}C chemical shifts of 13mer2AP

Res	C1'	C3'	C2	C5	C6	C8
G1	82.2	73.1	-	-	-	134.6
C2	83.9	76.5	-	95.5	140.1	-
T3	82.8	75.3	-	-	136.6	-
G4	81.4	76.5	-	-	-	135.5
C5	83.2	76.1	-	95.5	139.7	-
A6	81.7	76.8	151.5	-	-	138.7
X7	82.2	76.9	-	-	147.6	138.2
A8	81.9	76.6	152.0	-	-	138.1
C9	83.1	74.0	-	95.1	139.1	-
G10	81.9	76.7	-	-	-	135.5
T11	82.7	75.1	-	-	135.8	-
C12	85.1	76.5	-	96.0	141.1	-
G13	82.0	70.5	-	-	-	135.5
C14	83.7	74.9	-	96.8	140.4	-
G15	81.2	76.7	-	-	-	136.4
A16	82.1	76.9	152.7	-	-	138.4
C17	83.2	74.0	-	95.2	139.3	-
G18	82.1	75.3	-	-	-	135.5
T19	82.7	75.1	-	-	135.8	-
T20	82.5	74.5	-	-	137.2	-
T21	81.4	74.4	-	-	136.7	-
G22	81.2	76.6	-	-	-	135.5
C23	83.4	74.0	-	95.5	139.9	-
A24	82.1	76.1	151.5	-	-	138.8
G25	81.2	75.3	-	-	-	134.4
C26	84.0	68.6	-	95.4	139.9	-

10 Chemical shift differences for 13merRef and 13mer2AP

The chemical shift differences are calculated as X(13merRef)-X(13mer2AP).

Tabelle 13: ^1H chemical shift differences between 13merRef and 13mer2AP, for X7 H2 the difference the CSD between atoms H2 in 13merRef and H6 in 13mer2AP is given.

Res	H1'	H2'	H2"	H3'	H4'	H1/H3 H41/H42	H2/H5 H7	H6/H8
G1	0.00	0.00	0.00	-0.01	-0.02	-	-	0.00
C2	-0.01	0.00	0.00	-0.01	-0.01	-0.01/0.00	0.00	0.00
T3	-0.01	0.00	-0.01	-0.01	-0.01	0.00	-0.01	0.00
G4	-0.01	-0.03	-0.02	-0.02	-0.02	-0.01	-	-0.03
C5	-0.08	-0.08	-0.07	-0.04	-0.04	0.03/-0.01	-0.02	-0.04
A6	-0.17	0.01	-0.05	-0.02	-0.01	-	-0.16	-0.01
X7	0.29	0.06	0.14	0.04	0.05	-	-0.67	0.16
A8	-0.12	-0.06	-0.06	-0.03	-0.03	-	-0.13	-0.04
C9	0.00	-0.05	-0.02	-0.06	0.00	-0.03/0.02	-0.06	-0.04
G10	-0.01	-0.01	-0.02	-0.01	-0.02	-0.05	-	-0.02
T11	-0.01	-0.01	-0.01	-0.01	-0.01	-0.02	-0.03	-0.02
C12	-0.01	-0.01	-0.01	-0.02	-0.01	-0.02/-0.01	-0.01	-0.01
G13	-0.01	-0.02	-0.02	-0.02	-0.01	-	-	-0.01
C14	0.00	0.00	-0.01	-0.01	-0.02	-	0.01	0.00
G15	-0.01	-0.01	-0.02	-0.02	-0.01	-0.01	-	-0.01
A16	-0.02	-0.02	-0.02	-0.02	-0.02	-	-0.01	-0.02
C17	-0.04	-0.04	-0.02	-0.03	-0.02	-0.01/-0.02	-0.02	-0.02
G18	0.00	-0.03	0.00	-0.02	-0.01	0.03	-	-0.03
T19	-0.03	0.00	-0.03	-0.01	0.01	0.21	-0.04	-0.04
T20	0.09	-0.04	0.09	0.02	-0.03	0.53	-0.03	0.01
T21	0.11	-0.04	0.01	0.01	0.01	0.13	0.00	-0.02
G22	-0.04	0.00	-0.05	0.00	0.00	-0.05	-	0.00
C23	0.01	0.01	0.01	-0.01	0.01	-0.02/-0.02	-0.02	0.01
A24	0.00	-0.01	-0.01	-0.01	-0.01	-	-0.02	-0.01
G25	-0.01	-0.01	-0.01	-0.02	-0.01	0.00	-	0.00
C26	0.01	-0.02	-0.02	-0.01	0.00	-	0.07	0.03

Helical parameter

The averaged values for the 10 minimum-energy, violation-free structures for all helical parameters are listed in the Appendix in sections 11 and 12 for 13merRef and 13mer2AP, respectively. Their RMSD is given as the uncertainty.

11 Helical parameters for 13merRef

Tabelle 14: *Base pair parameters for 13merRef, translational*

base pair	Shear (Sx)/Å	Stretch (Sy)/Å	Stagger (Sz)/Å
G1–C26	-0,42 ± 0,01	-0,26 ± 0,00	-0,14 ± 0,04
C2–G25	0,42 ± 0,02	-0,26 ± 0,01	-0,13 ± 0,05
T3–A24	-0,07 ± 0,01	-0,26 ± 0,01	0,07 ± 0,04
G4–C23	-0,37 ± 0,01	-0,28 ± 0,00	0,10 ± 0,06
C5–G22	0,40 ± 0,01	-0,26 ± 0,01	-0,13 ± 0,05
A6–T21	0,02 ± 0,02	-0,26 ± 0,01	-0,04 ± 0,03
A7–T20	0,07 ± 0,01	-0,27 ± 0,01	-0,24 ± 0,02
A8–T19	0,03 ± 0,02	-0,28 ± 0,01	-0,06 ± 0,04
C9–G18	0,39 ± 0,01	-0,27 ± 0,00	-0,05 ± 0,07
G10–C17	-0,40 ± 0,01	-0,26 ± 0,00	0,08 ± 0,03
T11–A16	-0,07 ± 0,00	-0,27 ± 0,01	0,09 ± 0,07
C12–G15	0,40 ± 0,01	-0,27 ± 0,00	-0,11 ± 0,05
G13–C14	-0,40 ± 0,01	-0,26 ± 0,00	-0,03 ± 0,02
average	0.00 ± 0.33	-0.26 ± 0.01	-0.05 ± 0.11

Helical parameter

Tabelle 15: *Base pair parameters for 13merRef, rotational*

base pair	Buckle $(\chi)/°$	Propeller Twist $(\omega)/°$	Opening $(\sigma)/°$
G1–C26	-1,4 ± 1,5	2,4 ± 0,6	1,7 ± 0,1
C2–G25	-4,7 ± 1,9	3,8 ± 1,7	1,5 ± 0,2
T3–A24	-8,0 ± 1,7	-1,8 ± 1,7	-2,9 ± 0,1
G4–C23	-0,4 ± 0,5	5,4 ± 2,5	1,5 ± 0,1
C5–G22	5,3 ± 0,5	3,5 ± 1,7	1,6 ± 0,1
A6–T21	6,3 ± 0,4	-1,8 ± 0,4	-2,9 ± 0,2
A7–T20	4,9 ± 0,4	-8,2 ± 1,3	-2,5 ± 0,3
A8–T19	5,1 ± 0,7	-10,6 ± 1,0	-2,7 ± 0,3
C9–G18	4,3 ± 1,0	-6,0 ± 1,2	1,5 ± 0,1
G10–C17	-1,2 ± 1,3	1,2 ± 1,8	1,5 ± 0,1
T11–A16	2,6 ± 2,7	-3,4 ± 1,4	-3,0 ± 0,2
C12–G15	4,6 ± 1,6	-8,9 ± 0,6	2,0 ± 0,1
G13–C14	-0,6 ± 0,8	-0,6 ± 0,5	1,4 ± 0,1
average	1.3 ± 4.4	-1.9 ± 5.2	-0.1 ± 2.2

Tabelle 16: *Base pair step parameters for 13merRef, translational*

base pair step	Shift (Sx)/Å	Slide (Sy)/Å	Rise (Sz)/Å
G1–C2	-0,12 ± 0,03	-0,71 ± 0,11	3,40 ± 0,07
C2–T3	-0,30 ± 0,05	-1,35 ± 0,04	3,38 ± 0,04
T3–G4	0,44 ± 0,05	-0,57 ± 0,11	2,81 ± 0,06
G4–C5	0,10 ± 0,05	-0,25 ± 0,10	3,06 ± 0,05
C5–A6	-0,61 ± 0,11	-0,67 ± 0,06	3,01 ± 0,03
A6–A7	-0,36 ± 0,03	-0,87 ± 0,02	3,27 ± 0,01
A7–A8	-0,45 ± 0,07	-0,66 ± 0,06	3,11 ± 0,03
A8–C9	-0,02 ± 0,08	-0,72 ± 0,08	3,20 ± 0,01
C9–G10	0,17 ± 0,07	-0,98 ± 0,09	3,21 ± 0,05
G10–T11	-0,23 ± 0,03	-0,92 ± 0,04	3,14 ± 0,03
T11–C12	0,34 ± 0,10	-0,28 ± 0,09	3,00 ± 0,06
C12–G13	0,02 ± 0,04	-0,99 ± 0,05	3,01 ± 0,04
average	-0.08 ± 0.32	-0.75 ± 0.31	3.13 ± 0.17

11 Helical parameters for 13merRef

Tabelle 17: Base pair step parameters for 13merRef, rotational

base pair step	Tilt $(\tau)/°$	Roll $(\rho)/°$	Twist $(\Omega)/°$
G1–C2	-1,0 ± 0,4	-4,9 ± 1,0	41,1 ± 0,6
C2–T3	1,6 ± 0,3	-2,6 ± 1,3	30,4 ± 0,3
T3–G4	-0,6 ± 0,3	3,7 ± 0,9	35,2 ± 0,6
G4–C5	2,1 ± 0,1	2,1 ± 0,8	40,6 ± 0,6
C5–A6	-1,1 ± 0,2	4,6 ± 0,4	35,2 ± 0,4
A6–A7	0,3 ± 0,2	-2,4 ± 0,6	33,4 ± 0,2
A7–A8	-4,2 ± 0,3	1,3 ± 0,3	36,0 ± 0,5
A8–C9	-2,7 ± 0,5	-2,8 ± 0,6	37,6 ± 0,6
C9–G10	-2,3 ± 0,4	0,7 ± 1,2	33,3 ± 0,8
G10–T11	1,8 ± 0,3	-3,0 ± 0,6	35,3 ± 0,6
T11–C12	3,3 ± 0,5	2,7 ± 1,4	39,6 ± 0,7
C12–G13	-1,0 ± 0,5	10,0 ± 1,3	30,6 ± 0,5
average	-0.3 ± 2.2	0.8 ± 4.2	35.7 ± 3.5

Helical parameter

12 Helical parameters for 13mer2AP

Tabelle 18: *Base pair parameters for 13mer2AP, translational*

base pair	Shear (Sx)/Å	Stretch (Sy)/Å	Stagger (Sz)/Å
G1–C26	-0,40 ± 0,01	-0,26 ± 0,00	-0,11 ± 0,04
C2–G25	0,35 ± 0,02	-0,28 ± 0,01	-0,10 ± 0,05
T3–A24	-0,07 ± 0,00	-0,26 ± 0,00	-0,02 ± 0,05
G4–C23	-0,37 ± 0,01	-0,29 ± 0,00	0,33 ± 0,03
C5–G22	0,29 ± 0,03	-0,27 ± 0,01	-0,10 ± 0,05
A6–T21	0,01 ± 0,02	-0,27 ± 0,00	-0,02 ± 0,04
2AP7–T20	-0,02 ± 0,00	-0,32 ± 0,00	-0,23 ± 0,03
A8–T19	0,08 ± 0,01	-0,26 ± 0,00	0,12 ± 0,03
C9–G18	0,43 ± 0,02	-0,25 ± 0,01	-0,18 ± 0,06
G10–C17	-0,39 ± 0,02	-0,27 ± 0,00	-0,16 ± 0,03
T11–A16	-0,07 ± 0,01	-0,26 ± 0,00	-0,13 ± 0,05
C12–G15	0,41 ± 0,01	-0,27 ± 0,01	-0,17 ± 0,04
G13–C14	-0,37 ± 0,02	-0,26 ± 0,01	-0,05 ± 0,03
average	-0.01 ± 0.31	-0.27 ± 0.02	-0.06 ± 0.15

Tabelle 19: *Base pair parameters for 13mer2AP, rotational*

base pair	Buckle (χ)/°	Propeller Twist (ω)/°	Opening (σ)/°
G1–C26	-2,2 ± 1,2	2,1 ± 0,4	1,5 ± 0,1
C2–G25	-3,0 ± 1,4	6,0 ± 2,0	1,0 ± 0,1
T3–A24	-5,8 ± 1,5	-2,4 ± 1,3	-2,8 ± 0,1
G4–C23	5,7 ± 0,7	6,9 ± 1,6	1,8 ± 0,2
C5–G22	8,0 ± 0,6	1,9 ± 1,0	0,9 ± 0,1
A6–T21	6,4 ± 0,3	-4,9 ± 0,3	-2,8 ± 0,1
2AP7–T20	2,3 ± 0,3	-1,2 ± 0,3	-5,7 ± 0,1
A8–T19	5,3 ± 0,7	-2,5 ± 1,1	-3,0 ± 0,1
C9–G18	8,4 ± 0,7	-3,7 ± 1,1	1,8 ± 0,2
G10–C17	-0,5 ± 1,3	6,2 ± 1,5	1,3 ± 0,2
T11–A16	2,8 ± 2,2	0,8 ± 1,0	-2,9 ± 0,1
C12–G15	2,7 ± 1,4	-6,5 ± 1,1	2,1 ± 0,1
G13–C14	-2,1 ± 1,5	-1,0 ± 1,0	1,3 ± 0,1
average	2.1 ± 4.5	0.1 ± 4.3	-0.4 ± 2.6

12 Helical parameters for 13mer2AP

Tabelle 20: *Base pair step parameters for 13mer2AP, translational*

base pair step	Shift (Sx)/Å	Slide (Sy)/Å	Rise (Sz)/Å
G1–C2	-0,03 ± 0,07	-0,44 ± 0,11	3,28 ± 0,05
C2–T3	-0,04 ± 0,05	-1,25 ± 0,04	3,28 ± 0,05
T3–G4	0,28 ± 0,02	-0,85 ± 0,04	2,73 ± 0,06
G4–C5	0,06 ± 0,03	-0,29 ± 0,05	3,16 ± 0,03
C5–A6	-0,43 ± 0,03	-0,73 ± 0,05	3,07 ± 0,04
A6–2AP7	-0,58 ± 0,02	-1,07 ± 0,07	3,31 ± 0,02
2AP7–A8	-0,13 ± 0,03	-0,66 ± 0,04	3,14 ± 0,02
A8–C9	-0,10 ± 0,05	-0,98 ± 0,05	3,18 ± 0,02
C9–G10	0,11 ± 0,04	-1,10 ± 0,05	3,29 ± 0,05
G10–T11	0,06 ± 0,05	-0,87 ± 0,06	3,29 ± 0,03
T11–C12	0,31 ± 0,07	-0,39 ± 0,13	3,10 ± 0,04
C12–G13	-0,09 ± 0,09	-0,99 ± 0,08	3,01 ± 0,07
average	-0.05 ± 0.26	-0.80 ± 0.30	3.15 ± 0.17

Tabelle 21: *Base pair step parameters for 13mer2AP, rotational*

base pair step	Tilt (τ)/°	Roll (ρ)/°	Twist (Ω)/°
G1–C2	-1,1 ± 0,5	-2,3 ± 1,5	42,3 ± 0,8
C2–T3	2,3 ± 0,6	-1,1 ± 1,4	31,1 ± 0,4
T3–G4	-3,0 ± 0,3	-0,5 ± 0,5	34,0 ± 0,5
G4–C5	4,1 ± 0,3	0,7 ± 0,9	41,3 ± 0,4
C5–A6	-1,0 ± 0,2	2,5 ± 0,4	35,5 ± 0,5
A6–2AP7	-2,1 ± 0,1	0,4 ± 0,3	34,0 ± 0,4
2AP7–A8	-5,3 ± 0,3	-3,3 ± 0,3	37,6 ± 0,3
A8–C9	-0,3 ± 0,4	-3,0 ± 0,5	34,1 ± 0,5
C9–G10	-1,5 ± 0,3	3,9 ± 0,9	32,9 ± 0,5
G10–T11	3,3 ± 0,4	-5,5 ± 0,6	35,8 ± 0,3
T11–C12	3,0 ± 0,4	4,3 ± 1,6	39,1 ± 1,0
C12–G13	-0,8 ± 0,3	10,7 ± 1,6	30,6 ± 0,6
average	-0.2 ± 2.8	0.6 ± 4.3	35.7 ± 3.7

List of abbreviations

WC Watson-Crick

DNA DesoxyriboNucleic Acid

NMR Nuclear Magnetic Resonance

SNP Single Nnucleotide Polymorphism

DETEQ Detection by Electron Transfer controlled Emission Quenching

FIT Forced IntercalaTion probe

FRET Fluorescence Resonance Excitation Transfer

2AP 2-Aminopurin

HNF 2-Hydroxy-7-nitrofluorene

ABA ABAsic site

A Adenine

T Thymine

G Guanine

C Cytosine

TRIS Trishydroxymethylaminomethane

NOE Nuclear Overhauser Enhancement

Helical parameter

NOESY Nuclear Overhauser Enhancement Spectroscopy

ISPA Isolated Spin Pair Approximation

RDC Residual Dipolar Coupling

PAS Principle Axis System

SA Simulated Annealing

MD Molecular Dynamics

HPLC High Pressure Liquid Chromatography

WATERGATE Water suppression by GrAdient Tailored Excitation

DQF-COSY Double Quantum Filtered COrrelated Spectroscopy

TOCSY TOtal Correlation Spectroscopy

HMQC Heteronuclear Multiple Quantum Coherence Spectroscopy

DFT Density Functional Theory

TZVP Triple Zeta Valence plus Polarization

MEP Molecular Electrostatic Potential

P.E.-COSY Primitive Exclusive COrrelated Spectroscopy

RMSD Root Mean Square Deviation

CSD Chemical Shift Difference

Index of figures

1.1 DNA duplex sequence with chemical structure of the 2AP-T and A-T base pairs 5

1.2 DNA duplex sequence with chemical structure of the HNF fluorphore and abasic site 5

2.1 Structure and nomenclature of the Watson-Crick base pairs A:T and G:C . 8

2.2 Nomenclature and structure of 2''-deoxy-β-D-ribose 9

2.3 Dihedral angles in the sugar-phosphate backbone of DNA 9

2.4 Comparison of A- und B-form DNA 9

2.5 Helical parameters for the orientational description of base pairs 10

2.6 Kinetic scheme of imino proton exchange 12

2.7 Scheme of a general inversion recovery pulse sequence 18

2.8 Pulse shapes and excitation forms 19

2.9 Scheme of a general NOESY pulse sequence 21

2.10 The cross-relaxation network of N groups of nuclei 22

2.11 Two different approaches to obtain distances from NOE experiments . . . 26

2.12 Steric interaction Pf1-DNA and principle of RDC measurement 29

2.13 Orientation of the magnetic field and the internuclear vector in the macromolecular frame 31

2.14 Orientation of the internuclear vector in the eigenframe of the alignment tensor and orientational degeneracy of Residual Dipolar Couplings 33

Index of figures

3.1 Imino proton signal intensities for the temperature range 283-328 K. in 13mer2AP. 43

4.1 DNA duplex sequence with chemical structure of the 2AP-T and A-T base pairs 57

4.2 Sequential assignment in the sugar H1' base H6/H8 region of the NOESY-spectrum in D_2O for 13merRef and 13mer2AP 62

4.3 Absolute per-residue chemical shift differences of 13mer2AP to 13merRef 64

4.4 Structures of 13merRef and 13mer2AP 65

4.5 Overlay of the experimental and back-calculated NOESY-spectra 66

4.6 Plot of the experimental vs predicted Residual Dipolar Couplings for *13merRef* and *13mer2AP* 66

4.7 13merRef and 13mer2AP: Differences in translational helical parameters between base pair partners 67

4.8 13merRef and 13mer2AP: Differences in rotational helical parameter between base pair partners 67

4.9 13merRef and 13mer2AP: Differences in translational helical parameter between base pairs 68

4.10 13merRef and 13mer2AP: Differences in rotational helical parameter between base pairs 68

4.11 T20 and T3 imino proton intensities for *13mer2AP* at 298 K 71

4.12 Saturation transfer experiments in H_2O 72

4.13 Base pair lifetime determination for 13merRef and 13mer2AP 73

4.14 Melting curves via 13mer2AP absorbance and 2AP fluorescence yield. . . 74

4.15 Linear fits for the imino proton exchange times of T3,T11,G15 and G25 vs the inverse base catalyst concentration. 74

4.16 Comparison of the three central base pairs of the average structures for 13merRef and 13mer2AP 78

4.17 DNA duplex sequence with chemical structure of the HNF and abasic site 86

Index of figures

4.18 Sequential assignment in the sugar H1' base H6/H8 region of the NOESY-spectrum in D$_2$O for 13merHNF 88
4.19 Comparison of the imino proton region of 13merRef, 13merRefGC and 13merHNF 89
4.20 Absolute per-residue chemical shift differences of 13merHNF to the corresponding native structures 90
4.21 10 minimum-energy, violation-free structures of 13merHNF in face-up and face-down orientation................................ 91
4.22 Averaged, minimized structures of 13merHNF in face-up and face-down orientation... 91
4.23 Close-up view of the average structures of 13merHNF in face-up and face-down orientation 92
4.24 Overlay of the experimental and back-calculated NOESY-spectra for 13merHNF 92
4.25 Plot of the experimental vs predicted Residual Dipolar Couplings for 13merHNF 93
4.26 Absorption changes upon hybridisation, when lowering temperature from 85 to 25°C for 13merHNF 94
4.27 Temperature-dependence of the spectra in Fig. 4.26. 94

Index of tables

3.1	Experimentally determined Residual Dipolar Couplings for *13merRef*	48
3.2	Experimentally determined Residual Dipolar Couplings for *13mer2AP*	49
3.3	Overview of structural statistics for 13merRef and 13mer2AP calculations	50
3.4	Experimentally determined Residual Dipolar Couplings for *13merHNF*	55
3.5	Overview of structural statistics for 13merHNF calculations	56
4.1	Selected ^{1}H Chemical shift differences (CSD) for *13merRef* and *13mer2AP*	63
2	^{1}H chemical shifts of sugar protons of *13merHNF*	262
3	^{1}H chemical shifts of base protons of *13merHNF*	263
4	^{13}C chemical shifts of *13merHNF*	264
5	^{1}H chemical shifts of sugar protons of *13merRefGC*	265
6	^{1}H chemical shifts of base protons of *13merRefGC*	266
7	^{1}H chemical shifts of sugar protons of *13merRef*	267
8	^{1}H chemical shifts of base protons of *13merRef*	268
9	^{13}C chemical shifts of *13merRef*	269
10	^{1}H chemical shifts of sugar protons of *13mer2AP*	270
11	^{1}H chemical shifts of base protons of *13mer2AP*	271
12	^{13}C chemical shifts of *13mer2AP*	272
13	^{1}H chemical shift differences between *13merRef* and *13mer2AP*	273
14	Base pair parameters for *13merRef*, translational	275
15	Base pair parameters for *13merRef*, rotational	276
16	Base pair step parameters for *13merRef*, translational	276

Index of tables

17	Base pair step parameters for *13merRef*, rotational	277
18	Base pair parameters for *13mer2AP*, translational	278
19	Base pair parameters for *13mer2AP*, rotational	278
20	Base pair step parameters for *13mer2AP*, translational	279
21	Base pair step parameters for *13mer2AP*, rotational	279

Die VDM Verlagsservicegesellschaft sucht für wissenschaftliche Verlage abgeschlossene und herausragende

Dissertationen, Habilitationen, Diplomarbeiten, Master Theses, Magisterarbeiten usw.

für die kostenlose Publikation als Fachbuch.

Sie verfügen über eine Arbeit, die hohen inhaltlichen und formalen Ansprüchen genügt, und haben Interesse an einer honorarvergüteten Publikation?

Dann senden Sie bitte erste Informationen über sich und Ihre Arbeit per Email an *info@vdm-vsg.de*.

Sie erhalten kurzfristig unser Feedback!

VDM Verlagsservicegesellschaft mbH
Dudweiler Landstr. 99 Telefon +49 681 3720 174
D - 66123 Saarbrücken Fax   +49 681 3720 1749
www.vdm-vsg.de

Die VDM Verlagsservicegesellschaft mbH vertritt

Printed by Books on Demand GmbH, Norderstedt / Germany